AF361014

# Salutogenesis and Coping

# Salutogenesis and Coping: Ways to Overcome Stress and Conflicts

Editors

**Orna Braun-Lewensohn**
**Claude-Hélène Mayer**

MDPI • Basel • Beijing • Wuhan • Barcelona • Belgrade • Manchester • Tokyo • Cluj • Tianjin

*Editors*
Orna Braun-Lewensohn
Ben-Gurion University of the Negev
Israel

Claude-Hélène Mayer
University of Johannesburg
South Africa

*Editorial Office*
MDPI
St. Alban-Anlage 66
4052 Basel, Switzerland

This is a reprint of articles from the Special Issue published online in the open access journal *International Journal of Environmental Research and Public Health* (ISSN 1660-4601) (available at: https://www.mdpi.com/journal/ijerph/special_issues/Salutogenesis_Coping).

For citation purposes, cite each article independently as indicated on the article page online and as indicated below:

LastName, A.A.; LastName, B.B.; LastName, C.C. Article Title. *Journal Name* **Year**, *Article Number*, Page Range.

**ISBN 978-3-03943-446-6 (Hbk)**
**ISBN 978-3-03943-447-3 (PDF)**

Cover image courtesy of Schatz estate, Jerusalem.

# Contents

# About the Editors

**Orna Braun-Lewensohn** Orna Braun-Lewensohn is a Full Professor and the head of the Department of Interdisciplinary Studies at the Ben-Gurion University of the Negev (Israel). She also serves as a faculty member at the "Conflict Resolution and Conflict Management" Program. She received her Ph.D. at the Faculty of Psychology and Educational Sciences, Vrije Universiteit Brussels in 2007. Her major research interests include mental health outcomes and coping during or following stressful events. The focus of her research is personal as well as communal coping resources in different cultural groups. Her theoretical perspective is the salutogenic model of Antonovsky and the coping theory of Lazarus and Folkman. She is considered an expert in this field. In recent years, she was funded by: Ministry of Science and Technology—The Integration of Academic Minority Women into the Israeli Workforce: Ultra-Orthodox and Bedouin Women; The Israel National Institute for Health—Barriers and catalysts for the use of mental health services in light of the mental health reform; Israeli Science Foundation (ISF)—Salutogenesis as a universal construct and to develop a sensitive research tool for understanding a sense of coherence among indigenous cultures; Israeli Democracy Institute - Between the 'out' and the 'in': Integration of Ultra-Orthodox academics in the Israeli job market; Joint Tevet; Ministry of Internal Security; Peres Center for Peace, among others. She publishes extensively in journals such as: Current Psychiatry Reports, Anxiety, Stress, & Coping, Community Mental Health Journal, Journal of Adolescence, Journal of Positive Psychology, and Social Indicators Research.

**Claude-Hélène Mayer** (Dr. habil., Ph.D., Ph.D.) is a Professor in Industrial and Organizational Psychology at the Department of Industrial Psychology and People Management at the University of Johannesburg, an Adjunct Professor at the European University Viadrina in Frankfurt (Oder), Germany, and a Senior Research Associate at Rhodes University, Grahamstown, South Africa. She holds a Ph.D. in Psychology (University of Pretoria, South Africa), a Ph.D. in Management (Rhodes University, South Africa), a doctorate (Georg-August University, Germany) in Political Sciences (socio-cultural anthropology and intercultural didactics), and a habilitation (European University Viadrina, Germany) in Psychology with a focus on work, organizational, and cultural psychology. She has published several monographs, text collections, accredited journal articles, and Special Issues on transcultural mental health, sense of coherence, shame, culture and health, transcultural conflict management and mediation, women in leadership in culturally diverse work contexts, constellation work, coaching, and psychobiography.

International Journal of
*Environmental Research and Public Health*

*Editorial*

# Salutogenesis and Coping: Ways to Overcome Stress and Conflict

**Orna Braun-Lewensohn [1,*] and Claude-Hélène Mayer [2,3]**

[1]  Department of Interdisciplinary Studies, Conflict Management & Resolution Program,
   Ben-Gurion University of the Negev, Beer-Sheva 8410501, Israel
[2]  Department of Industrial Psychology and People Management, Auckland Park Campus,
   University of Johannesburg, Johannesburg 2006, South Africa; claudemayer@gmx.net
[3]  Institut für Therapeutische Kommunikation und Sprachgebrauch, Europa Universität Viadrina,
   Logenstrasse 11, 15230 Frankfurt (Oder), Germany
*  Correspondence: ornabl@bgu.ac.il

Received: 8 September 2020; Accepted: 10 September 2020; Published: 13 September 2020

**Abstract:** This Special Issue aims to explore the concepts of stress, coping resources, and coping strategies, which are rooted in several theories, such as the stress and coping theory and the salutogenesis theory, and to understand how their core constructs are manifested in various ethnic and cultural groups around the world. This Special Issue includes 13 articles on salutogenesis and coping from different disciplinary, socio-cultural, historical, political, and economic perspectives. These articles address salutogenesis on the individual, organizational, and societal levels. The empirical studies are based in different societal and national contexts and refer to different ethnic groups within those contexts. Other studies examine international leaders in industry from a global perspective and present a systemic review of the literature concerning individuals in specific professions, such as nursing. The studies in the current Special Issue set the ground for continuing research toward even more comprehensive theoretical grounds; studies that incorporate several theoretical backgrounds and explore a broad theoretical model that may help us to understand successful adaptation in various contexts. In summary, results of studies that incorporate these theories may promote our understanding of the effects of coping resources and strategies, including acculturation strategies used among minority groups for positive adaptation.

**Keywords:** salutogenesis; stress; coping; conflict

---

## 1. Introduction

The stress appraisal and coping theory [1], views coping as an interactional process between an individual and his/her environment, which can be defined as the effort exerted by the individual to deal with demands from the environment, in order to make those demands more tolerable and reduce stress and conflict. This means that the characteristics of an individual and the way that he or she appraises a situation are important elements for that individual's well-being in the aftermath of a stressful or conflictual encounter. Moreover, in the cognitive process of appraisal, one of the components that the individual assesses is the resources s/he has to deal with the situation.

To this end, sense of coherence (SOC), which is the central component of the salutogenic model, can be perceived as a secondary appraisal that facilitates the exploration of resources available to the individual to deal with the stressful situation. The salutogenic model looks for functions of positive qualities rather than healing from sickness [2,3]. Its main construct, SOC, is an enduring tendency to see the world as more or less comprehensible, manageable, and meaningful [4]. In accordance with salutogenesis, a person with a strong SOC is more likely to evaluate a stimulus as neutral [2]. Therefore, an individual with a strong SOC is less likely than one with a weak SOC to perceive stressful situations as threatening and, therefore, as anxiety-provoking. SOC determines the ability of individuals to use resources that are available to them to promote their well-being [5]. Moreover, SOC includes components that consolidate resilience and expand subjective mental health [2].

Coping strategies are the behavioral component of the process and can be defined as the actual effort made in the attempt to render a perceived stressor or conflict more tolerable and to minimize the distress induced by the situation. Most models of coping assume that individuals who cope more effectively with stressful and conflictual life events will exhibit lower levels of anxiety or depression [1]. Studies have shown that emotion-focused strategies of coping tend to be associated with more psychological problems, whereas, problem-focused strategies or active coping tend to be linked to more well-being [6].

We thought that looking at the stress appraisal and coping theory of Lazarus and Folkman [1] and the salutogenesis model of Antonovsky together might provide us with a more comprehensive understanding of the resources that facilitate certain coping strategies and the behaviors that are more or less adaptive in different situations of stress and conflict. Through the lens of a more integrative model, several issues can be highlighted. First, different types of events can be examined to determine whether different resources, coping, and responses are exhibited. Second, cultural contexts can be taken into consideration to understand the cognitive, behavioral, and emotional processes of individuals in the course of these events. Finally, we can consider a more comprehensive set of outcomes that includes positive (and not only pathological) outcomes.

## 2. The Aim of this Special Issue

This Special Issue aims to explore the concepts of stress, coping resources, and coping strategies, which are rooted in several theories, such as the stress and coping theory of Lazarus and Folkman [1], and the salutogenesis theory of Antonovsky [7], and to understand how their core constructs are manifested in various ethnic and cultural groups around the world.

These theories suggest that their main concepts, namely, several ways of coping, hope, personal and collective SOC, and others, are universal and, therefore, predict that, in all cultures, they could be considered as potential protectors against stress. However, to date, studies involving a non-Western population have reported ambiguous results.

In this Special Issue, we aim to address these concerns comprehensively by inviting researchers from around the world to present their studies based on special research methods and mixed research methods. There studies will enable a fundamental understanding of positive adaptation in stressful and conflictual situations among various cultural and ethnic groups and in different contexts around the world.

## 3. The Contributions in this Special Issue

This Special Issue includes 13 articles on salutogenesis and coping from different disciplinary, socio-cultural, historical, political, and economic perspectives. These articles address salutogenesis on individual, organizational, and societal levels. The empirical studies are based in different societal and national contexts, including Israel, Spain, Norway, and the Democratic Republic of Congo, and refer to different ethnic groups within those countries. Other studies examine international leaders in industry from a global perspective and present a systemic review of the literature concerning individuals in specific professions, such as nursing.

We decided to organize this Special Issue around several themes: first, the age of the participants (from youngest to adults); second, special populations such as minority groups, volunteers, health workers etc.; and third, the settings on which the studies focused, for example, workplaces. The Special Issue opens with a paper on adolescents, the youngest group examined in this volume, and moves on to a paper on a minority student population. Other papers focused on minority groups highlight refugees from the civil war in Syria and educated ultra-orthodox Jews in the workplace. The last prominent theme of the current Special Issue is the workplace, a focal point of many of the articles in this volume. Two articles focus on professionals in the special context of political violence, while some others focus on work in health-care settings. Following these articles, one paper focuses on SOC and coping among employees in the Democratic Republic of Congo. Finally, the last article in this volume explores SOC among international leaders and compassionate love as a coping mechanism.

Here is a brief introduction to the articles that make up this Special Issue.

A highly interesting study explores the associations between sex, age, socio-economic status, stress, SOC, and health among adolescents in Norway. The authors, Unni Karin Moksnes and Geir Arild Espnes [8], investigate SOC and stress interrelationships and point out that SOC is a major coping resource in the context of depression and mental well-being.

Sarah Abu-Kaf and Enas Khalaf [9], present a study on experiences of acculturative stress among Arab students at Israeli institutions of higher learning. The authors combined the theories of coping and salutogenesis and report gender differences in the use of different coping strategies and in levels of depressive symptoms. Moreover, they report that SOC differentially mediates the relationships between acculturative stress and depressive symptoms.

Antony Bernard and Aden Paul Flotman [10], explore the identity work of a group of eight consulting psychology doctoral students. The students wrote self-reflective essays about becoming a consulting psychologist and findings describe how students cope with performance.

Orna Braun-Lewensohn, Sarah Abu-Kaf and Khaled Al-Said [11], present findings on the coping resources and mental health of women in refugee camps. These authors explore personal and the community SOC and their influence on perceived danger and coping. The authors also demonstrate that SOC is crucial for good adaptation. These results are discussed in light of salutogenic theory.

In her article on coping strategies of college-educated, ultra-orthodox Jews in the general Israeli workforce, Tehila Kalagy [12], speaks about societal transitions and the changes in the values of ethnic groups and in workplaces. This article contributes to minority research, as well as our understanding of the professional integration and adaptability of members of minority groups in the workplace and how those individuals cope with the challenges they face.

Tal Litvak-Hirsch and Alon Lazar [13] explore long-term mindfulness training and its contribution to personal and professional coping among teachers living in a conflict zone. These authors present their findings from a qualitative study conducted in the Western Negev region of Israel. Interviewees reported that their coping skills had been heightened as result of being able to put aside intrusive thoughts and feelings that used to paralyze them and focus on active coping, centered on what they needed to do promptly. The interviewees also reported increased compassion and self-acceptance of emotions and behaviors. This article presents an important contribution to stress management in war zones through mindfulness training.

In their article, Dorit Segal-Engelchin, Netta Achdut, Ephrat Huss and Orly Sarid [14], focus on CB-ART (cognitive behavioral and art-based) intervention during the 2014 Gaza conflict. The authors present findings regarding the ability of interventions to decrease stress and trauma among individuals working in medical professions. Specifically, they describe how arts-based methods supported coping and built resources to deal with stress and trauma.

The next article sheds light on the situation in South Africa and refers to the experiences of volunteers in the health-care context and their well-being. Antoni Barnard and Aleksandra Furtak [15], argue that volunteers in South Africa need psychological resilience from a salutogenic perspective. What really keeps them healthy is an inner drive and a calling in the context of the work orientation, which can be increased when organizations invest in developmental interventions.

Natura Colomer Pérez, Elena Chover-Sierra, Vicente Gea-Caballero and Joan J Paredes-Carbonell [16], address people's health-assets mapping processes and design-dynamization strategies for health promotion. The authors present a salutogenic model of health and a health-assets model and report findings from the nursing context in Spain. Their results show that SOC can be strengthened through the use of salutogenic and asset-based approaches.

The article by Giuseppe Michele Masanotti, Silvia Paolucci, Elia Abbafati, Claudio Serratore and Michaela Caricato [17], provides a systematic review of SOC among nurses. They report that low SOC is a predictor of depressive state, burnout, and job dissatisfaction among female nurses and that, therefore, SOC could be a health-promoting resource.

Shir Daphna-Tekoah, Talia Megadasi Brikman, Eric Scheier and Uri Balla [18], close the section on health professions with a timely manuscript on COVID-19. They used a unique methodology involving a listening guide and narrative analysis to understand the physical and psychological needs of heath professional during the pandemic, in order to build and provide suitable support programs for those professionals.

Jeremy Mitonga-Monga and Claude-Hélène Mayer [19], present empirical research findings on coping, SOC, burnout, and work engagement in the context of the Democratic Republic of Congo (DRC), where a void in research on salutogenesis and coping still exists. The authors examined the moderating effect of coping in the relationships between SOC, burnout, and work engagement and found that there is a positive relationship between coping and SOC; however, SOC is negatively related to work engagement and burnout. The authors provide recommendations for future theory and practice to increase engagement, performance, and productivity based on increased SOC and coping mechanisms.

International leaders need new skills in the rapidly changing world of work, as well as new resources to cope with and manage stress. In the last article in this issue, Claude-Hélène Mayer and Rudolph M. Oosthuizen [20], present findings from an international study showing that SOC, compassionate love, and coping interrelate are important resources for staying healthy.

## 4. Conclusions

*The Way Forward*

This Special Issue presents the latest studies on salutogenesis and coping in specific cultural and transcultural contexts. These studies present particular insights into specific socio-cultural contexts from qualitative and quantitative empirical, theoretical, and conceptual stances. The articles will lead to deeper discourse, new critical thinking, and expanded contextual knowledge, and will build a foundation for future research and applied interventions with regard to salutogenesis and coping.

The studies in the current Special Issue set the ground for continuing research toward even more comprehensive theoretical grounds: studies which incorporate several theoretical backgrounds and explore a broad theoretical model that may help us to understand successful adaptation in various contexts. We suggest that future studies in the field should incorporate several theories into one model: theories of stress appraisal and coping [1], salutogenesis [4,7], and acculturation [21,22], which are fundamental to the understanding of successful adaptation in various situations. Each of these theoretical foundations will contribute its own driven variables to a model that will encompass the socio-ecological surroundings of the participants. Such studies will enable examination of how different demographic and contextual variables, cognitive appraisals, coping resources, and coping and acculturation strategies relate to each other and to psychological adaptation, on one hand, and various psychological problems, on the other. A comprehensive and coherent model of relations among the variables that relies on the above-mentioned three well-established theories could advance our theoretical and practical knowledge of how people cope and adapt in various contexts and cultures. In summary, the results of studies that incorporate these theories may promote the understanding of the effect of coping resources, and strategies, in addition to acculturation strategies (among minority groups) for positive adaptation. Practically, such research has the potential to help parents, educators, leaders, and policymakers to become better aware of the difficulties experienced by individuals who are confronted with meaningful challenges and stressors. This awareness can assist the establishment of research-based, theory-driven prevention and intervention programs to promote adjustment and adaptation in numerous contexts and cultures.

**Author Contributions:** All authors have read and agreed to the published version of the manuscript.

**Funding:** This research received no external funding.

**Conflicts of Interest:** The authors declare no conflict of interest.

## References


1. Lazarus, R.S.; Folkman, S. *Stress, Appraisal, and Coping*; Springer: New York, NY, USA, 1984.
2. Mayer, C.H. *The Meaning of Sense of Coherence in Transcultural Management*; Waxmann Verlag: Münster, Germany, 2011; Volume 563.
3. Mittelmark, M.B.; Bauer, G.F. The meanings of salutogenesis. In *The Handbook of Salutogenesis*; Springer: New York, NY, USA, 2017; pp. 7–13.
4. Antonovsky, A. *Unraveling the Mystery of Health*; Jossey-Bass: San Francisco, CA, USA, 1987.
5. Eriksson, M. The sense of coherence in the salutogenic model of health. In *The Handbook of Salutogenesis*; Mittelmark, M.B., Sagy, S., Eriksson, M., Bauer, G., Pelikan, J.M., Lindström, B., Espnes, G.A., Eds.; Springer: New York, NY, USA, 2017; pp. 91–96.
6. Braun-Lewensohn, O. Coping and social support in children exposed to mass trauma. *Curr. Psychiatr. Rep.* **2015**, *17*, 46–56. [CrossRef] [PubMed]
7. Antonovsky, A. *Health, Stress, and Coping: New Perspectives on Mental and Physical Well-Being*; Jossey-Bass: San Francisco, CA, USA, 1979.
8. Moksnes, U.K.; Espnes, G.A. Sense of Coherence in Association with Stress Experience and Health in Adolescents. *Int. J. Environ. Res. Public Health* **2020**, *17*, 3003. [CrossRef] [PubMed]
9. Abu-Kaf, S.; Khalaf, E. Acculturative Stress among Arab Students in Israel: The Roles of Sense of Coherence and Coping Strategies. *Int. J. Environ. Res. Public Health* **2020**, *17*, 5106. [CrossRef] [PubMed]
10. Bernard, A.; Flotman, A.P. Coping Dynamic of Consulting Psychology Doctoral Students Transitioning a Professional Role Identity: A Systems Psychodynamic Perspective. *Int. J. Environ. Res. Public Health* **2020**, *17*, 5492. [CrossRef] [PubMed]
11. Braun-Lewensohn, O.; Abu-Kaf, S.; Al-Said, K. Women in Refugee Camps: Which Coping Resources Help them to Adapt? *Int. J. Environ. Res. Public Health* **2019**, *16*, 3990. [CrossRef] [PubMed]
12. Kalagy, T. Enclave in Transition: Ways of Coping of Academics from Ultra-Orthodox (Haredim) Minority Group with Challenges of Integration into the Workforce. *Int. J. Environ. Res. Public Health* **2020**, *17*, 2373. [CrossRef] [PubMed]
13. Litvak-Hirsch, T.; Lazar, A. The Contribution of Long-Term Mindfulness Training on Personal and Professional Coping for Teachers Living in a Conflict Zone: A Qualitative Perspective. *Int. J. Environ. Res. Public Health* **2020**, *17*, 4096. [CrossRef] [PubMed]
14. Segal-Engelchin, D.; Achdut, N.; Huss, E.; Sarid, O. CB-Art Interventions Implemented with Mental Health Professionals Working in a Shared War Reality: Transforming Negative Images and Enhancing Coping Resources. *Int. J. Environ. Res. Public Health* **2020**, *17*, 2287. [CrossRef] [PubMed]
15. Barnard, A.; Furtak, A. Psychological Resilience of Volunteers in a South African Health Care Context: A Salutogenic Approach and Hermeneutic Phenomenological Inquiry. *Int. J. Environ. Res. Public Health* **2020**, *17*, 2922. [CrossRef] [PubMed]
16. Pérez, N.C.; Chover-Sierra, E.; Gea-Caballero, V.; Paredes-Carbonell, J.J. Health Assets, Vocation and Zest for Healthcare Work. A Salutogenic Approach to Active Coping among Certified Nursing Assistant Students. *Int. J. Environ. Res. Public Health* **2020**, *17*, 3586. [CrossRef] [PubMed]
17. Masanotti, G.M.; Paolucci, S.; Abbafati, E.; Serratore, C.; Caricato, M. Sense of Coherence in Nurses: A Systematic Review. *Int. J. Environ. Res. Public Health* **2020**, *17*, 1861. [CrossRef] [PubMed]
18. Daphna-Tekoah, S.; Brikman, T.M.; Scheier, E.; Balla, U. Listening to Hospital Personnel's Narratives during the COVID-19 Outbreak. *Int. J. Environ. Res. Public Health* **2020**, *17*, 6413. [CrossRef] [PubMed]
19. Mitonga-Monga, J.; Mayer, C.-H. Sense of Coherence, Burnout, and Work Engagement: The Moderating Effect of Coping in the Democratic Republic of Congo. *Int. J. Environ. Res. Public Health* **2020**, *17*, 4127. [CrossRef] [PubMed]
20. Mayer, C.-H.; Oosthuizen, R.M. Sense of Coherence, Compassionate Love and Coping in International Leaders during the Transition into the Fourth Industrial Revolution. *Int. J. Environ. Res. Public Health* **2020**, *17*, 2829. [CrossRef] [PubMed]

21. Berry, J.W. Immigration, acculturation, and adaptation. *Appl. Psychol.* **1997**, *46*, 5–34. [CrossRef]
22. Berry, J.W. Mobility and Acculturation. In *The Psychology of Global Mobility*; Carr, S.C., Ed.; Springer: New York, NY, USA, 2010; pp. 193–210.

International Journal of
*Environmental Research and Public Health*

*Article*

# Sense of Coherence in Association with Stress Experience and Health in Adolescents

**Unni Karin Moksnes [1,2,*] and Geir Arild Espnes [1]**

[1]  Department of Public Health and Nursing, NTNU Center for Health Promotion Research, Norwegian University of Science and Technology, 7030 Trondheim, Norway; geir.arild.espnes@ntnu.no
[2]  Faculty of Nursing and Health Sciences, Nord University, 8026 Bodø, Norway
*  Correspondence: unni.moksnes@ntnu.no

Received: 17 March 2020; Accepted: 18 April 2020; Published: 26 April 2020

**Abstract:** This study investigated the associations between sex, age, socio-economic status, stress, sense of coherence (SOC), and health (mental wellbeing, depressive symptoms, self-rated health, and subjective health complaints) in Norwegian adolescents aged 13–19 years. Furthermore, the study investigated the potential protective or compensatory role from SOC on the association between stress and health. *Methods:* The study was based on a cross-sectional sample of 1233 adolescents. Data were analyzed with descriptive, comparative, and multiple linear regression analyses. *Results:* Girls reported significantly higher scores on depressive symptoms and subjective health complaints than boys. Stress was significantly and positively associated with depressive symptoms. SOC associated significantly with all outcome variables; and especially with mental wellbeing and depressive symptoms. Significant interaction effects of sex in combination with stress and SOC were found in association with depression and mental wellbeing. Associations were strongest for girls. *Conclusion:* The findings provided support for the significant role of SOC as a coping resource, especially in relation to adolescents' mental health; weaker associations were found with subjective health complains and self-rated health. The findings also mainly supported a compensatory role of SOC on the association between stress and health during adolescence.

**Keywords:** subjective health complaints; self-rated health; mental health; stress; sense of coherence; salutogenesis; moderator

---

## 1. Introduction

A fair opportunity for every young person to reach their full health potential is a democratic goal for most societies, regardless of demographic, social, economic, educational, and cultural factors [1,2]. Hence, in order to promote positive development in adolescents it is important to investigate how adolescents evaluate their health, and what factors have the greatest impact on their health, as assessed through self-reports. This was also interesting in reference to the fact that young people especially during this period of life experience changes and transitions, which might influence their health and well-being throughout the life course [3,4].

In general, in Norwegian and most other Western societies, children and adolescents growing up today are characterized by good health and a high quality of life. However, self-reported mental health problems have increased in recent years, both globally and nationally and account for a large proportion of negative health outcomes in young people, in all societies [5–7]. In Norway, it is estimated that approximately one in five adolescents have mental health problems affecting their daily life and seven percent have symptoms that meet the requirements for a psychiatric diagnosis [8]. Mental health problems seem to be especially evident in girls, where the proportion of girls aged 15–20 years who are given a psychiatric diagnosis (most common problems are depression, anxiety, eating disorders, and behavioral disorders), has increased from five to seven percent per year, from 2011 to 2016. [8].

Adolescents typically have low rates of serious medical illnesses, but studies show an increase in reports of subjective health complains (SHC), especially among girls, during the adolescent years [9–12]. These complaints refer to mental and physical 'unexplained symptoms', often related to stress experience [9–12]. A well-used indicator to assess the overall perception of health status, is to ask people to self-rate their health (SRH) [13,14]. Previous studies suggest that adolescents' perception of health seem to be relatively stable during the adolescent years [13–16]. However, sex differences in SRH are often reported to increase with age, where especially girls seem to report more negative evaluations of SRH than boys [9–11,17–19]. There is evidence to show that this health deterioration, along with an increase in SHC, relates to a broad spectrum of medical, physical, psychological, and psychosocial factors, where an increased experience of multiple independent and cumulative stressors is recognized as one important factor [10,17,20,21]. Research shows that stress levels increase from preadolescence to adolescence, where girls report higher stressor load and seem to be more vulnerable to the negative psychological effects of stress than boys [4,20]. In order to promote positive functioning, health, and wellbeing in the adolescent population, it is important to gain a better understanding of how stress relates to adolescents' overall experience of health, as well as investigating the role of potential protective factors in this context. The concept of sense of coherence (SOC) is central in the exploration of what coping resources are crucial for the individual's capacity to cope with stressors in daily life and create health (salutogenesis) as a complementary approach to the traditional focus on risks for disease (pathogenesis) [22,23]. SOC is described as a personal coping resource and life orientation, which is recognized as the ability to perceive life as comprehensible, manageable, and meaningful, and the perception of having resources needed to cope with normative and non-normative stressors in daily life [22,23].

SOC is a central resource for the protection and promotion of health [24]. A strong SOC is associated with a positive mental health and subjective well-being and a lower severity of symptoms of anxiety and depression [22,24–26]. Through the last years, a discussion has evolved regarding the weak associations between SOC and physical health [27]. This has been explained by the fact that SOC mainly comprises the individual's mental, social, and spiritual resources for coping with life challenges [24]. Studies in adolescent samples have, however, shown positive associations between SOC and perceived positive health [28–30], and negative associations between SOC and SHC [17,31,32]. Where adolescents have been examined for 'normal' life stressors, such as academic, school, or peer pressure as well as family conflicts, it has been shown that those with stronger SOC report lower stress levels [26,32–34].

It is unclear whether SOC has a compensatory or protective role on the association between perceived stress and health. A compensatory model proposes that SOC operates as a resource, irrespective of stress levels (compensation), while a protective model claims that SOC is activated in the face of adversity (buffer effect). In adult samples, SOC seems to have both a protective and compensatory role in association with different health outcomes [24,35]. Studies conducted in adolescents focusing on daily life stressors have shown that SOC has a weak-to-moderate stress protective role in relation to SHC [21,31,36]. In studies based on Norwegian adolescent samples, support for a stress compensatory role of SOC has mainly been found in relation to both SHC [17], life satisfaction [37], and symptoms of anxiety and depression [26]. These studies have similarities with the present study by investing the role of stress and SOC in relation to mental and physical symptoms. However, the present study extends these studies by investigating the health outcomes more broadly, including subjective-, physical-, and mental health, as well as investigating the potential moderating role of sex and SOC on the association between stress and health in a sample of Norwegian adolescents age 13–19 years in rural areas in mid-Norway. The present study also included socio-economic status that are relevant to investigate in relation to adolescents' health and wellbeing [1,2].

The aims of the study were to investigate in adolescents:

1. Sex differences in self-reported health (SHC, SRH, mental wellbeing, and symptoms of depression);
2. The relation between stress, SOC, and health; and potential sex differences in these associations;

3.   The potential protective or compensatory role of SOC on the association between stress and health.

## 2. Method

### 2.1. Participants

This cross-sectional study was based on data from adolescents in public lower- and upper-secondary schools, in five municipalities from inland and coastal rural areas in the county of Trøndelag, located in Central Norway. The schools offer vocational and academic study tracks that are representative of Norwegian upper secondary schools. In the data collection from 2016, 1906 students were invited to participate in the study, with N = 1282 responding on a questionnaire (a response rate of 67%). Non-responses were caused by students not being at school at the time of data collection, non-willingness to participate or because some classes did not have the chance to participate as the teachers could not administer the questionnaire. No detailed information was available on non-responders. Adolescents <13 or >19 years (n = 49) were excluded, resulting in n = 1233 (64%) being included in the study sample (Table 1).

**Table 1.** Demographic characteristics of the sample.

| Variables | Total n (%) | |
|---|---|---|
| **Gender** | | |
| Boys | 580 (47.0) | |
| Girls | 644 (52.2) | |
| Missing | 9 (0.7) | |
| **Age** | | |
| 13–14 years | 381 (30.9) | |
| 15–16 years | 453 (36.7) | |
| 17–19 years | 399 (32.3) | |
| **Family economy** | | |
| Bad economy all the time | 113 (9.2) | |
| More or less bad economy | 243 (19.7) | |
| Neither had bad or good economy | 264 (21.4) | |
| More or less good economy | 327 (26.5) | |
| Good economy all the time | 254 (20.6) | |
| Missing | 32 (2.6) | |
| **Parents' education** | Mother | Father |
| Primary and lower secondary school | 37 (3.0) | 69 (5.6) |
| Upper secondary school | 283 (23.0) | 366 (29.7) |
| University up to 4 years | 303 (24.6) | 197 (16.0) |
| University more than 4 years | 221 (17.9) | 161 (13.1) |
| Unknown | 365 (29.6) | 393 (31.9) |
| Missing | 24 (1.9) | 47 (3.8) |
| **Parents' job status** | Mother | Father |
| Fulltime job | 798 (64.7) | 1018 (82.6) |
| Part-time job | 238 (19.3) | 86 (7.0) |
| Unemployed / on leave | 47 (3.8) | 28 (2.3) |
| Staying at home | 83 (6.7) | 32 (2.6) |
| Other | 41 (3.3) | 37 (3.0) |
| Missing | 26 (2.1) | 32 (2.6) |
| Total | 1233 (100) | |

*2.2. Procedure*

Data collection was approved by the Regional Committee for Medical Research Ethics (approval number 2016/1165). Prior to data collection, a written information letter was sent to all students and to parents of those ≤15 years, underscoring that participation was voluntary and anonymous, that participants were free to withdraw from the study, and that the collected information was treated with confidentiality. According to research ethical guidelines, written consent was required from adolescents and their parents when adolescents were ≤15 years. Adolescents ≥16 years gave consent by answering the questionnaire. Questionnaire administration was completed with help from teachers in whole class groups during one regular school session (of the teachers' choice) of 45 min, in 2016.

*2.3. Measures*

Self-rated health (SRH) was assessed by one item, "How is your health now?" The response options were: 1–'bad', 2–'not so good', 3–'good', 4–'very good', and 5–'extremely good'. Assessment of health using one item was previously found to be satisfactory for use in other studies on adolescents' health [13,14].

Subjective health complaints (SHC) was measured by 12 items comprising physical symptoms (e.g., stomachache, headache, pain in the back/arms/legs, and cold) and mental symptoms (e.g., bad mood, felt lonely, nervous, sad, or irritable). Participants responded on a four-point scale ranging from 1–'not bothered' to 4–'very much bothered', where higher sum scores indicated higher symptom load. Cronbach's α for the instrument was 0.86.

Sense of coherence (SOC) was assessed with the 13-item Orientation to Life Questionnaire consisting of 13-items rated on a seven-point scale; higher sum scores indicated stronger SOC. The questionnaire has been extensively validated and used cross-culturally, both in adult and adolescent samples [38,39]. In the present study, Cronbach's α was 0.82.

Adolescent stress was measured by use of the Norwegian 30-item version of the Adolescent Stress Questionnaire (ASQ-N). Each item was rated on a five-point Likert scale ranging from 1–'not at all stressful' or 'irrelevant to me' to 5–'very stressful', where a higher sum score indicated higher stress level. The scale was validated for use in Norwegian adolescents [40] and adolescents in other European countries [41–43]. Cronbach's α for the instrument in the present study was 0.94.

Mental well-being (MWB) was assessed with the 14-item version of Warwick–Edinburgh Mental Well-Being Scale (WEMWBS) [44]. The respondents were asked how they had felt about seven positively worded statements over the past two weeks. The values ranged from 1–'None of the time' to 5–'All of the time', where higher sum scores indicated higher levels of mental well-being (range 14–70). The WEMWBS was validated in the general population [44,45], clinical samples [46], and in adolescents [47–49]. Cronbach's α for the scale in the present study was 0.91.

Symptoms of depression was measured using a non-clinical depression scale appropriate for measuring non-clinical depressive attributes [3]. The scale consisted of a 15-item questionnaire measuring respondents' levels of current depressive moods. Item choice was informed by reference to commonly experienced depressive features outlined in the Diagnostic and Statistical Manual–Fourth Edition [50], and to the Zung Self Rating Depression Scale [51]. The items were rated on a 5-point Likert scale ranging from 1–'never' to 5–'always', where higher scores indicated a higher symptom load. The scale was used in previous studies in the adolescent population [26,40] Cronbach's α for the instrument in the present study was 0.94.

Socioeconomic status (SES) was measured in terms of mother's and father's education, employment status, and adolescents' perception of their family's economic situation. Mother's and father's education were assessed separately using one item: "'What is your parents' highest education?"; 1–'Primary and lower secondary school', 2–'Upper secondary school', 3–'University up to 4 years', 4–'University, more than 4 years', 5–'Don't know'. Mother's and father's employment status was assessed separately with the item "'What is your parents' employment status?"; 1–'stay at home', 2–'unemployed', 3–'part time job', 4–'full time job', 5–'other'. Adolescents' perception of family economy was assessed by one item:

"How has the family economy been during the last two years?"; 1–'We have had bad economy the whole time', to 5–'We have had good economy the whole time'.

*2.4. Statistical Analyses*

Statistical analyses were conducted using SPSS, version 22.0 BM SPSS, Armonk, NY, USA. Descriptive statistics included frequencies, means, and standard deviations. T-tests were calculated to test sex mean differences on the scales in the study. To evaluate the strength of the sex mean differences, effect sizes were calculated following Cohen's [52] guidelines for small (0.20), medium (0.50), and large (0.80+) effect sizes. Bivariate correlations between the continuous variables of age, SES, stress, SOC, and health (MWB, depression, SRH, SHC) was tested using Pearson's product-moment correlation. Multiple linear regression analysis was applied to investigate associations between sex, age, SES, stress, SOC, and the outcome of each of SRH, SHC, MWB, and depressive symptoms. The interaction effects including combinations of sex, stress, and SOC were also tested. An assumption for conducting linear regression analysis is to have continuous variables. As stated by Wu and Leung [53], Likert scales are often treated as interval scales when included in regression analyses, when strictly speaking, it is an ordinal scale. Meanwhile, a study by Tacoby [54] also showed that the decisions used in measurement levels depended on the researcher's interpretation of the differences among the observational categories into which the empirical objects are divided. When considering the dependent and independent variables of stress and SOC for use in the present study, the assumption of continuous variables was met as the variables were constructed as sum scores. The SES variables including mother's and father's education level and employment status were originally scaled at the ordinal level. In the analyses, the variables were therefore constructed as summed scores representing parents' education and parents' employment status. In the survey, the values 'I don't know' and 'other' were included in the assessment of SES variables to ensure valid responses from the participants. In the regression analyses, these values were excluded, due to the assumption of including only continuous variables. Model assumptions for linear regression analysis were tested, and no indications of multicollinearity (VIF < 0.10 and tolerance > 0.02, correlations < 0.80) were found. The assumptions of linearity, homoscedasticity, and independent residuals were also met, where the Durbin Watson test were close to 2 for all models and the residuals were normally distributed through an inspection of the scatterplot [52]. The independent variables were included in the following order: (1) sex and age; (2) SES, (3) stress; (4) SOC; (5) sex × stress, and sex × SOC, and SOC × stress. The last step of the four regression models is presented in the results section; statistical significance was set to $p \leq 0.05$.

## 3. Results

*3.1. Mean Scores and Correlations of the Included Scales*

The distribution of sex, age, and socio-economic status (SES) is presented in Table 1. When looking at sex, 580 (47%) were girls and 644 (52.2%) were boys; 9 did not report sex. Mean age was 16.62 years (SD = 1.61 years) for the total sample; for boys it was 16.68 years (SD = 1.60 years), and for girls it was 16.55 years (SD = 1.61 years). Table 2 presents an overview of the sex mean differences on the included scales. Boys scored significantly higher on SOC, MWB, and SRH, whereas girls scored significantly higher on SHC and depressive symptoms, showing weak-to-moderate strong mean differences. The correlation analysis is displayed in Table 3. The main variables of MWB, depressive symptoms, SHC, SRH, stress, and SOC showed moderate-to-strong correlations in expected directions; the strongest correlations were between SOC, depression, and MWB. The SES variables moreover showed weak to moderate strong correlations with the other variables.

**Table 2.** Sex mean differences on stress, sense of coherence, mental wellbeing, symptoms of depression, self-rated health, and subjective health complaints.

| | Mental Wellbeing (n = 728) | Symptoms of Depression (n = 729) | Self-Rated Health (n = 1209) | Subjective Health Complaints (n = 759) | Sense of Coherence (n = 715) | Stress (n = 730) |
|---|---|---|---|---|---|---|
| | *Mean* (SD) | *Mean* (SD) | *Mean* (SD) | *Mean* (SD) | *Mean* (SD) | *Mean* (SD) |
| Girls | 46.41 (9.75) | 35.95 (12.88) | 3.36 (1.27) | 23.75 (7.70) | 57.95 (13.26) | 66.18 (15.59) |
| Boys | 49.82 (9.48) | 28.03 (10.71) | 3.14 (1.44) | 20.61 (8.70) | 63.96 (12.51) | 58.08 (14.15) |
| Total | 48.08 (9.77) | 32.06 (12.45) | 3.25 (1.36) | 22.16 (8.33) | 60.88 (13.20) | 62.29 (15.50) |
| Range | 13–70 | 14–73 | 1–5 | 12–60 | 18–91 | 40–116 |
| *t*-value | −4.76 *** | 9.00 *** | 2.91 ** | 5.34 *** | −6.20 *** | 7.29 *** |
| Cohen's *d* | 0.35 | 0.67 | 0.16 | 0.38 | 0.47 | 0.54 |

Note. * $p \leq 0.05$; ** $p \leq 0.01$; *** $p \leq 0.001$.

**Table 3.** Correlations between the study variables.

| | MWB | D | SHC | SRH | S | SOC | Age | PE | PVS | FE |
|---|---|---|---|---|---|---|---|---|---|---|
| Mental wellbeing (MWB) | - | −0.58 ** | −0.24 ** | 0.41 ** | −0.33 ** | 0.61 ** | −0.09 * | 0.01 | −0.10 ** | 0.24 ** |
| Depression | | - | 0.44 ** | −0.36 ** | 0.60 ** | −0.75 ** | 0.14 ** | −0.08 * | 0.13 ** | −0.22 ** |
| Subjective health symptoms (SHC) | | | - | −0.25 ** | 0.33 ** | −0.39 ** | 0.11 ** | −0.08 * | 0.04 | −0.13 ** |
| Self-rated health (SRH) | | | | - | −0.19 ** | 0.37 ** | −0.24 ** | 0.05 | 0.03 | 0.69 ** |
| Stress | | | | | - | −0.51 ** | 0.15 ** | −0.10 ** | 0.03 | −0.10 ** |
| Sense of Coherence (SOC) | | | | | | - | −0.12 ** | 0.07 | −0.11 ** | 0.25 ** |
| Age | | | | | | | - | −0.25 ** | −0.11 ** | −0.21 ** |
| Parents' education (PE) | | | | | | | | - | 0.13 ** | 0.01 |
| Parents' vocational status (PVS) | | | | | | | | | - | 0.00 |
| Family economy (FE) | | | | | | | | | | - |

Note. * $p \leq 0.05$; ** $p \leq 0.01$.

### 3.2. Regression Analyses for Variables Associated with Mental Wellbeing (MWB) and Depressive Symptoms

Table 4 presents the results of the multiple linear regression analyses investigating the associations between sex, age, SES, stress, SOC, and the dependent variables depressive symptoms and MWB. When looking at the two models, sex was significantly related with depressive symptoms, where girls reported higher scores than boys; no significant sex differences were found on MWB. Age showed a non-significant association with MWB and a weak positive and significant association with depressive symptoms, indicating that adolescents seem to have a stable level of MWB and a weak increase in symptoms of depression across age groups. Of the SES variables, perception of stronger family economy showed a significant positive and weak association with MWB. Parents' employment status also showed a significant and positive association with depressive symptoms. Stress was significantly positively associated with depressive symptoms (22% explained variance), but not with MWB, after being controlled for the other variables. A strong positive relation was found between SOC and MWB (20% explained variance), whereas a significant strong and inverse relation was found between SOC and depressive symptoms (24% explained variance), controlled for the other variables. Significant interaction effects were found between sex × stress on MWB, and of sex × SOC on depressive symptoms, where the associations were strongest for girls. A significant interaction effect was also found between stress × SOC on depression, indicating that the strength of the relation between stress and depressive symptoms depended on the level of SOC. The total explained variance in the two regression models was 41% in the model with MWB and 68% in the model with depressive symptoms.

**Table 4.** Summary of the hierarchical regression analysis for variables associated with mental wellbeing and depressive symptoms.

| | Mental Wellbeing (n = 494) | | | | | Symptoms of Depression (n = 494) | | | | |
|---|---|---|---|---|---|---|---|---|---|---|
| | **B** | *SE B* | β | **95% CI** | **F** | **B** | *SE B* | β | **95% CI** | **F** |
| Constant | 47.88 | 4.37 | | | 34.54 *** | 34.76 | 4.31 | | | 103.45 *** |
| Sex | 0.36 | 0.72 | 0.02 | −1.04–1.77 | | −3.13 | 0.71 | −0.12 *** | −4.52−−1.75 | |
| Age | 0.23 | 0.20 | 0.04 | −0.63–0.17 | | 0.41 | 0.20 | 0.05 * | 0.02–0.81 | |
| Parents' education | 0.05 | 0.22 | 0.01 | −0.37–0.48 | | 0.27 | 0.21 | 0.03 | −0.15–0.69 | |
| Parents' employment status | 0.00 | 0.32 | 0.00 | −0.62–0.63 | | −1.00 | 0.32 | −0.09 ** | −1.62−−0.38 | |
| Family economy | 1.08 | 0.42 | 0.10 ** | 0.26–1.90 | | −0.68 | 0.41 | −0.05 | −1.49–0.13 | |
| Stress | −0.07 | 0.04 | −0.11 | −0.14–0.00 | | 0.15 | 0.04 | 0.19 *** | 0.08–0.22 | |
| SOC | 0.42 | 0.04 | 0.59 *** | 0.34–0.50 | | -0.65 | 0.04 | −0.68 *** | −0.73−−0.57 | |
| Stress × sex | 0.12 | 0.05 | 0.12 * | 0.01–0.22 | | −0.04 | 0.05 | −0.03 | −0.14–0.07 | |
| SOC × sex | −0.04 | 0.06 | −0.03 | −0.16–0.09 | | 0.14 | 0.06 | 0.09 * | 0.02–0.26 | |
| Stress × SOC | 0.00 | 0.00 | 0.02 | −0.00–0.00 | | −0.01 | 0.00 | −0.08 ** | −0.01–0.00 | |

Note. * $p \leq 0.05$; ** $p \leq 0.01$; *** $p \leq 0.001$. Sex: value 0—girls; value 1—boys. Cases deleted listwise. Adjusted $R^2$ = 0.41 for model with mental wellbeing and $R^2$ = 0.68 for model with depression.

## 3.3. Regression Analyses for Variables Associated with Self-Rated Health (SRH) and Subjective Health Complaints (SHC)

When looking at the results from the regression analyses with SRH and SHC as outcome variables (Table 5), sex was significantly associated with SHC, where girls scored higher than boys. Age showed a weak, significant inverse association with SRH, but was not significantly associated with SHC. Adolescents' perception of stronger family economy associated significantly with higher scores on SRH; the other associations including SES were non-significant. Stress was not significantly associated with either SRH or SHC. Stronger SOC was significantly associated with higher levels of SRH and lower levels of SHC. A significant interaction effect was found between stress × SOC on SRH; the other interaction effects were non-significant. The regression models totally explained 21% of the variance in SRH and 22% of the variance in SHC.

**Table 5.** Summary of the hierarchical regression analysis for variables associated with self-rated health and subjective health symptoms.

| | Self-Rated Health (n = 493) | | | | | Subjective Health Complaints (n = 493) | | | | |
|---|---|---|---|---|---|---|---|---|---|---|
| | **B** | *SE B* | β | **95% CI** | **F** | **B** | *SE B* | β | **95% CI** | **F** |
| Constant | 4.57 | 0.47 | | | 10.05 *** | 26.53 | 4.24 | | | 14.00 *** |
| Sex | 0.02 | 0.08 | 0.01 | −0.13–0.17 | | −2.07 | 0.70 | −0.12 ** | −3.44−−0.71 | |
| Age | −0.08 | 0.02 | −0.14 ** | −0.12−−0.03 | | 0.12 | 0.20 | 0.03 | −0.27–0.51 | |
| Parents' education | 0.02 | 0.02 | 0.04 | −0.02–0.07 | | −0.15 | 0.21 | −0.03 | −0.56–0.26 | |
| Parents' vocational status | 0.02 | 0.03 | 0.02 | −0.05–0.08 | | −0.55 | 0.31 | −0.08 | −1.16–0.06 | |
| Family economy | 0.14 | 0.05 | 0.13 ** | 0.05–0.22 | | −0.13 | 0.41 | 0.01 | −0.93–0.67 | |
| Stress | 0.00 | 0.00 | 0.03 | −0.01–0.01 | | 0.05 | 0.03 | 0.09 | −0.02–0.12 | |
| SOC | 0.03 | 0.00 | 0.37 *** | 0.02–0.03 | | −0.19 | 0.04 | −0.31 *** | −0.27−−0.11 | |
| Stress × sex | 0.01 | 0.01 | 0.05 | −0.01–0.02 | | 0.05 | 0.05 | 0.06 | −0.05–0.16 | |
| SOC × sex | −0.00 | 0.01 | −0.02 | −0.02–0.01 | | 0.01 | 0.06 | 0.01 | −0.11–0.13 | |
| Stress × SOC | 0.00 | 0.00 | 0.14 ** | 0.00–0.00 | | −0.00 | 0.00 | −0.05 | −0.01–0.00 | |

Note. * $p \leq 0.05$; ** $p \leq 0.01$; *** $p \leq 0.001$. Sex: value 0—girls; value 1—boys. Cases deleted listwise. Adjusted $R^2$ = 0.21 for model with self-rated health and $R^2$ = 0.22 for model with subjective health complaints.

## 4. Discussion

This study investigated the role of sex, age, SES, stress, and SOC in association with four outcome variables—subjective health complaints (SHC), self-rated health (SRH), mental wellbeing (MWB), and depressive symptoms in Norwegian adolescents.

The sex differences found in SHC were in line with previous findings showing that girls generally report more health complaints than boys [9–12]. Sex differences in depressive symptoms are well-established in the research literature, showing that girls report higher levels of depressive symptoms than boys during the adolescent years [5–8,55]. The focus in discussions has been placed on whether the symptoms represent real changes in mental health or whether especially girls' report of higher levels of depressive symptoms and other mental health problems partly result from gender role differences, and a shift in how symptoms are perceived and reported by informants [6]; however, this might not be regarded as a key explanatory factor. In Norway, the Norwegian public health report states that the causes of the increased report of mental health problems in adolescents are complex and might be explained by a range of psychological, biological, and psychosocial factors in the different situations that adolescents partake in, as well as broader socioeconomic and cultural influences in society [1,2,8]. This points to the fact that the causes of the increased reported rates of mental health problems needs to be further investigated.

The results showed that higher stress level associated significantly with higher levels of depressive symptoms and with lower MWB, especially in girls. The associations between stress and each of SRH and SHC were non-significant. Although exposure to stressful events is a normal part of adolescent life, exposure to multiple independent and cumulative stressors plays a substantial role in the development of mental health problems, where girls seem to be more vulnerable to the negative health effects of stress than boys [4,56]. The perceived importance of the stressor and the individual's evaluations of the ability to cope with the stressor, are fundamental for the impact of the stressor and for the health outcomes of stress. However, one should be aware of possible reciprocal associations; just as stress experience might lead to more mental health problems, it is equally possible that mental health problems can lead to more vulnerability to perceived situations and experiences as stressful, leading to spiraling negative effects.

The findings showed support for SOC as strongly associated with adolescents' perception of depressive symptoms and especially MWB, and weaker associations were found with SHC and SRH. Furthermore, a significant but weak moderating role of sex on the relationship between SOC and depressive symptoms was found, showing that SOC seemed to be a relevant coping resource especially for girls' experience of depressive symptoms. When considering the interaction effects of stress by SOC, the results mainly supported a compensatory role of SOC in relation to MWB and SHC, whereas weak but significant support for a protective/buffering role of SOC was found in relation to depressive symptoms and SRH. The results thus indicated that SOC seemed to be a stronger coping resource for adolescents' mental health, compared with SHC and SRH, despite experience of stressors [24,26,27]. Antonovsky assumed that the individual is constantly exposed to stressors in daily life that might reduce health temporarily, but in the long term, this also has the potential to strengthen the individual and help cope with stress. Through the identification and use of different resistance resources, the individual develops a strong SOC that helps one to mobilize resources to cope with stressors and manage tension successfully, which promotes movement on the positive end of the ease/dis-ease continuum [22–24].

Although no causal conclusions could be drawn, the results provide insight into the importance of stress experience and SOC, especially in association with adolescents' report of mental health, controlled for sex, age, and SES. The findings thus support the importance of strengthening SOC in adolescents, among an array of other possible personal and social coping resources (e.g., self-esteem, self-efficacy, and resilience). Interestingly, the study showed a stronger association between stress and MWB and between SOC and depressive symptoms for girls, which shows that stress and SOC might affect girls' and boys' mental health differently, during adolescence.

Working on promoting adolescents' coping resources is important for strengthening their ability to cope with life stressors and natural ups and downs, which is important for their overall health and wellbeing. This requires cross-sectorial action that should be integrated in central developmental contexts where adolescents and adults meet on a regular basis (e.g., family, school, peers, and neighborhood) [24,34]. Although health is influenced by different areas of the adolescents' lives, school is one important setting. In Norway, the new interdisciplinary theme of "public health and coping" has been implemented in both elementary and secondary school as part of the compulsory curriculum. This strategy presents an opportunity for implementing universal health promoting strategies focusing on coping with normative stressors in daily life and strengthening adolescents' coping resources through socio-emotional learning and promotion of health literacy, which might also contribute to facilitating SOC [22,29].

*Strengths and Limitations*

The strengths of this study were the use of validated instruments, the relatively large sample size, and high response rate. However, the cross-sectional design did not allow us to make conclusions regarding causality and it is possible that the variables might be reciprocally related. A longitudinal design would have been preferable in order to draw conclusions about the relative strength of the variables in predicting health outcomes.

The data were based on self-reports from adolescents and should be evaluated with reference to potential self-reporting bias. Self-reporting requires that adolescents can understand and reflect around aspects related to health and illness (e.g., social desirability and over- and under-reporting). This might especially be relevant for the youngest adolescents, with reference to potential challenges regarding reflections on abstract concepts. The sample size could contribute to protection from the influences of potential bias related to sample selection and self-reports. The study was based on public lower- and upper-secondary schools in rural areas of mid-Norway; the findings might therefore not generalize to schools in urban areas and larger cities, and private schools. Regarding the recruitment of adolescents and administration of questionnaires, the teachers were strongly encouraged by the principal to administer the questionnaire to the students, however, administration was based on the teachers' decision depending on time needed for educational activities. The study did not have any data on students who did not participate in the study or the parents' mental health status, which was also a limitation of the present study.

## 5. Conclusions

The present study showed that girls reported significantly higher levels of depressive symptoms and SHC than boys, after controlling for sex, age, SES, stress, and SOC. Stress associated with significantly higher levels of depressive symptoms, where the association between stress and depression was significantly stronger in girls. The results showed that SOC is a stronger coping resource in association with mental health (especially for girls) than with SHC and SRH. The findings also support a compensatory role of SOC on the association between stress and health during adolescence.

**Author Contributions:** Conceptualization, U.K.M. and G.A.E.; methodology, U.K.M. and G.A.E.; software, U.K.M. and G.A.E.; validation, U.K.M. and G.A.E.; formal analysis, U.K.M. and G.A.E.; investigation, U.K.M. and G.A.E.; resources, U.K.M. and G.A.E.; data curation, U.K.M. and G.A.E.; writing—original draft preparation, U.K.M.; writing—review and editing, G.A.E.; visualization, U.K.M. and G.A.E.; supervision, G.A.E.; project administration, G.A.E., funding acquisition, G.A.E. All authors have read and agreed to the published version of the manuscript.

**Funding:** This research received no specific grant from any funding agency in the public, commercial, or not-for-profit sectors.

**Acknowledgments:** We would like to express our gratitude to the adolescents who participated and generously contributed their time, the teachers for their valuable help administering the questionnaire. Also thank you to Professor Geir Arild Espnes and Jan Erik Ingebrigtsen for leading the data collection.

**Conflicts of Interest:** The authors declare no conflict of interest.

## References

1.  World Health Organization. Growing up unequal: Gender and socioeconomic differences in young people's health and well-being. In *Health Behaviour in School-Aged Children (HBSC) Study 2016*; International Report from the 2013/2014 Survey; Regional office for Europe: Copenhagen, Denmark, 2016.
2.  Patton, G.C.; Sawyer, S.M.; Santelli, J.S.; Ross, D.A.; Afifi, R.; Allen, N.B.; Arora, M.; Azzopardi, P.; Baldwin, W.; Bonell, C.; et al. Our future: Lancet commission on adolescent health and wellbeing. *Lancet* **2016**, *387*, 2423–2478. [CrossRef]
3.  Byrne, D.G.; Davenport, S.C.; Mazanov, J. Profiles of adolescent stress: The development of the adolescent stress questionnaire (ASQ). *J. Adolesc.* **2007**, *30*, 393–416. [CrossRef] [PubMed]
4.  Charbonneau, A.M.; Mezulis, A.H.; Hyde, J.S. Stress and emotional reactivity as explanations for gender differences in adolescents' depressive symptoms. *J. Youth Adolesc.* **2009**, *38*, 1050–1058. [CrossRef] [PubMed]
5.  Collishaw, S. Annual Research Review: Secular trends in child and adolescent mental health. *J. Child Psychol. Psychiatry* **2014**, *56*, 370–393. [CrossRef]
6.  Bor, W.; Dean, A.J.; Najman, J.; Hayatbakhsh, R. Are child and adolescent mental health problems increasing in the 21st century? A systematic review. *Aust. N. Z. J. Psychiatry* **2014**, *48*, 606–616. [CrossRef]
7.  Von Soest, T.; Wichstrøm, L. Secular trends in depressive symptoms among Norwegian adolescents from 1992 to 2010. *J. Abnorm. Child Psychol.* **2014**, *42*, 403–415. [CrossRef]
8.  Reneflot, A.; Aarø, L.E.; Aase, H.; Reichborn-Kjennerud, T.; Tambs, K.; Øverland, S. *Mental Health in Norway*; 2018 Report; Norwegian Institute of Public Health: Oslo, Norway, 2018.
9.  Torsheim, T.; Ravens-Sieberer, U.; Hetland, J.; Välimaa, R.; Danielson, M.; Overpeck, M. Cross-national variation of gender differences in adolescent subjective health in Europe and North America. *Soc. Sci. Med.* **2006**, *62*, 815–827. [CrossRef]
10. Wiklund, M.; Malmgren-Olsson, E.B.; Öhman, A.; Bergström, E.; Fjellman-Wiklund, A. Subjective health complaints in older adolescents are related to perceived stress, anxiety and gender—A cross-sectional school study in Northern Sweden. *BMC Public Health.* **2012**, *12*, 993–1006. [CrossRef]
11. Ravens-Sieberer, U.; Torsheim, T.; Hetland, J.; Volleberg, W.; Cavallo, F.; Jericek, H.; Alikasifoglu, M.; Välimaa, R.; Ottova, V.; Erhart, M. Subjective health, symptom load and quality of life of children and adolescents in Europe. *Int. J. Public Health* **2009**, *54*, 151–159. [CrossRef]
12. Gobina, I.; Välimaa, R.; Tynjälä, J.; Villberg, J.; Villerusa, A.; Iannotti, R.J.; Godeau, E.; Gabhainn, G.E.; Andersen, A.; Holstein, B.E. The medicine use and corresponding subjective health complaints among adolescents, a cross-national survey. *Pharmacoepidemiol. Drug Saf.* **2011**, *20*, 424–431. [CrossRef]
13. Breidablik, H.J.; Meland, E.; Lydersen, S. Self-rated health in adolescence: A multifactorial composite. *Scand. J. Public Health* **2008**, *36*, 12–20. [CrossRef] [PubMed]
14. Breidablik, H.J.; Meland, E.; Lydersen, S. Self-rated health during adolescence: Stability and predictors of change (Young-HUNT study, Norway). *Eur. J. Public Health* **2008**, *19*, 73–78. [CrossRef] [PubMed]
15. Boardman, J.D. Self-rated health among U.S. adolescents. *J. Adolesc. Health* **2006**, *38*, 401–408. [CrossRef] [PubMed]
16. Fosse, N.E.; Haas, S.A. Validity and stability of self-reported health among adolescents in a longitudinal, nationally representative survey. *Pediatrics* **2009**, *123*, 496–501. [CrossRef] [PubMed]
17. Moksnes, U.K.; Rannestad, T.; Byrne, D.G.; Espnes, G.A. The association between stress, sense of coherence and subjective health complaints in adolescents: Sense of coherence as a potential moderator. *Stress Health.* **2010**, *27*, e157–e165. [CrossRef]
18. Jerden, L.; Burell, G.; Stenlund, H.; Weinehall, L.; Bergström, E. Gender differences and predictors of self-rated health development among Swedish adolescents. *J. Adolesc. Health.* **2011**, *48*, 143–150. [CrossRef]
19. Cavallo, F.; Dalmasso, P.; Jordan, V.O.; Brooks, F.; Mazur, J.; Välimaa, R.; Gobina, I. Trends in self-rated health in European and North-American adolescents from 2002 to 2010 in 32 countries. *Eur. J. Public Health* **2015**, *25*, 13–15. [CrossRef]
20. Sundblad, G.B.; Jansson, A.; Saartok, T.; Renström, P.; Engström, L.M. Self-rated pain and perceived health in relation to stress and physical activity among school-students: A 3-year follow-up. *Pain* **2008**, *136*, 239–249. [CrossRef]

21. Torsheim, T.; Aaroe, L.E.; Wold, B. Sense of coherence and school-related stress as predictors of subjective health complaints in early adolescence: Interactive, indirect or direct relationships? *Soc. Sci. Med.* **2001**, *53*, 603–614. [CrossRef]

22. Braun-Lewensohn, O.; Idan, O.; Lindström, B.; Margalit, M. Salutogenesis: Sense of coherence in adolescence. In *The Handbook of Salutogenesis*; Mittelmark, M., Sagy, S., Eriksson, M., Bauer, G.F., Pelikan, J.F., Lindström, B., Espnes, G.A., Eds.; Springer Nature: Basel, Switzerland, 2016.

23. Eriksson, M. The sense of coherence in the salutogenic model of health. In *The Handbook of Salutogenesis*; Mittelmark, M., Sagy, S., Eriksson, M., Bauer, G.F., Pelikan, J.F., Lindström, B., Espnes, G.A., Eds.; Springer Nature: Basel, Switzerland, 2016; pp. 91–96.

24. Eriksson, M.; Lindström, B. Antonovsky's sense of coherence scale and the relation with health: A systematic review. *J. Epidemiol. Community Health* **2006**, *60*, 376–381. [CrossRef]

25. Moksnes, U.K.; Løhre, A.; Espnes, G.A. The association between sense of coherence and life satisfaction in adolescents. *Qual. Life Res.* **2013**, *22*, 1331–1338. [CrossRef] [PubMed]

26. Moksnes, U.K.; Espnes, G.A.; Haugan, G. Stress, sense of coherence and emotional symptoms in adolescents. *Psychol. Health* **2014**, *29*, 32–49. [CrossRef] [PubMed]

27. Flensborg-Madsen, T.; Ventegodt, S.; Merrick, J. Sense of coherence and physical health. A review of previous findings. *Sci. World J.* **2005**, *25*, 665–673. [CrossRef] [PubMed]

28. Apers, S.; Luyckx, K.; Rassart, J.; Goossens, E.; Budts, W.; Moons, P. Sense of coherence is a predictor of perceived health in adolescents with congenital heart disease: A cross-lagged prospective study. *Int. J. Nurs. Stud.* **2013**, *50*, 776–785. [CrossRef]

29. Garcia-Moya, I.; Rivera, F.; Moreno, C. School context and health in adolescence: The role of sense of coherence. *Scan. J. Psychol.* **2013**, *54*, 243–249. [CrossRef]

30. Honkinen, P.L.; Suominen, S.B.; Välimaa, R.S.; Helenius, H.Y.; Rautava, P.T. Factors associated with perceived health among 12-year-old school-children. Relevance of physical exercise and sense of coherence. *Scand. J. Public Health* **2005**, *33*, 35–41. [CrossRef]

31. Modin, B.; Ostberg, V.; Toivanen, S.; Sundell, K. Psychosocial working conditions, school sense of coherence and subjective health complaints. A multilevel analysis of ninth grade pupils in the Stockholm area. *J. Adolesc.* **2011**, *34*, 129–139. [CrossRef]

32. Simonsson, B.; Nilsson, K.W.; Leppert, J.; Diwan, V.K. Psychosomatic complaints and sense of coherence among adolescents in a county in Sweden: A cross-sectional school survey. *Biopsychosoc. Med.* **2008**, *2*, 1–8. [CrossRef]

33. Ristakari, T.; Sourander, S.; Rønning, J.A.; Nikolakaros, G.; Helenius, H. Life events, self-reported psychopathology and sense of coherence among young men—A population based study. *Nordic. J. Psychiatry* **2008**, *62*, 464–471. [CrossRef]

34. Nielsen, A.M.; Hansson, K. Associations between adolescents' health, stress and sense of coherence. *Stress Health* **2007**, *23*, 331–341. [CrossRef]

35. Richardson, C.G.; Ratner, P. Sense of coherence as a moderator of the effects of stressful life events on health. *J. Epidemiol. Community Health* **2005**, *59*, 979–984. [CrossRef] [PubMed]

36. Moksnes, U.K.; Espnes, G.A. Stress, sense of coherence and subjective health in adolescents aged 13–18 years. *Scand. J. Public Health* **2017**, *45*, 397–403. [CrossRef] [PubMed]

37. Moksnes, U.K.; Haugan, G. Stressor experience negatively affects life satisfaction in adolescents the positive role of sense of coherence. *Qual. Life Res.* **2015**, *24*, 2473–2481. [CrossRef] [PubMed]

38. Moksnes, U.K.; Haugan, G. Validation of the Orientation to Life Questionnaire in Norwegian adolescents, construct validity across samples. *Soc. Ind. Res.* **2013**, *119*, 1105–1120. [CrossRef]

39. Eriksson, M.; Mittelmark, M.B. The Sense of Coherence and Its Measurement. In *The Handbook of Salutogenesis*; Mittelmark, M., Sagy, S., Eriksson, M., Bauer, G.F., Pelikan, J.F., Lindström, B., Espnes, G.A., Eds.; Springer Nature: Basel, Switzerland, 2016; pp. 97–106.

40. Moksnes, U.K.; Espnes, G.A. Evaluation of the Norwegian version of the Adolescent Stress Questionnaire (ASQ-N): Factorial validity across samples. *Scand. J. Psychol.* **2011**, *52*, 601–608. [CrossRef]

41. De Vriendt, T.; Clays, E.; Moreno, L.; Bergman, P.; Vicente-Rodriguez, G.; Nagy, E.; Dietrich, S.; Manios, Y.; De Henauw, S.; HELENA Study Group. Reliability and validity of the Adolescent Stress Questionnaire in a sample of European adolescents—The HELENA study. *BMC Public Health.* **2011**, *11*, 717. [CrossRef]

42. Darviri, C.; Legaki, P.E.; Chatzioannidou, P.; Gnardellis, C.; Kraniotou, C.; Tigani, X.; Alexopoulos, E.C. Adolescent Stress Questionnaire: Reliability and validity of the Greek version and its description in a sample of high school (lyceum) students. *J. Adolesc.* **2014**, *37*, 1373–1377. [CrossRef]

43. McKay, M.T.; Percy, A.; Byrne, D.G. Support for the multidimensional adolescent stress questionnaire in a sample of adolescents in the United Kingdom. *Stress Health* **2014**, *32*, 12–19. [CrossRef]

44. Tennant, R.; Hiller, L.; Fishwick, R.; Platt, S.; Joseph, S.; Weich, S.; Parkinson, J.; Secker, J.; Stewart-Brown, S. The Warwick-Edinburgh Mental Well-Being Scale (WEMWBS): Development and UK validation. *Health Qual. Life Outcomes* **2007**, *5*, 63. [CrossRef]

45. Stewart-Brown, S.; Tennant, A.; Tennant, R.; Platt, S.; Parkonson, J.; Weich, S. Internal construct validity of the Warwick-Edinburgh Mental Well-being Scale (WEMWBS): A Rasch analysis using data from the Scottish Health Education Population Survey. *Health Qual. Life Outcomes* **2009**, *7*, 15. [CrossRef]

46. Smith, O.R.F.; Alves, D.E.; Knapstad, M.; Haug, E.; Aarø, L.E. Measuring mental well-being in Norway: Validation of the Warwick-Edinburgh Mental Well-being Scale (WEMWBS). *BMC Psychiatry* **2017**, *17*, 182. [CrossRef] [PubMed]

47. Clarke, A.; Friede, T.; Putz, R.; Ashdown, J.; Martin, S.; Blake, A.; Adi, Y.; Parkinson, J.; Flynn, P.; Platt, S.; et al. Warwick-Edinburgh Mental Well-being Scale (WEMWBS): Validated for teenage school students in England and Scotland. A mixed methods assessment. *BMC Public Health* **2011**, *11*, 487. [CrossRef] [PubMed]

48. McKay, M.T.; Andretta, J.R. Evidence for the Psychometric Validity, Internal Consistency and Measurement Invariance of Warwick Edinburgh Mental Well-being Scale Scores in Scottish and Irish Adolescents. *Psychiatry Res.* **2017**, *255*, 382–386. [CrossRef] [PubMed]

49. Ringdal, R.; Eilertsen, M.E.B.; Bjørnsen, H.N.; Espnes, G.A.; Moksnes, U.K. Validation of two versions of the Warwick-Edinburgh Mental Well-Being Scale among Norwegian adolescents. *Scand. J. Public Health* **2018**, *46*, 718–725. [CrossRef]

50. American Psychiatric Association. *Diagnostic and Statistical Manual of Mental Disorders DSM-IV-TR*, 4th ed.; American Psychiatric Association: Washington, DC, USA, 2000.

51. Zung, W.W.K. A self-rating depression scale. *Arch. Gen. Psychiatry* **1965**, *12*, 63–70. [CrossRef]

52. Field, A. *Discovering Statistics Using SPSS*, 2nd ed.; Sage Publications: London, UK, 1988.

53. Wu, U.; Leung, S.O. Can likert scales be treated as interval scales?—A simulation study. *J. Soc. Serv. Res.* **2015**, *56*, 203–396. [CrossRef]

54. Tacoby, W.G. Levels of measurement and political research: An optimistic view. *Midw. Polit. Sci. Ass.* **1999**, *43*, 271–301.

55. Thapar, A.; Collishaw, S.; Pine, D.; Thapar, A.K. Depression in adolescence. *Lancet* **2012**, *379*, 1056–1067. [CrossRef]

56. Grant, E.K.; McMahon, S.D.; Duffy, S.N.; Taylor, J.J.; Comas, B.E. Stressors and mental health problems in childhood and adolescence. In *The Handbook of Stress Science: Biology, Psychology, and Health*; Contrada, R., Baum, A., Eds.; Springer Publisher Company: New York, NY, USA, 2010; pp. 359–372.

International Journal of
*Environmental Research
and Public Health*

*Article*

# Acculturative Stress among Arab Students in Israel: The Roles of Sense of Coherence and Coping Strategies

**Sarah Abu-Kaf * and Enas Khalaf**

Conflict Management and Resolution Program, Department of Multidisciplinary Studies, Ben-Gurion University of the Negev, Beer-Sheva 8410501, Israel; enas.kh88@gmail.com
* Correspondence: aks@bgu.ac.il; Tel.: +972-50-473-9984

Received: 18 May 2020; Accepted: 10 July 2020; Published: 15 July 2020

**Abstract:** Background: In Israeli colleges and universities, many Arab students experience acculturative stress. Such stress arises from the need to learn new cultural rules, manage the overarching conflict inherent in maintaining elements of their culture of origin (i.e., Arab culture) while incorporating elements of the host culture (i.e., Jewish academic culture), and deal with experiences of prejudice and discrimination present in the host culture. Methods: This study investigated the association between acculturative stress and depressive symptoms among 170 Arab undergraduates from northern and central Israel. It also explored the roles of sense of coherence and coping strategies in the relationship between acculturative stress and depressive symptoms. Participants completed questionnaires on acculturative stress, depressive symptoms, sense of coherence, coping strategies, and demographics. Results: The findings reveal gender differences in the use of different coping strategies and in levels of depressive symptoms. However, academic-year differences were found only in levels of sense of coherence and depressive symptoms. Specifically, female students expressed higher levels of both active and avoidant coping. Moreover, female students and those in their first and second years of university studies reported higher levels of depressive symptoms. Among the male students, acculturative stress was related to depressive symptoms indirectly via sense of coherence and active coping. In contrast, among the female students, acculturative stress was related to depressive symptoms both directly and indirectly via sense of coherence and avoidant coping. Among first- and second-year students, acculturative stress was related to depressive symptoms indirectly via sense of coherence and avoidant coping. However, among third- and fourth-year students, acculturative stress was related to depressive symptoms both directly and indirectly via sense of coherence. Conclusions: This article underscores the significance of gender and academic-year differences in pathways involving acculturative stress.

**Keywords:** acculturative stress; depression; students; Arab; coping strategies; sense of coherence

---

## 1. Introduction

### 1.1. Acculturative Stress

Redfield, Linton, and Herskovits [1] defined acculturation as a "phenomenon which results when groups of individuals having different cultures come into continuous first-hand contact with subsequent changes in the original cultural patterns of each other or both groups" (p. 149). The acculturation process can be positive, improving one's life chances in the new culture. However, it also imposes stress on the individual due to the challenging nature of change and adaptation to new cultural and social expectations. Acculturative stress is usually experienced by those who are in the process of acculturating to the dominant society by adapting the dominant culture's language and norms [2].

Berry and colleagues [2] described acculturative stress as "a reduction in health status (including psychological, somatic, and social aspects) of individuals who are undergoing acculturation, and for which there is evidence that these health phenomena are related systematically to acculturation phenomena" (p. 491). Drawing on broader stress and adaptation theory (e.g., [3]), Berry [4] claimed that acculturative stress is a stress reaction to life events that are rooted in the experience of acculturation. Individuals experience change events that challenge their cultural understandings about how to live. Such stress arises from multiple aspects of the acculturation process, such as learning new and sometimes confusing cultural rules and expectations, dealing with experiences of prejudice and discrimination, and managing the overarching conflict inherent in maintaining elements of the old culture while incorporating elements of the new [4–6]. The aspects of acculturative stress that are salient to college students may relate to language proficiency, unfamiliarity with prevailing cultural practices, cultural self-consciousness, the experience of conflicting value systems, and experiences of discrimination (e.g., [7–9]). In previous research, acculturative stress has repeatedly been found to be associated with mental-health problems such as anxiety and depression, feelings of alienation, identity confusion, and heightened levels of psychosomatic symptoms [2,10–12]. Among international college students and college students from cultural-ethnic minorities, acculturative stress has been associated with a number of psychological challenges, including depression [13,14].

### 1.2. Protective Factors: Sense of Coherence and Coping Strategies

General models of stress posit that perceiving a situation as being threatening or beyond one's coping abilities causes stress and leads to negative affect [3,15]. Resilience studies (e.g., [16]) have highlighted the importance of identifying protective factors that reduce the negative effects of stressful events and encourage positive outcomes. Sense of coherence and coping strategies have been identified as the main stress-buffering variables.

### 1.2.1. Sense of Coherence

Sense of coherence (SOC) refers to a permanent attitude according to which individuals view and understand life and is a measure of the capacity to assess and use available resistance resources to maintain and improve health when faced with stressful situations [17]. In other words, SOC can be defined as a way of viewing life and the ability to manage the stressors that are faced in life [18].

According to Antonovsky [17], SOC is an important resource that enables people to manage stress, to evaluate their external and internal resources, and to identify and use those resources, in order to promote effective coping and adjustment. SOC explains why individuals experiencing stressful or challenging events in their lives are capable of dealing with them [17]. This sense develops during childhood and early adulthood and stabilizes around the age of 30 (in the period of early adulthood [19]). However, Eriksson [20] claimed that SOC tends to increase with age over one's whole lifespan. Researchers have argued that SOC is a construct that develops differently, according to environmental characteristics and life experiences [21]. SOC integrates three components: comprehensibility, manageability, and meaningfulness. Comprehensibility refers to the individual's ability to perceive life events as comprehensible and consistent, and to reasonably predict what will happen in the future. Manageability refers to the ability of the individual to understand that the resources at one's disposal are sufficient to cope with life's difficulties. Meaningfulness is the extent to which an individual feels that life makes sense emotionally [19]. Meaningfulness motivates an individual to seek resolutions to events or situations that are considered stressful [22]. Antonovsky claimed that an ability to define stressors as irrelevant, neutral, or even as a challenge indicates that a person has a strong SOC, while considering a stressor as endangering one's well-being is indicative of low SOC [19]. Previous literature has stated that SOC has important positive effects on reactions to stress, as well as problem-solving and emotional coping in general, particularly among individuals who are members of ethnic and cultural minorities [23–26]. Previous research has found that an individual with a strong SOC is more likely to feel less stress and to have more social support that can be called upon in his or her efforts to cope with stress [19,27]. Previous cross-sectional studies have reported a significant inverse correlation between SOC and depression [28]. Researchers have also found that a strong SOC is associated with fewer depressive symptoms [29]. According to a review conducted by Eriksson and Lindström [30], SOC is strongly, negatively related to perceived depression; stronger SOC is associated with fewer symptoms of perceived depression.

Previous research on SOC in Eastern-collectivistic contexts such as the Arab minority in Israel has revealed two important findings. First, researchers found lower levels of SOC among the Arab minority [23,31]. Second, over time, changes in SOC levels were observed (higher levels) and SOC became a strong predictor of stress reactions [24]. However, this important resource has not been previously examined among the Arab student population and it is important to evaluate the levels and implications of SOC on depressive symptoms among Arab students.

### 1.2.2. Coping Strategies

Coping processes are complex responses that occur when an individual attempts to remove a source of stress or a perceived threat from his or her environment. The reaction to an event has been found to be as important as the event itself [32]. Coping strategies include cognitive or behavioral efforts to manage situations appraised as taxing or exceeding a person's resources [3]. A common characteristic of many coping taxonomies has been the distinction between strategies that are active and oriented toward confronting the problem (i.e., active coping strategies) and strategies that entail an effort to reduce tension by avoiding dealing with the problem (i.e., avoidant coping strategies [33]). Research on the effects of coping strategies on adjustment has found that active coping strategies are more effective and that they moderate the adverse influence of negative life events on psychological functioning [34,35]. In contrast, avoidant coping strategies tend to be associated with psychological distress [33,35–37].

Berry [4] noted that, when acculturative stress is not managed well, it will increase and its effect will be even more negative. In addition, if such stressors become overwhelming, the immediate effects can be significantly negative and damaging, even to the point of personal crises, anxiety, and depression. When acculturative problems (stressors) arise, but are successfully managed, stress levels are similarly low and the immediate effects are positive [38].

### 1.2.3. Coping Strategies in a Collectivistic Cultural Context

Sociocultural groups appear to generate not only consensual belief systems concerning the origin and meaning of stressors, but also beliefs concerning the most appropriate means to cope with stressors. Empirical investigations of coping strategies across cultures have yielded mixed findings. However, overall, there is significant support for the idea that individuals from collectivistic cultural contexts are more likely to use avoidant coping strategies [39,40], whereas individuals from individualistic cultural contexts are more likely to use active and problem-focused coping strategies [40,41]. Among Arab students in Israel (specifically Bedouin Arabs), levels of active coping strategies are similar to those found in Israeli Jewish society [42]. However, Arab students have also reported using more avoidant coping strategies than Jewish students [42]. This study provides support for the important role of avoidant coping strategies. Bedouin Arab students tend to be more depressed because they tend to use avoidant coping strategies more often. That is, they attempt to reduce tension by avoiding dealing with problems (i.e., behavioral disengagement, self-distraction, denial, and self blame [42]).

### 1.3. Gender Differences in Acculturative Stress, SOC, Coping Strategies, and Depression

Gender has a variety of effects on the acculturation process. There is substantial evidence that women may be at greater risk for problems in the acculturative process than men (e.g., [43,44]). Female immigrants reported higher levels of acculturative stress than men across multiple domains including homesickness, social isolation, employment barriers, discrimination, and civic disengagement [44]. Attempts by women to take on the new roles available in the "new" society may bring them into conflict with their heritage culture, placing them at risk for acculturative stress and negative outcomes [43,45]. In addition, the majority of previous studies have reported higher SOC scores among men [20]. Gender differences have also been found in relation to coping strategies. In general, women tend to exploit social support, affective release, emotional regulation, and emotionally focused and "tend-and-befriend" strategies (which may be considered a form of active coping( [46,47]. In contrast, the coping efforts of men are directed toward "fight-or-flight" responses (i.e., gaining control over the situation and invoking disengagement responses) [47,48]. Gender differences have also been found in levels of depression. For example, women are twice as likely as men to have higher scores on self-reported depression symptom measures [49]. However, previous research among students from a subgroup belonging to the Arab minority in Israel (i.e., Bedouin Arabs) revealed no gender differences in mean levels of depressive symptoms [42,50]. It would be interesting to examine the generalizability of these findings among students from the larger Arab minority.

*1.4. Academic-Year Differences in Acculturative Stress, SOC, Coping Strategies and Depression*

Previous research suggests that students in the earlier years of their college educations are at higher risk for experiencing psychological distress and depression than students in the later years of their degree programs [51]. One possible explanation is that the period of transition from high school to college is stressful and this stress may be related to elevated rates of depression among these students [51]. Research on the effects of academic-year differences on acculturative stress among Chinese nursing students in Australia found higher levels of acculturative stress in the third year as compared to the second year. However, that study reported no significant differences in acculturative stress between first-year students and second-year students [52]. He et al. [52] compared the SOC levels among first-, second-, and third-year students and found no significant differences between the three groups. It is important to note that most the studies have focused on a single academic year [53] or have not reported comparisons of different years of study [54]. It is important to test whether students in the earlier years of college express different levels of acculturative stress and depressive symptoms. In addition, special attention should be paid to their coping resources and strategies, as compared to the resources and strategies found among students who are further along in their studies.

*1.5. Acculturative Stress Among Arab Students in Israel*

The Arab minority in Israel comprises about 21% of the entire population [55]. During the last decade (2008–2018), there was an 80% increase in the number of Arab students in academic institutions of higher learning in Israel [56]. Arab culture differs significantly from Jewish Israeli culture in terms of its emphasis on collectivistic ideals [57]. Jewish culture, being more individualistic and less authoritarian, emphasizes separation, independence, personal development, and achievement [58]. In addition, Arabs also differ from the Jewish majority in terms of language, religion, and other cultural factors [59]. This large cultural distance between the Arab minority and the Jewish majority is expected to increase the acculturative stress experienced by Arab individuals. Discrimination, prejudice, and negative stereotypes and attitudes of the host culture toward the minority group also increase acculturative stress [60]. Arabs in Israel are a largely underprivileged minority with a history of disadvantage in income, education, and employment [61]. They live in segregated residential areas [61]. Despite enjoying full citizenship status, the Arab minority is subject to various forms of discrimination that may contribute to social and economic disparities between them and the Jewish majority [62,63]. These experiences of discrimination are expected to contribute to acculturative stress among Arab students. These students doubt the readiness of the majority to welcome them and tend to feel that they are discriminated against [63]. Members of Arab society, who share more traditional and collectivist values [64], have to adjust to unfamiliar values and codes of behavior and fit in with the majority of students in a more Western-individualistic cultural milieu, which differs substantially from their native culture. To the best of our knowledge, there has been no previous research on acculturative stress within Arab society or specifically among students attending higher academic institutions within this society. Thus, the current study will address acculturative stress and its associations with depressive symptoms among female and male Arab students at different stages of their academic studies.

### 1.6. The Current Study

The main goals of the current study were to explore the associations between acculturative stress, SOC, coping strategies, and depressive symptoms among Arab students from northern and central Israel. Special attention was paid to the roles of those variables in the associations between acculturative stress and depressive symptoms. The current study also investigated gender differences in the levels and roles of SOC and the use of different coping strategies within the Arab minority in Israel. The effects of academic year (Year 1 + 2 vs. Year 3 + 4) on the levels of the study variables and the roles of SOC, active coping, and avoidant coping in the association between acculturative stress and depressive symptoms were also examined.

### 1.7. Hypotheses

The following hypotheses were tested:

1. There are gender differences in acculturative stress, SOC, coping strategies, and depressive symptoms [20,42–44,50,65,66].
2. There are academic-year differences in acculturative stress and depressive symptoms [51,52].
3. There is a positive association between acculturative stress and depressive symptoms [13,14].
4. SOC and coping strategies mediate the relationships between acculturative stress and depressive symptoms [42,67,68]. There are gender and academic-year differences in the pathways between acculturative stress and depressive symptoms (Figure 1).

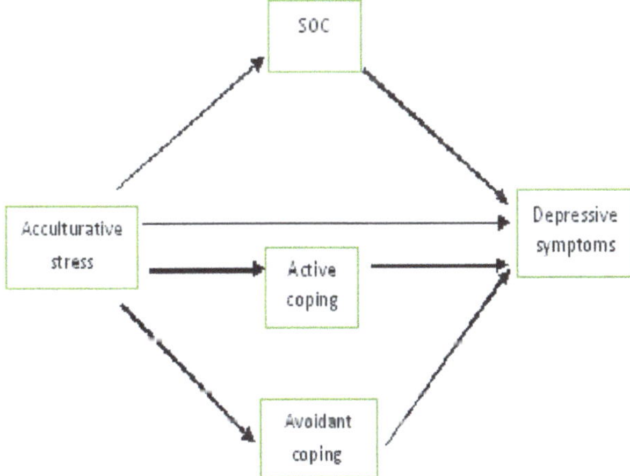


**Figure 1.** The hypothesized relationships between acculturative stress, avoidant coping, active coping, SOC, and depressive symptoms. SOC: Sense of coherence.

## 2. Materials and Methods

### 2.1. Participants and Procedure

We employed a cross-sectional research design. One hundred seventy individuals participated in the study, 103 female Arab students and 67 male Arab students from northern and central Israel who were studying at institutions of higher education (i.e., Ben-Gurion University of the Negev, the Hebrew University of Jerusalem, Tel Aviv University, and Haifa University).

The study was approved by the Department's Human Subjects Ethics Committee (Approved Ethics Form No. 0015-009). Participants were approached and included in the study using convenience sampling. All participants were from the northern and central regions of Israel and currently enrolled in an institution of higher learning. Students were recruited with the cooperation of close friends of the second author of this study who study in Israeli universities and by circulating the questionnaire through the social media. Since we used social-media platforms (mainly Facebook), we could not determine the response rate. The process of data collection took three weeks. Arab students were encouraged to complete the questionnaires via a link. The first page of the linked document included an informed-consent form and only students who checked a box to indicate their agreement to participate were directed to complete the full set of questionnaires. The purpose of the research study was also presented to the participants on the first page of that document. The participants were explicitly requested to refrain from providing identifying information. After the participants finished answering the questionnaires, they were presented with a final page that included a full description of the purpose of the study, contact information for the researchers, and a list of references related to the research topic. Arabic versions of the questionnaires were used. Each participant was reimbursed with a coupon for coffee and cake.

*2.2. Measures*

2.2.1. Acculturative Stress—Societal, Attitudinal, Familial, and Environmental—Revised–Short Form (SAFE-Short)

This 24-item measure is designed to assess negative stressors experienced by minority individuals. It captures both stress experienced within one's own group and stress experienced when engaging with the mainstream culture [7]. The items are statements that describe situations that may cause stress (e.g., "It bothers me that I cannot be with my family." and "My family members and I have different expectations about my future.") The items are rated on a Likert scale of 1–5 (1 = not stressful; 2 = somewhat stressful; 3 = stressful; 4 = very stressful; 5 = extremely stressful). The mean score of the scale ranged from 1 to 5. This scale has been found to be reliable across different ethnic and cultural groups [69]. In the current study, the internal consistency coefficient of the scale was 0.88.

2.2.2. Sense of Coherence (SOC) Questionnaire

This questionnaire consists of 13 items that measure a respondent's perception of life as comprehensible, manageable, and meaningful [18]. The items were rated using a 7-point Likert scale that had an anchoring phrase at each end. High scores indicated a strong SOC. The scale included such items as "Doing the things you do every day is" with answers ranging from 1 (a source of pain and boredom) to 7 (a source of deep pleasure and satisfaction). The mean score of the scale ranged from 1 to 7. In this study, we used the Arabic version of the SOC questionnaire. This version has been used in several previous research projects and has been found to be reliable and suitable for use among the Bedouin Arab population [23,70,71]. In the current study, the internal consistency reliability for the scale was 0.80.

2.2.3. Active and Avoidant Coping—The Coping Orientations to Problems Experienced Inventory (COPE)

The COPE-Short Form is a 28-item questionnaire used to assess different dimensions of active or avoidant coping strategies [72]. Participants rate each coping statement in terms of how frequently they use each strategy to manage stressful events, on a scale of 1 (never) to 5 (always). The subscales were aggregated to form two composite scales: active coping (14 items related to planning, instrumental support, emotional support, positive reframing, problem-solving, and humor) and avoidant coping (14 items related to self-blame, behavioral disengagement, self-distraction, substance use, and denial). Coping statements include items such as "I spent more time alone"; "I blamed myself"; "I tried to forget the whole thing"; "I've been getting emotional support from others"; and "I've been getting comfort from someone." The mean scores of the active-coping and avoidant-coping strategy subscales ranged from 1 to 5. In the current study, the Arabic version of the instrument [73] was used. The internal consistency coefficients of the active and avoidant coping scales of the instrument were 0.83 and 0.71, respectively.

### 2.2.4. Depressive Symptoms—The Center for Epidemiological Studies Depression Scale (CES-D)

The CES-D Scale is a 20-item inventory of symptoms of depression [74]. Respondents report how frequently symptoms have been experienced during the past month, using a 4-point Likert scale that ranges from 0 (rarely or none of the time; less than once a day) to 3 (most or all of the time; 5–7 days a week). Items include: "I was bothered by things that usually don't bother me" and "I felt depressed." The total score of the scale ranged from 0 to 60. Score above the stricter diagnostic cut-off point of 23 indicate severe levels of depressive symptoms [75]. In our previous studies, the Cronbach's internal consistency alpha coefficients for the Arabic version were around 0.90 [50]. In the current study, the internal consistency coefficient of the scale was 0.93.

### 2.2.5. Demographics

Participants were asked to report their gender, age, marital status, institutional affiliation and academic year, parents' levels of education, and household income. The variable academic year was recoded to new variable; students in their first and second academic year were included in the group Year 1 + 2 and students in their third and fourth academic year were included in the group Year 3 + 4.

### *2.3. Data Analysis*

The collected data were analyzed using SPSS (IBM SPSS Statistics 26.0, Chicago, IL, USA) and structural equation modeling (SEM), which was carried out using SPSS AMOS 26 software [76]. Three sets of analyses were conducted. First, to test the gender and academic-year differences, we conducted two independent-sample *t*-tests with five dependent variables: acculturative stress, SOC, active coping, avoidant coping, and depressive symptoms. The second hypothesis was tested using Pearson correlations, which involved acculturative stress, depression, SOC, and coping strategies. We used SEM to test the direct and direct effects of acculturative stress on depressive symptoms via SOC and coping strategies among female and male Arab students, as well as among Year 1 + 2 and Year 3 + 4 Arab students.

## 3. Results

### *3.1. The Study Population*

One hundred seventy individuals participated in the study, 103 female Arab students and 67 male Arab students. The participants' had a mean age of 21.88 years (*SD* = 2.54). The majority of the students were single and the children of parents who had 12 or fewer years of education (82.4%). A complete description of the demographic characteristics of the study population is presented in Table 1.

**Table 1.** Frequencies and percentages of demographic variables.

| Demographic Variable | Frequency | Percentage (%) |
|---|---|---|
| Gender | | |
| Women | 103 | 69.5 |
| Men | 67 | 30.5 |
| Marital status | | |
| Single | 140 | 82.4 |
| Engaged | 15 | 8.8 |
| Married | 14 | 8.2 |
| Divorced | 1 | 0.6 |
| Widow/er | | |
| Academic year | | |
| First | 48 | 28.2 |
| Second | 50 | 29.4 |
| Third | 44 | 25.9 |
| Fourth | 28 | 16.5 |
| Household income | | |
| Much more than average | 6 | 3.5 |
| More than average | 38 | 22.4 |
| Similar to average | 55 | 32.4 |
| Less than average | 39 | 22.9 |
| Much less than average | 32 | 18.8 |

### 3.2. Gender Differences in Levels of Acculturative Stress, SOC, Coping Strategies, and Depression

Our first hypothesis was that gender differences would be related to levels of acculturative stress, coping strategies, and depression. To test that hypothesis, we conducted independent-sample *t*-tests with five dependent variables (i.e., acculturative stress, SOC, active coping, avoidant coping, and depressive symptoms). This analysis revealed significant gender differences in the use of active and avoidant coping strategies, as well as depressive symptoms. In addition, non-significant gender differences were found in the levels of acculturative stress and SOC.

As shown in Table 2, female Arab students reported higher levels of avoidant coping, active coping, and depressive symptoms, as compared to male students. Concerning depressive symptoms, forty-five (44%) of the female participants had mean CES-D scores that exceeded the stricter diagnostic cut-off point of 23. In contrast, 22 (33%) of the male participants had CES-D scores above the stricter diagnostic cut-off point.

**Table 2.** Gender differences in acculturative stress, SOC, active coping, avoidant coping, and depressive symptoms.

| Variable | Male<br>$n = 67$ | | Female<br>$n = 103$ | | *t*-Value | Hedges' *g* |
|---|---|---|---|---|---|---|
| | *M* | *SD* | *M* | *SD* | | |
| Acculturative stress (0–5) | 1.80 | 0.86 | 1.84 | 0.78 | −0.27 | 0.04 |
| SOC (1–7) | 4.28 | 0.90 | 4.08 | 0.97 | 1.35 | 0.21 |
| Coping strategies | | | | | | |
| Active coping (1–5) | 2.40 | 0.42 | 2.60 | 0.47 | −2.88 * | 0.45 |
| Avoidant coping (1–5) | 2.0 | 0.41 | 2.20 | 0.45 | −3.08 ** | 0.48 |
| Depressive symptoms (0–60) | 18.64 | 10.86 | 23.47 | 13.41 | −2.29 * | 0.31 |

* $p < 0.05$; ** $p < 0.01$; SOC, sense of coherence.

### 3.3. Academic-Year Differences in Levels of Acculturative Stress, SOC, Coping Strategies, and Depression

Our second hypothesis was that academic year differences would be related to levels of acculturative stress and depression. To test that hypothesis, we conducted independent-sample *t*-tests with five dependent variables (i.e., acculturative stress, SOC, active coping, avoidant coping, and depressive symptoms). This analysis revealed significant academic year differences only in depressive symptoms. In addition, non-significant gender differences were found in the levels of acculturative stress. As shown in Table 3, Year 1 + 2 Arab students reported higher levels of depressive symptoms as compared to Year 3 + 4 students.

**Table 3.** Academic-year differences in acculturative stress, SOC, active coping, avoidant coping, and depressive symptoms.

| Variable | Year 1 + 2 Arab Students $n = 98$ | | Year 3 + 4 Arab Students $n = 72$ | | *t*-Value | Hedges' *g* |
|---|---|---|---|---|---|---|
| | *M* | *SD* | *M* | *SD* | | |
| Acculturative stress (0–5) | 1.81 | 0.81 | 1.83 | 0.82 | −0.15 | 0.02 |
| SOC (1–7) | 4.02 | 0.93 | 4.34 | 0.93 | −2.28 * | 0.35 |
| Coping strategies | | | | | | |
| Active coping (1–5) | 2.48 | 0.48 | 2.52 | 0.44 | −0.45 | 0.07 |
| Avoidant coping (1–5) | 2.16 | 0.49 | 2.03 | 0.37 | 1.91 | 0.30 |
| Depressive symptoms (0–60) | 23.42 | 13.05 | 19.04 | 11.74 | 2.25 * | 0.35 |

* $p < 0.05$; SOC, sense of coherence.

### 3.4. Relationships Between the Study Variables Among the Students

We computed Pearson's correlations between the study variables. As shown in Table 4, a positive association was found between acculturative stress and depressive symptoms. In addition, avoidant coping was found to be positively associated with acculturative stress. In other words, higher levels of acculturative stress were related to both greater use of avoidant coping strategies and more depressive symptoms. In addition, negative associations were found between acculturative stress and SOC and active coping. Higher levels of acculturative stress were related to lower levels of SOC and less use of active coping strategies among Arab students.

**Table 4.** Pearson's correlations between the study variables.

| Variables | 1 | 2 | 3 | 4 |
|---|---|---|---|---|
| 1. Acculturative stress | | | | |
| 2. SOC | −0.37 *** | | | |
| 3. Avoidant coping | 0.18 * | −0.41 *** | | |
| 4. Active coping | −0.17 * | 0.20 * | 0.23 ** | |
| 5. Depressive symptoms | 0.40 *** | −0.69 *** | 0.44 *** | −0.24 ** |

* $p < 0.05$; ** $p < 0.01$; *** $p < 0.001$; SOC, sense of coherence.

### 3.5. Direct and Indirect Relationships between Acculturative Stress and Depression among Male and Female Arab Students

Multiple-group SEM analysis was performed with SPSS AMOS software [76], using the maximum-likelihood estimation to test how well the data fit the hypothesized model. AMOS generates a variety of indices for evaluating fit; models with chi-square/degrees of freedom ratios of less than two considered acceptable. We also employed the non-normed fit index (NNFI) [77], the Tucker–Lewis index (TLI), the comparative fit index (CFI) [78], and the root mean square error of approximation (RMSEA). Index values between 0.00 and 0.08 are generally deemed acceptable [79]. The fit indices of the hypothesized model were as follows: CFI = 0.99, NNFI = 0.91, RMSEA = 0.07, CMIN/*df* = 1.83, and $p < 0.05$. Thus, the hypothesized model fit the data well.

The full models explained 48% and 56% of the variance of depressive symptoms among female and male students, respectively. All of the coefficients reported in the text and in Figure 2 are standardized. As shown in Figure 2, among the female participants, acculturative stress had a significant direct effect ($\beta = 0.15$) and an indirect effect ($\beta = 0.20$) on depressive symptoms via SOC and avoidant coping. However, among the male participants, we observed a stronger indirect effect of acculturative stress on depression ($\beta = 0.40$) via SOC and active coping. Among the men, we did not find any significant direct effect of acculturative stress on depressive symptoms ($\beta = 0.13$).

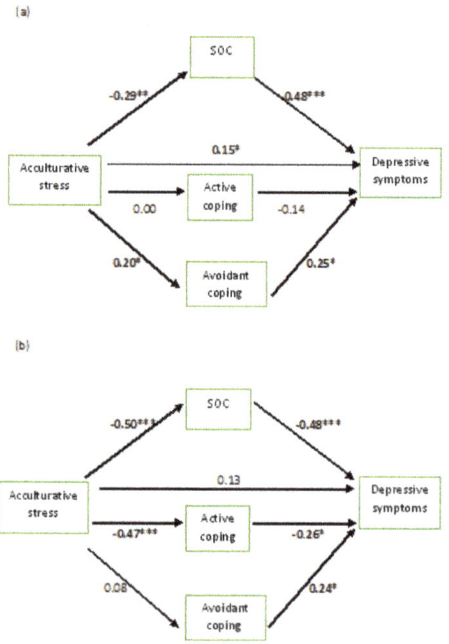


**Figure 2.** Direct and indirect effects of acculturative stress on depressive symptoms through SOC, avoidant coping, and active coping among: (**a**) female Arab students; and (**b**) male Arab students. *Note.* All of the coefficients in the figures are standardized. * $p < 0.05$; ** $p < 0.01$; *** $p < 0.001$; SOC, sense of coherence.

Among the women, acculturative stress had significant direct effects on SOC ($\beta = -0.29$) and avoidant coping ($\beta = 0.20$). Moreover, we observed significant direct effects of SOC ($\beta = -0.48$) and avoidant coping ($\beta = 0.25$) on depressive symptoms. In comparison, among the men, acculturative stress had strong direct effects on SOC ($\beta = -0.50$) and active coping ($\beta = -0.47$). We also observed significant direct effects of SOC ($\beta = -0.48$) and active coping ($\beta = -0.26$) on depressive symptoms among the men. The results from this analysis underscore the significance of the indirect effect of acculturative stress on depressive symptoms through SOC and coping strategies among female and male Arab students. However, our results support the direct effect of acculturative stress on depression among female students, but not among male students.

*3.6. Direct and Indirect Relationships between Acculturative Stress and Depression among Year 1 + 2 and Year 3 + 4 Arab Students*

Multiple-group SEM analysis was performed with SPSS AMOS software to compare the effects of the different variables on depressive symptoms in the two groups of Year 1 + 2 and Year 3 + 4

Arab Students. The fit indices of the hypothesized model were as follows: CFI = 0.98, NNFI = 0.93, RMSEA = 0.08, CMIN/*df* = 1.94, and *p* < 0.05. Thus, the hypothesized model fit the data well.

The full models explained 45% and 53% of the variance of depressive symptoms among Year 1 + 2 students and Year 3 + 4 students, respectively. All of the coefficients reported in the text and in Figure 2 are standardized. As shown in Figure 3, among the Year 1 + 2 participants, acculturative stress had an indirect effect (β = 0.27) on depressive symptoms via SOC and avoidant coping. However, among the Year 3 + 4 participants, we observed a direct effect of acculturative stress on depression (β = 0.23) and an indirect effect of acculturative stress on depression (β = 0.22) via SOC. Among the Year 1 + 2 students, we did not find any significant direct effect of acculturative stress on depressive symptoms (β = 0.11).

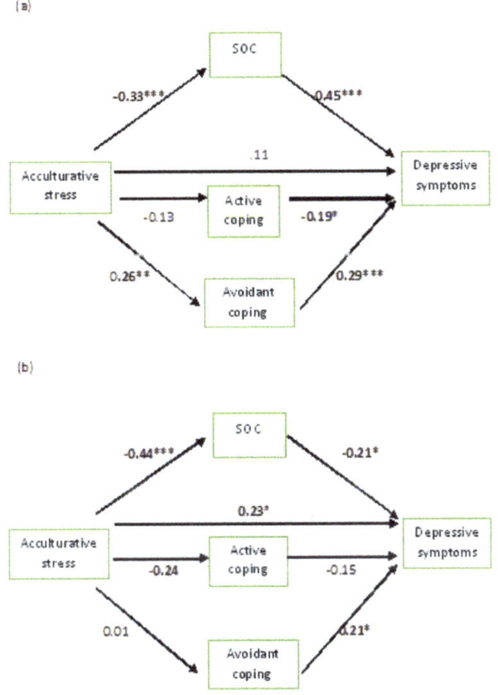


**Figure 3.** Direct and indirect effects of acculturative stress on depressive symptoms through SOC, avoidant coping, and active coping among: (**a**) Year 1 + 2 Arab students; and (**b**) Year 3 + 4 Arab students. *Note.* All of the coefficients in the figures are standardized. * *p* < 0.05; ** *p* < 0.01; *** *p* < 0.001; SOC, sense of coherence.

Among the Year 1 + 2 students, acculturative stress had significant direct effects on SOC (β = −0.33) and avoidant coping (β = 0.26). Moreover, we observed significant direct effects of SOC (β = −0.46), active coping (β = −0.19), and avoidant coping (β = 0.29) on depressive symptoms among those students. In comparison, among the Year 3 + 4 students, acculturative stress had strong direct effects on SOC (β = −0.44) and active coping (β = −0.24). We also observed significant direct effects of SOC (β = −0.42) and avoidant coping (β = 0.21) on depressive symptoms among the Year 3 + 4 students.

The results of these analyses underscore the significance of the indirect effect of acculturative stress on depressive symptoms, through SOC and avoidant-coping strategies, among Year 1 + 2 Arab students. However, our results support the direct effect and the indirect effect of acculturative stress on depressive symptoms through SOC only among Year 3 + 4 Arab students.

An examination of the total (direct and indirect) effects of acculturative stress revealed a meaningful picture. Among the female participants, acculturative stress affected depressive symptoms directly and indirectly through SOC and avoidant coping. However, among Arab male participants, acculturative stress affected depressive symptoms only indirectly, through SOC and active coping. Moreover, among Year 1 + 2 students, acculturative stress was related to depressive symptoms indirectly via SOC and avoidant coping. However, among the Year 3 + 4 students, acculturative stress was related to depressive symptoms both directly and indirectly via SOC.

## 4. Discussion

The main aim of this study was to examine the association between acculturative stress and depressive symptoms. We were interested in exploring the roles of SOC and avoidant and active coping strategies in this relationship among male and female Arab college students from northern and central Israel who are at different stages of their studies in Israeli institutions of higher education. Important findings emerged from this study. Gender differences were found in both active and avoidant coping, as well as in depressive symptoms. Academic-year differences were found in SOC and depressive symptoms. In addition, higher levels of acculturative stress were found to be related to lower levels of SOC, less use of active coping, more use of avoidant coping, and higher levels of depressive symptoms. Among the female Arab students, acculturative stress affected depressive symptoms both directly and indirectly through SOC and avoidant coping. However, among Arab males, acculturative stress only affected depressive symptoms indirectly, through SOC and active coping. Among Year 1 + 2 students, acculturative stress affected depressive symptoms indirectly via SOC and avoidant coping. However, among the Year 3 + 4 students, acculturative stress affected depressive symptoms both directly and indirectly via SOC.

### 4.1. Gender Differences in Coping Strategies

Gender differences were found in both active and avoidant coping, with females scoring higher in both active and avoidant coping than their male counterparts. This is consistent with previous literature that has noted that coping mechanisms may be gender-specific [80] and that females tend to face higher levels of stress, which are associated with the use of more coping resources [66]. Abu-Kaf and Braun-Lewensohn [42] confirmed that female Bedouin Arab students report greater use of both active coping and avoidant coping, as compared to male Bedouin Arab students. Bedouin Arab females reported using a variety of coping strategies as they confront problems/stressors (i.e., social support, emotional regulation, affective release, and emotion-focused strategies). They also attempt to reduce tension by avoiding dealing with problems (i.e., behavioral disengagement, self-distraction, denial, and self-blame). This finding may be related to the understanding that females suffer disproportionately from life stressors. Understanding that difference is essential for the discussion of gender differences in personal methods of coping and efforts to overcome stressful situations [3]. In other words, the greater number of life stressors that Arab females experience may be associated with the use of more coping mechanisms, whether avoidant or active [66].

### 4.2. Gender Differences in Depressive Symptoms

In terms of depressive symptoms, the current study revealed a higher monthly prevalence rate, as well as higher scores (mean) among female Arab students, as compared to male Arab students. These finding provide more support for previous research, which found gender differences in depression. For example, women are twice as likely as men to have higher scores on self-reported measures of depressive symptoms [49]. It is important to mention that mean depression scores and the monthly prevalence rate among male Arab students were higher than those usually observed among male and female student samples in Israel. Previous research found that the reported rates of severe levels of depressive symptoms range from 12.5% to 16.7% among male Jewish students and range from 15.1% to 17% among Jewish female students [50]. In the current study, about one-third of the Arab male students scored above the stricter diagnostic cut-off point of 23, indicating severe levels of depressive symptoms. Mental-health problems have been found to be more prevalent among Arab students, particularly among female Arab students [42,81]. Depressive symptoms among students have been found to affect learning and memory processes, leading to lower levels of academic achievement [82], poor attendance, failure to complete the academic degree [83], dropping out of the academic institution [84], and even suicidal ideation [85].

### 4.3. Academic-Year Differences in Depressive Symptoms

Our results revealed differences in depressive symptoms among Year 1 + 2 students as compared to Year 3 + 4 students. Year 1 + 2 students reported higher levels of depressive symptoms than Year 3 + 4 students. This finding supports previous research which addressed the vulnerability of students in the earlier years of their college educations to psychological distress, specifically depressive symptoms [51]. Previous studies have demonstrated that the period of transition from high school to college is very stressful for many students, in general, and for student from cultural-ethnic minority groups, in particular [50,62].

### 4.4. Academic-Year Differences in SOC

Another interesting finding is related to the higher levels of SOC observed among Year 3 + 4 students, as compared to Year 1 + 2 students. This finding does not support the findings of a previous study that found no academic-year differences in SOC among Chinese nursing students in Australia. Our finding may be explained by the fact that more advanced Arab students have more experience, greater knowledge about the academic environment, and larger social networks on campus than students who are in the earlier stages of their education [86]. This may affect their perception of college life and the academic environment as structured and predictable, as well as increase their capacity to assess and use available resources to face and cope with stressful situations [21].

### 4.5. Associations Between Acculturative Stress and Depressive Symptoms

Our findings revealed that acculturative stress is positively related to depressive symptoms. This finding supports previous research that has found acculturative stress to be associated with mental-health problems, including anxiety and depression [2,10–12]. Among college students (international students and students from a cultural-ethnic minority), acculturative stress has been associated with a number of psychological challenges, including depression [13,14]. This finding may be related to the associations that have been found between acculturative stress and feelings of alienation, being discriminated against, and feeling that one does not belong, which may contribute to feelings of disconnection and depression [87].

### 4.6. The Direct and Indirect Effects of SOC and Coping Strategies on the Association Between Acculturative Stress and Depressive Symptoms

Among both males and females, acculturative stress had an indirect effect on depressive symptoms via SOC. This confirms the result of no gender bias in the levels of SOC in this study. Many cross-sectional studies have reported a significant inverse correlation between SOC and depression. According to Sairenchi, Haruyama, Ishikawa, Wada, Kimura, and Muto [88], SOC can predict the onset of depression in Japanese workers. Workers with low levels of SOC might suffer from depression more than workers with high SOC; no gender differences were found in these associations. Therefore, examining the levels of SOC among students of both genders may be useful for identifying Arab students at high risk of future depressive symptoms.

Among female Arab students, the effect of acculturative stress was amplified through the indirect effects of avoidant coping. Among the male students, the effect of acculturative stress was amplified through the indirect effects of active coping. Our findings underscore previous claims that passive/avoidant coping is strongly related to general psychological distress and depression [42] and that active coping is inversely related to depression [89].

The fact that cultures vary in the degree to which gender roles are emphasized could contribute to this difference. Williams and Best [90] confirmed that men and women in traditional-collectivistic cultures (i.e., Arab culture, in this case) tend to emphasize gender-role differences, whereas those in more Western-individualistic cultures tend to minimize them. Presumably, such values could affect the development of gender differences, in general, and the direct and indirect effects of acculturative stress on depressive symptoms, in particular. Thus, Arab males, who are seen in their patriarchal society as the main providers and guardians of their families [57], may be driven to use active coping strategies as they confront challenges. Among Arab males, acculturative stress may lead to more depressive symptoms by affecting (i.e., decreasing) their use of active coping strategies. Arab women are expected to adopt more avoidant/passive coping strategies, in order to fit in their social roles and fulfill their obligations in more traditional cultural contexts [57]. Among these women, acculturative stress may lead to more depressive symptoms by affecting (i.e., increasing) their use of avoidant coping strategies.

Among both Year 1 + 2 and Year 3 + 4 Arab students, acculturative stress had an indirect effect on depressive symptoms via SOC. This finding confirms the findings of previous studies conducted among Bedouin Arabs, which found that, over time, SOC becomes a strong predictor of stress reactions [24]. Among Year 1 + 2 Arab students, acculturative stress had an indirect effect on depressive symptoms via the use of avoidant-coping strategies. Similar to what was observed for female Arab students, the role of avoidant coping in the association between acculturative stress and depressive symptoms underscores previous findings that indicated that passive/avoidant coping is strongly related to general psychological distress and depression [42].

The present study suggests that Arab individuals of both genders and across different stages in their academic studies who face relatively high levels of acculturative stress tend to exhibit higher levels of depressive symptoms, but that the effect of acculturative stress on depression develops via similar, as well as different pathways among female and male, and among more junior (Year 1 + 2) and more advanced (Year 3 + 4) Arab students. The female Arab students tend to be more depressed as a direct result of their higher levels of acculturative stress, which arise from the need to learn new and sometimes confusing cultural rules and expectations, deal with experiences of prejudice and discrimination, and manage the overarching conflict between maintaining elements of their old culture while incorporating elements of the new. These depression levels are also an indirect result of their lower levels of SOC, as well as their stronger tendency to use avoidant coping strategies (i.e., to attempt to reduce tension by avoiding dealing with problems through behavioral disengagement, self-distraction, denial, and self-blame). In contrast, male Arab students experiencing high levels of acculturative stress tend to be more depressed as an indirect result of their lower levels of SOC and their limited use of active coping strategies (i.e., planning, instrumental support, emotional support, positive reframing, problem-solving, and humor). Year 1 + 2 students experiencing high levels of acculturative stress tend to be more depressed as an indirect result of their lower levels of SOC and their increased use of avoidant coping strategies (like the female students, they tend to increase their use of behavioral disengagement, self-distraction, denial, and self-blame). However, Year 3 + 4 Arab students tend to be more depressed as a direct result of their higher levels of acculturative stress and also as an indirect result of their lower levels of SOC.

The current study contributes to the understanding of gender and academic-year differences in coping resources and the use of different strategies to deal with acculturative stressors in the context of higher education. This research suggests that female and male students, as well as students at different stages of their degree programs, from a more traditional, Eastern-collectivistic cultural background who exhibit high levels of acculturative stress tend to have different and distinct pathways to emotional distress. However, because the research design is cross-sectional, we cannot completely exclude the possibility that depression may lead to greater acculturative stress and lower SOC, as well as lowered motivation to deal actively with stressors and high levels of avoidance and withdrawal.

### 4.7. Limitations and Directions for Future Research

Although the fact that this work may be the first study of acculturative stress and depressive symptoms among Arab students in Israel underscores its value, it also has several limitations that should be taken into account when considering the research findings. There are also areas that warrant further attention in future research. First, the study design was cross-sectional and, therefore, we cannot make any claims of causality. In addition, we did not evaluate past depressive experiences among those students, an important variable that may affect the prospective levels of acculturative stress and depressive symptoms during their academic studies. Thus, future research studies should employ a longitudinal design and aim to test prior levels of depressive symptoms, as well as the prospective relationship between acculturative stress and depressive symptoms in the academic setting. Second, our data were based on self-report measures, which have particular limitations, such as restrictive rating scales, and can be limited by the introspective ability of participants, inaccurate interpretations of particular questions, and biased responses. In future research, the use of other methods of data collection (especially diaries and interviews) would be beneficial and important for the evaluation of the validity of the obtained findings. Third, the study involved participants from only four of the many institutions of higher education in Israel: Ben-Gurion University of the Negev, Tel Aviv University, The Hebrew University of Jerusalem, and Haifa University. Therefore, future research should include students from wider range of Israeli academic institutions and the generalizability of the current finding should be evaluated. Another avenue for future research might be to extend the present model by examining cultural factors such as self-construal, collectivism level, and somatization, which may be important indicators of psychological distress in Arab cultural contexts [50,91,92].

## 5. Conclusions

The importance of the current study lies in its examination of the roles of coping resources and coping strategies in the association between acculturative stress and depressive symptoms among female and male, as well as more junior and more advanced, Arab college students. The present study highlights the importance of SOC, as well as active coping and avoidant coping, in the relationship between acculturative stress and depressive symptoms. Gender and academic-year similarity was found with regard to the role of SOC, whereas differences were found in the direct effects and the roles of active and avoidant coping. Among the male students, active coping played a significant role in the association between acculturative stress and depressive symptoms. In contrast, among female and Year 1 + 2 students, avoidant coping played a significant role. This knowledge is expected to help the employees of academic departments understand the different aspects of the coping and distress that characterize Arab students. Male and female students and students at different stages of their degree programs may need different interventions to help them adjust to academic life, the acculturation process, and the stresses of this process. Psychological counseling and guidance programs should be tailored to the specific needs of Arab students, with special attention given to the roles of gender, academic year, coping resources, and coping strategies.

**Author Contributions:** Conceptualization, S.A.-K. and E.K.; methodology, S.A.-K. and E.K.; software, S.A.-K.; formal analysis, S.A.-K. and E.K.; investigation, S.A.-K. and E.K.; resources, S.A.-K.; data curation, E.K.; writing—original draft preparation, S.A.-K.; writing—review and editing, S.A.-K.; visualization, S.A.-K.; supervision, S.A.-K.; and project administration, E.K. All authors have read and agreed to the published version of the manuscript.

**Funding:** This research received no external funding.

**Conflicts of Interest:** The authors declare no conflict of interest.

## References

1. Redfield, R.; Linton, R.; Herskovits, M.J. Memorandum for the study of acculturation. *Am. Anthropol.* **1936**, *38*, 149–152. [CrossRef]

2.    Berry, J.W.; Kim, U.; Minde, T.; Mok, D. Comparative studies of acculturative stress. *Int. Migr. Rev.* **1987**, *21*, 491–511. [CrossRef]

3.    Lazarus, R.S.; Folkman, S. *Stress, Appraisal, and Coping*; Springer: New York, NY, USA, 1984.

4.    Berry, J.W. Contexts of acculturation. In *The Cambridge Handbook of Acculturation Psychology*; Sam, D.L., Berry, J.W., Eds.; Cambridge University Press: New York, NY, USA, 2006; pp. 27–42.

5.    Berry, J.W. Immigration, acculturation, and adaptation. *Appl. Psychol. Int. Rev.* **1997**, *46*, 5–68. [CrossRef]

6.    Suárez-Orozco, C.; Suárez-Orozco, M. *Children of Immigration*; Harvard University Press: Cambridge, MA, USA, 2001.

7.    Mena, F.J.; Padilla, A.M.; Maldonado, M. Acculturative stress and specific coping strategies among immigrant and later generation college students [Special issue]. *Hispanic J. Behav. Sci.* **1987**, *9*, 207–225. [CrossRef]

8.    Sullivan, C.; Kashubeck-West, S. The interplay of international students' acculturative stress, social support, and acculturation modes. *J. Int. Stud.* **2015**, *5*, 1–11.

9.    Torres, V. Influences on ethnic identity development of Latino college students in the first 2 years of college. *J. Coll. Student Dev.* **2003**, *44*, 532–547. [CrossRef]

10.   Sirin, S.R.; Sin, E.; Clingain, C.; Rogers-Sirin, L. Acculturative stress and mental health: Implications for immigrant-origin youth. *Pediatr. Clin.* **2019**, *66*, 641–653.

11.   Torres, L. Predicting levels of Latino depression: Acculturation, acculturative stress, and coping. *Cult. Div. Ethn. Minor. Psychol.* **2010**, *16*, 256–263. [CrossRef]

12.   Williams, C.L.; Berry, J.W. Primary prevention of acculturative stress among refugees: Application of psychological theory and practice. *Am. Psychol.* **1991**, *46*, 632. [CrossRef]

13.   Lee, M.; Nezu, A.M.; Nezu, C.M. Acculturative stress, social problem solving, and depressive symptoms among Korean American immigrants. *Transcult. Psychiatr.* **2018**, *55*, 710–729. [CrossRef]

14.   Sam, D.L. Acculturation: Conceptual background and core components. In *The Cambridge Handbook of Acculturation Psychology*; Sam, D.L., Berry, J.W., Eds.; Cambridge University Press: Cambridge, UK, 2006. [CrossRef]

15.   Cohen, S.; Wills, T.A. Stress, social support, and the buffering hypothesis. *Psychol. Bull.* **1985**, *98*, 310–357. [CrossRef] [PubMed]

16.   Masten, A.; Coatsworth, J. The development of competence in favorable and unfavorable environments: Lessons from research on successful children. *Am. Psychol.* **1998**, *53*, 205–220. [CrossRef]

17.   Antonovsky, A. *Health, Stress, and Coping*; Jossey-Bass: San Francisco, CA, USA, 1979.

18.   Antonovsky, A. The structure and properties of the sense of coherence scale. *Soc. Sci. Med.* **1993**, *36*, 725–733. [CrossRef]

19.   Antonovsky, A. *Unraveling the Mystery of Health: How People Manage. Stress and Stay Well*; Jossey-Bass: San Francisco, CA, USA, 1987.

20.   Eriksson, M. *Unravelling the Mystery of Salutogenesis: The Evidence Base of the Salutogenic Research as Measured by Antonovsky's Sense of Coherence Scale*; Folkhälsan Research Centre: Turku, Finland, 2007.

21.   Sagy, S.; Antonovsky, H. Factors related to the development of the sense of coherence (SOC) in adolescence. A retrospective study. *Pol. Psychol. Bull.* **1999**, *30*, 255–262.

22.   Mosley-Hänninen, P. *Contextualising the Salutogenic Perspective on Adolescent Health and the Sense of Coherence in Families: A Study among Adolescents and Their Families in the Swedish Speaking Finland*; Laurea University of Applied Sciences: Vantaa, Finland, 2009; Abstract. Master of Health care, Dana: 09. 11. 2009. Available online: https://core.ac.uk/download/pdf/37985723.pdf (accessed on 21 January 2020).

23.   Abu-Kaf, S.; Braun-Lewensohn, O.; Kalagy, T. Youth in the midst of escalated political violence: Sense of coherence and hope among Jewish and Bedouin Arab adolescents. *Child. Adolesc. Psychiatr. Ment. Health* **2017**, *11*, 42. [CrossRef]

24.   Abu-Kaf, S.; Braun-Lewensohn, O. Coping resources and stress reactions among Bedouin Arab adolescents during three military operations. *Psychiatr. Res.* **2019**, *273*, 559–566. [CrossRef] [PubMed]

25.   Braun-Lewensohn, O.; Abu-Kaf, S.; Al-Said, K.; Huss, E. Analysis of the differential relationship between the perception of one's life and coping resources among three generations of Bedouin women. *Int. J. Environ. Res. Public Health* **2019**, *16*, 804. [CrossRef]

26.   Braun-Lewensohn, O.; Sagy, S.; Roth, G. Coping strategies as mediators of the relationship between sense of coherence and stress reactions: Israeli adolescents under missile attacks. *Anxiety Stress Coping* **2011**, *24*, 327–341. [CrossRef]

27. Peker, K.; Bermek, G.; Uysal, O. Factors related to sense of coherence among dental students at Istanbul University. *J. Dent. Educ.* **2012**, *76*, 774–782. [CrossRef]
28. Blom, E.C.H.; Serlachius, E.; Larsson, J.O.; Theorell, T.; Ingvar, M. Low sense of coherence (SOC) is a mirror of general anxiety and persistent depressive symptoms in adolescent girls—A cross-sectional study of a clinical and a non-clinical cohort. *Health Qual. Life Outcomes* **2010**, *8*, 58. [CrossRef]
29. Dezutter, J.; Wiesmann, U.; Apers, S.; Luyckx, K. Sense of coherence, depressive feelings and life satisfaction in older persons: A closer look at the role of integrity and despair. *Aging Ment. Health* **2013**, *17*, 839–843. [CrossRef]
30. Eriksson, M.; Lindström, B. Antonovsky's sense of coherence scale and the relation with health: A systematic review. *J. Epidemiol. Community Health* **2006**, *60*, 376–381. [CrossRef]
31. Braun-Lewensohn, O. Coping resources and stress reactions among three cultural groups one year after a natural disaster. *Clin. Soc. Work J.* **2014**, *42*, 366–374. [CrossRef]
32. Robin-Walton, L.A. Comparison of Perceived Stress Levels and Coping Styles of Junior and Senior Students in Nursing and Social Work Programs. Doctoral Dissertation, Marshall University, Huntington, WV, USA, 2002.
33. Billing, A.; Moos, R.H. Coping, stress, and social resources among adults with unipolar depression. *J. Pers. Soc. Psychol.* **1984**, *46*, 877–891. [CrossRef]
34. Mitchell, R.E.; Cronkite, R.C.; Moos, R.H. Stress, coping, and depression among married couples. *J. Abnorm. Psychol.* **1983**, *92*, 433–448. [CrossRef] [PubMed]
35. Moskowitz, J.T.; Hult, J.R.; Bussolari, C.; Acree, M. What works in coping with HIV? A meta-analysis with implications for coping with serious illness. *Psychol. Bull.* **2009**, *135*, 121–141. [CrossRef]
36. Aldwin, C.M.; Revenson, T.A. Does coping help? A reexamination of the relation between coping and mental health. *J. Pers. Soc. Psychol.* **1987**, *53*, 337–348. [CrossRef] [PubMed]
37. Holahan, C.J.; Moos, R.H. Risk, resistance, and psychological distress: A longitudinal analysis with adults and children. *J. Abnorm. Psychol.* **1987**, *96*, 3. [CrossRef] [PubMed]
38. Sam, D.L.; Berry, J.W. Acculturation: When individuals and groups of different cultural backgrounds meet. *Perspect. Psychol. Sci.* **2010**, *5*, 472–481. [CrossRef]
39. Chang, E.C. A look at the coping strategies and styles of Asian Americans: Similar and different? In *Coping with Stress: Effective People and Processes*; Snyder, C.R., Ed.; Oxford University Press: New York, NY, USA, 2001; pp. 222–239.
40. Taylor, S.E.; Sherman, D.K.; Jarcho, J.; Takagi, K.; Dunagan, M.S. Culture and social support: Who seeks it and why? *J. Pers. Soc. Psychol.* **2004**, *87*, 354–362. [CrossRef]
41. Essau, C.; Trommsdorff, G. Coping with university-related problems: A cross-cultural comparison. *J. Cross-Cult. Psychol.* **1996**, *27*, 315–328. [CrossRef]
42. Abu-Kaf, S.; Braun-Lewensohn, O. Paths to depression among two different cultural contexts: Comparing Bedouin Arab and Jewish students. *J. Cross-Cult. Psychol.* **2015**, *46*, 612–630. [CrossRef]
43. Castillo, L.G.; Navarro, R.L.; Walker, J.E.O.; Schwartz, S.J.; Zamboanga, B.L.; Whitbourne, S.K.; Weisskirch, R.S.; Kim, S.Y.; Park, I.J.; Vazsonyi, A.T.; et al. Gender Matters: The Influence of Acculturation and Acculturative Stress on Latino College Student Depressive Symptomatology. *J. Lat. Psychol.* **2015**, *3*, 40–55. [CrossRef]
44. Park, C.; Spruill, T.M.; Butler, M.J.; Kwon, S.C.; Redeker, N.S.; Gharzeddine, R.; Whittemore, R. Gender differences in acculturative stress and habitual sleep duration in Korean American immigrants. *J. Immigr. Minor. Health* **2020**, *22*, 736–745. [CrossRef] [PubMed]
45. Moghaddam, F.M.; Taylor, D.M.; Ditto, B.; Jacobs, K.; Bianchi, E. Psychological distress and perceived discrimination: A study of women from India. *Int. J. Intercult. Relations* **2002**, *26*, 381–390. [CrossRef]
46. Hilt, L.M.; McLaughlin, K.A.; Nolen-Hoeksema, S. Examination of the response styles theory in a community sample of young adolescents. *J. Abnorm. Child. Psychol.* **2010**, *38*, 545–556. [CrossRef]
47. Taylor, S.E.; Klein, L.C.; Lewis, B.P.; Gruenewald, T.L.; Gurung, R.A.R.; Updegraff, J.A. Biobehavioral responses to stress in females: Tend-and-befriend, not fight-or-flight. *Psychol. Rev.* **2000**, *107*, 411–429. [CrossRef] [PubMed]
48. Hampel, P.; Petermann, F. Age and gender effects on coping in children and adolescents. *J. Youth Adolesc.* **2005**, *34*, 73–83. [CrossRef]
49. Nolen-Hoeksema, S. Gender differences in depression. *Curr. Dir. Psychol. Sci.* **2001**, *10*, 173–176. [CrossRef]

50. Abu-Kaf, S.; Shahar, G. Depression and somatic symptoms among two ethnic groups in Israel: Testing three theoretical models. *Isr. J. Psychiatr. Rel. Sci.* **2017**, *54*, 32–39.

51. Hwang, W.C.; Ting, J.Y. Disaggregating the effects of acculturation and acculturative stress on the mental health of Asian Americans. *Cult. Divers. Ethn. Minor. Psychol.* **2008**, *14*, 147. [CrossRef] [PubMed]

52. He, F.X.; Lopez, V.; Leigh, M.C. Perceived acculturative stress and sense of coherence in Chinese nursing students in Australia. *Nurse Educ. Today* **2012**, *32*, 345–350. [CrossRef] [PubMed]

53. Baykan, Z.; Naçar, M. Depression, anxiety, and stress among last-year students at Erciyes University Medical School. *Acad. Psychiatr.* **2012**, *36*, 64–65. [CrossRef] [PubMed]

54. Dahlin, M.; Joneborg, N.; Runeson, B. Stress and depression among medical students: A cross-sectional study. *Med. Educ.* **2005**, *39*, 594–604. [CrossRef] [PubMed]

55. Central Bureau of Statistics (Israel). Statistical Abstract of Israel. 2020. Available online: https://www.cbs.gov.il/en/mediarelease/Pages/2020/Israe-Independence-Day-2020.aspx (accessed on 15 March 2020).

56. Israeli Council for Higher Education. Doubling the Number of Arab Students. 2019. Available online: https://che.org.il/en/doubling-number-arab-students/ (accessed on 20 March 2020).

57. Haj Yahia, M. The Palestinian family in Israel. *Ment. Health Palest. Citiz. Isr.* **2019**, 97–120.

58. Oyserman, D.; Coon, H.M.; Kemmelmeier, M. Rethinking individualism and collectivism: Evaluation of theoretical assumptions and meta-analysis. *Psychol. Bull.* **2002**, *128*, 3–72. [CrossRef] [PubMed]

59. Al-Krenawi, A.; Graham, G.R.; Al-Bedah, E.A.; Kadri, H.M.; Sehwail, M.A. Cross-national comparison of Middle Eastern university students: Help-seeking behaviors, attitudes toward helping professionals, and cultural beliefs about mental health problems. *Community Ment. Health J.* **2009**, *45*, 26–36. [CrossRef] [PubMed]

60. Mahalingam, R. (Ed.) *Cultural Psychology of Immigrants*; Lawrence Erlbaum Associates: Mahwah, NJ, USA, 2006.

61. Haberfeld, Y.; Cohen, Y. Gender, ethnic, and national earnings gaps in Israel: The role of rising inequality. *Soc. Sci. Res.* **2007**, *36*, 654–672. [CrossRef]

62. Abu-Kaf, S. Mental health issues among Palestinian women in Israel. *Ment. Health Palest. Citiz. Isr.* **2019**, 121–148.

63. Baron-Epel, O.; Kaplan, G.; Moran, M. Perceived discrimination and health related quality of life among Arabs, immigrants and veteran Jews in Israel. *Public Health* **2010**, *10*, 282.

64. Dwairy, M. Individuation among Bedouin versus urban Arab adolescents: Ethnic and gender differences. *Cult. Div. Ethn. Minor. Psychol.* **2004**, *10*, 340–350. [CrossRef] [PubMed]

65. Carballo, M. *Scientific Consultation on the Social and Health Impact of Migration: Priorities for Research*; International Organization for Migration: Geneva, Switzerland, 1994.

66. Tamres, L.K.; Janicki, D.; Helgeson, V.S. Sex differences in coping behavior: A meta-analytic review and an examination of relative coping. *Pers. Soc. Psychol. Rev.* **2002**, *6*, 2–30. [CrossRef]

67. Katsiaficas, D.; Suárez-Orozco, C.; Sirin, S.; Gupta, T. Mediators of the Relationship Between Acculturative Stress and Internalization Symptoms for Immigrant Origin Youth. *Cult. Divers. Ethn. Minor. Psychol.* **2013**, *19*, 27–37. [CrossRef] [PubMed]

68. Vasquez, P. Religious Coping and Social Support as Mediators and/or Moderators and Acculturative Stress in a Latino Community Sample. 2010. Available online: https://epublications.marquette.edu/theses_open/71/ (accessed on 10 April 2020).

69. Joiner, T.E., Jr.; Walker, R.L. Construct validity of a measure of acculturative stress in African Americans. *Psychol. Assess.* **2002**, *14*, 462–466. [CrossRef] [PubMed]

70. Braun-Lewensohn, O.; Sagy, S. Coping resources as explanatory factors of stress reactions during missile attacks: Comparing Jewish and Arab adolescents in Israel. *Community Ment. Health J.* **2011**, *47*, 300–310. [CrossRef]

71. Daoud, N.; Braun-Lewensohn, O.; Eriksson, M.; Sagy, S. Sense of coherence and depressive symptoms among low-income Bedouin women in the Negev. *Isr. J. Ment. Health.* **2014**, *23*, 307–311. [CrossRef]

72. Carver, C.S. You want to measure coping but your protocol's too long: Consider the Brief COPE. *Int. J. Behav. Med.* **1997**, *4*, 92–100. [CrossRef]

73. Qouta, S.; Punamäki, R.L.; Sarraj, E.E. Prison experiences and coping styles among Palestinian men. *Peace Confl.* **1997**, *3*, 19–36. [CrossRef]

74. Radloff, L.S. The CES-D scale: A self-report depression scale for research in the general population. *Appl. Psychol. Meas.* **1977**, *1*, 385–401. [CrossRef]

75. Roberts, R.E.; Andrews, J.A.; Lewinsohn, P.M.; Hops, H. Assessment of depression in adolescents using the Center for Epidemiologic Studies Depression Scale. *J. Consult. Clin. Psychol.* **1990**, *2*, 122–128. [CrossRef]

76. Arbuckle, J.L. IBM SPSS Amos 26 User's Guide: Marketing Department; IBM SPSS, 2019. Available online: File:///C:/Users/User/Downloads/IBM_SPSS_Amos_User_Guide.pdf (accessed on 2 April 2020).

77. Bentler, P.M.; Bonett, D.G. Significance tests and goodness of fit in the analysis of covariance structure. *Psychol. Bull.* **1980**, *88*, 588–606. [CrossRef]

78. Bentler, P. Comparative fit indexes in structural models. *Psychol. Bull.* **1990**, *107*, 238–246. [CrossRef]

79. Hu, L.; Bentler, P.M. Fit indices in covariance structure modeling: Sensitivity to under parameterized model misspecification. *Psychol. Method* **1998**, *3*, 424–453. [CrossRef]

80. Folkman, S.; Lazarus, R.S. An analysis of coping in a middle-aged community sample. *J. Health Soc. Behav.* **1980**, *21*, 219–239. [CrossRef]

81. Al Khatib, S.A. Satisfaction with life, self-esteem, gender and marital status as predictors of depressive symptoms among United Arab Emirates college students. *Int. J. Psychol. Couns.* **2013**, *5*, 53–61.

82. Hysenbegasi, A.; Hass, S.L.; Rowland, C.R. The impact of depression on the academic productivity of university students. *J. Ment. Health Policy Econ.* **2005**, *8*, 145.

83. Hammen, C.; Rudolph, K.; Weisz, J.; Rao, U.; Burge, D. The context of depression in clinic-referred youth: Neglected areas in treatment. *J. Am. Acad. Child. Adolesc. Psychiatr.* **1999**, *38*, 64–71. [CrossRef] [PubMed]

84. King, N.J.; Bernstein, G.A. School refusal in children and adolescents: A review of the past 10 years. *J. Am. Acad. Child. Adolesc. Psychiatr.* **2001**, *40*, 197–205. [CrossRef] [PubMed]

85. Garlow, S.J.; Rosenberg, J.; Moore, J.D.; Haas, A.P.; Koestner, B.; Hendin, H.; Nemeroff, C.B. Depression, desperation, and suicidal ideation in college students: Results from the American Foundation for Suicide Prevention College Screening Project at Emory University. *Depress. Anxiety* **2008**, *25*, 482–488. [CrossRef] [PubMed]

86. Geiger, B. Female Arab students' experience of acculturation and cultural diversity upon accessing higher education in the Northern Galilee-Israel. *Int. J. High. Educ.* **2013**, *2*, 91. [CrossRef]

87. Tafoya, M. The Relationship of Acculturation and Acculturative Stress in Latina/o Youths' Psychosocial Functioning. Master's Thesis in Psychology, Utah State University, Logan, UT, USA, 2011. All Graduate Theses and Dissertations, 1116.

88. Sairenchi, T.; Haruyama, Y.; Ishikawa, Y.; Wada, K.; Kimura, K.; Muto, T. Sense of coherence as a predictor of onset of depression among Japanese workers: A cohort study. *BMC Public Health* **2011**, *11*, 205. [CrossRef] [PubMed]

89. Bardwell, W.A.; Ancoli-Israel, S.; Dimsdale, J.E. Types of coping strategies are associated with increased depressive symptoms in patients with obstructive sleep apnea. *Sleep* **2001**, *24*, 905–909. [CrossRef] [PubMed]

90. Williams, J.E.; Best, D. *Sex. and Psyche: Gender and Self Viewed Cross-Culturally*; Sage Publications: Thousand Oaks, CA, USA, 1990.

91. Abu-Kaf, S.; Shahar, G.; Noyman-Veksler, G.; Priel, B. Role of perceived social support in depressive and somatic symptoms experienced by Bedouin Arab and Jewish Israeli undergraduates. *Transcult. Psychiatr.* **2018**, *56*, 359–378. [CrossRef] [PubMed]

92. Kleinman, A. Culture and depression. *N. Eng. J. Med.* **2004**, *351*, 951–953. [CrossRef] [PubMed]

International Journal of
*Environmental Research
and Public Health*

*Article*

# Coping Dynamics of Consulting Psychology Doctoral Students Transitioning a Professional Role Identity: A Systems Psychodynamic Perspective

Antoni Barnard * and Aden-Paul Flotman

Department of Industrial & Organisational Psychology, UNISA, Pretoria 0003, South Africa; flotma@unisa.ac.za
* Correspondence: barnaha@unisa.ac.za; Tel.: +27-82-375-2696

Received: 1 June 2020; Accepted: 23 July 2020; Published: 30 July 2020

**Abstract:** To remain relevant and valuable, the psychology profession in South Africa continues to transform and evolve in response to the changing needs of society. Some psychologists embark on development opportunities to advance their professional qualifications and skills. In doing so, they experience identity tensions inherent to professional identity development and transformation. Understanding how psychologists cope with professional identity transition will enable them to develop a self-efficacious service offering and broaden the reach of psychology in the South African context. The aim of this study was to explore the identity work of a group of eight consulting psychology doctoral students to develop a system psychodynamic understanding of their coping dynamics while transitioning to a professional role identity. Students' self-reflective essays about becoming a consulting psychologist constituted the data protocols for the study and were analysed through hermeneutic phenomenological analysis. Findings describe how students cope with performance and survival anxieties through anti-task behaviour and immature as well as sophisticated psychodynamic defences. The study contributes to the exploration of the coping concept and its manifestation, by proposing defensive coping as a natural dynamic phenomenon in the process of adapting to a transforming professional identity.

**Keywords:** consulting psychology; coping; defences; identity work; identity tension; professional identity; system psychodynamic

---

## 1. Introduction

Twenty-five years post-apartheid the profession of psychology in South Africa has been beset by identity tensions regarding its relevance and value to a continuously transforming South African society [1–3]. The identity dilemmas of the psychology profession seem evident in the disagreement among its practitioners about changing the Regulations Defining the Scope of the Profession of Psychology. The original regulations defined a scope of practice for psychologists registered with the Health Professions Council of South Africa (HPCSA) to distinguish the different categories of counselling: industrial and organisational (IO), clinical, educational and research. The original regulations were however amended in 2011 without an adequate consultation process. Unlike the original widely accepted demarcated scope of practice, the amendments narrowed down practice opportunities for some of the psychology categories and seemed more favourable to others. The process was criticized for being forced upon the profession, for creating power disparities between the different categories of psychology and for marginalising certain practitioners, preventing them from contributing to sectors in society in dire need of mental health and well-being [1,4]. A large section of the profession was therefore dissatisfied and angry, resulting in a court order on 14 November 2016 by the High Court of South Africa (Western Cape Division, Case No: 12420/13) declaring the amended regulations invalid [5]. A notice by the Department of Health not to proceed with any amendments that were recently published [6] declared the original Regulations Defining the Scope of the Profession of Psychology of 2008 to remain in force.

Disputes about the amended scope of practice reflected underlying identity tensions within the psychology fraternity. On the positive side, the discussion also aligned with evolving frames of thought about the profession and its impact on society at large and about opportunities for continued professional development and multidisciplinary collaboration. One such opportunity was created in a doctoral programme specialising in consulting psychology. The programme is offered in collaboration by the Departments of Psychology and IO Psychology at UNISA. Students admitted to the programme include registered and practising psychologists across the five different HPCSA registration categories. These educational, clinical, IO, counselling or research psychologists entering the field of consulting psychology invariably experience a role identity transition [7] or an altering of their professional identity [8].

When transitioning to a new role, people naturally experience identity tensions between previously known meanings of the self and new role expectations [9]. They attempt to resolve these tensions by engaging in identity work [10]. Identity work entails the restructuring, reframing and development of identity meanings that constitute the self in specific social contexts [11–13]. It refers to the intrapersonal process in which individuals engage to form, repair, maintain, strengthen or revise their identities [14]. Identity work is therefore relevant to intrapersonal coping, because it enables individuals to cope with difficult work-life demands [10,15] and adjust to important work-life transitions [16,17]. Breakwell introduced the idea that identity work is based on intrapsychic and interpersonal coping strategies [9]. Observation of identity work is therefore expected to reveal the intrapersonal coping strategies that individuals apply to resolve the identity tensions they experience consequent to role transitions.

The aim of this study was to develop an understanding of the coping dynamics that consulting psychology doctoral students manifest when transitioning their professional role identity. Understanding how the students cope with the identity tensions they experience will enable them to develop a self-efficacious service offering to their clients and broaden the reach of psychology in the South African context. Identity work is a natural and involuntary [7] or subconscious [18] process, which in the context of this study happens in both the individual student and in the bigger systemic context of the psychology fraternity. It was therefore deemed useful to explore coping with transitioning a professional identity by applying a systems psychodynamic perspective.

## 2. Literature Review

### 2.1. System Psychodynamics

Systems psychodynamics is rooted in psychoanalysis, object relations, systems theory and the Tavistock Human Institute of Human Relations [19,20]. Although Freud was the father of psychoanalysis, it was Klein's object relations theory, group relations and open systems theory that contributed significantly to the systems psychodynamic paradigm [21,22]. The systems psychodynamic approach developed as a suitable stance to explore beneath-the-surface behaviour in organisations [23] and to the study of organisational change dynamics [24,25]. The central tenets of system psychodynamics lie in the semantic co-occurrence of the words systems and psychodynamic. Open systems principles are firstly applied in understanding behaviour, and individual behaviour is regarded as a function of systemic dynamics as much as it is a representation of (mirroring) systemic behaviour. Secondly, psychodynamic theory represents a fundamental focus on unconscious behavioural dynamics and includes the spectrum of psychoanalytic perspectives on individual and social experiences and mental processes. This includes un understanding of anxiety as the basis of group and unconscious systemic behaviour [26,27] and the involuntary use of defence mechanisms to cope with anxiety [28].

Central to psychodynamic thinking is the assumption that part of the mental life of individuals is hidden and affects them in ways of which they are not always aware—the unconscious life [29].

Furthermore, in system psychodynamics, anxiety is regarded as the driving force of all relational dynamics in a system, as any change is seen to arouse anxiety naturally [25]. Anxiety manifests itself as personal anxiety, task-related anxiety or role anxiety but always in a systems context and representative of larger anxieties in the system [26]. Typically, transitions in the work environment arouse performance and survival anxiety [30]. In response to this anxiety, defence mechanisms are used unconsciously [20,31] to remain in control and to experience a sense of safety, security and acceptance [27,32].

Defences are fundamental to coping in psychodynamic theory [33–35]. In systems psychodynamics, defences manifest themselves as basic assumption behaviour, which includes dependence, fight-flight, pairing [21], one-ness, we-ness and me-ness [20]; in primary defences such as splitting, introjection, suppression, denial, projection and projective identification [22,28,36] and in more sophisticated defences such as rationalisation and intellectualisation [37]. Defences operate on a continuum, ranging from primitive, debilitating impairments to more sophisticated competence-enhancing adaptations [25,38]. These defence mechanisms have a protective function and are neither good nor bad [39]. Individuals develop defences as coping strategies from an early age to deal with reality and to maintain a functional sense of self [28].

There is a longstanding link between coping theory and psychodynamics [40], yet contemporary perspectives on coping strategies typically focus on cognitive-behavioural perspectives such as emotion-focussed, problem-focussed, social-support coping and religious coping [41]. Such coping strategies have been distinguished as involving conscious, purposeful effort [33]. This paper takes the stance that coping is a process of psychological adjustment [42,43] that includes conscious and unconscious coping as means of adaptation [33,34,44]. Applying a system psychodynamic stance to understand coping is of value, since it provides a more holistic and systemic understanding of coping behaviour [45] and uncovers the unconscious as a creative source of knowledge [23,46]. When experiences are explored from a systems psychodynamic stance, one can enhance awareness, understanding and learning of both conscious and unconscious dynamics [47,48]. Such knowledge is deemed essential in managing and facilitating real behavioural change [25] and prevents anxiety and defences from becoming destructive in the workplace [49].

### 2.2. Systems Psychodynamics and Identity Work

Systems psychodynamics approaches identity firstly from a group relations stance, relating the individual's role identity to a systemic role identity [22]. From this perspective, taking up a new role entails a psychosocial dynamic that emerges from the interface between the person and the formal role as defined by a particular context [50]. As such, viewed from a psychodynamic perspective, identity work involves cognitive, emotional and social processes that are social or systemic in nature (i.e., cannot be done in isolation) [12,13], and such identity work is often stimulated by anxiety [51].

System psychodynamics further focus on understanding unconscious experience. In doing identity work, a person therefore consciously, but also unconsciously, negotiates personal beliefs, needs and aspirations in adjusting to latent role demands. Projective processes that relate to psychodynamic defence dynamics, in the form of projective identification, transference and counter-transference, are benign parts to be worked with by individuals in the process of identity work [52]. Psychodynamically, identity work is therefore essential in everyday coping with tensions in the self and maintaining well-being [44].

A psychodynamic lens to identity enriches our understanding of identity work beyond current identity theory [44]. System psychodynamic literature emphasises that creating reflective spaces is essential to facilitate constructive role transition [49,53]. In this regard and relevant to the consulting psychology programme context of this study, the concept of identity workspaces refers to institutions that provide a holding environment for individuals to do their identity work [54]. This holding environment serves as a social context that not only reduces distressing emotions but also actively facilitates sense-making, either aimed at identity stabilisation or identity transition [55].

A significant portion of identity work is emotional, dynamic and complex and lies beneath the surface of conscious behaviour [18,44]. Taking a systems psychodynamic stance stresses that this intrapsychic dimension of identity work must also be worked with. Thus, by "focusing on the ongoing dynamic interaction of individual and social, cognitive and emotional, conscious and unconscious factors, the system psychodynamic perspective is particularly well suited to enriching our understanding of identity and identification" [54].

## 3. Materials and Methods

This paper is part of a larger project studying the experiences of master's and doctoral students to improve academic curricula and the throughput of postgraduate students. The next sections give an outline of the research methodology that directed the study, an explication of the research setting, the data protocols and participants, as well as the analytic strategy applied.

### 3.1. Research Methodology

A qualitative inquiry was deemed most appropriate for the exploratory and descriptive nature of the study. The study followed a hermeneutic phenomenological approach in its overall design. In hermeneutic phenomenology knowledge generation equates to the co-construction of meaning between the researcher and the researched [56]. It allows for ascribing meaning to participants' experiences through critical interpretation, inevitably influenced by researcher experience and theoretical preconceptions [57–59]. Fundamentally, from a hermeneutic phenomenological perspective, researchers' critical interpretation of participants' phenomenological experiences is imperative for rigorous scientific research [60]. To achieve the stated objective, researchers' interpretations were influenced in particular by applying a systems psychodynamic lens or meta-theoretical orientation during data analysis. System psychodynamics refers to an extensive body of scholarship [46] that provides a sound meta-theory to develop truthful, useful and credible insights and hypotheses [48]. Both researchers (a black man and a white woman) hold PhDs in psychology and are part of faculty in an open-distance, e-learning tertiary institution. They are registered psychologists with the HPCSA with a keen research interest in socio-analytic research methodologies, hermeneutically informed by system psychodynamics. They are also both involved in the teaching of the course work component of the Consulting Psychology doctoral programme.

### 3.2. Research Setting

The doctoral programme in Consulting Psychology offered at a tertiary open distance and e-learning (ODeL) institution in South Africa constituted the setting for this research. The institution's two main campuses are located in the Gauteng region, with satellite campuses across South Africa. Gauteng is the smallest province yet the most densely populated (742.6 per square metre) in South Africa, accommodating 1.349 million of the 55.9 million South Africans (https://www.southafricanmi.com/sa-by-numbers.html). National and international student enrolments at this ODeL institution varied between 354,743 and 381,483 in the 2015 to 2018 registration periods. The Consulting Psychology doctoral programme has enrolled eight to 12 students per annum since its inception in 2005. The first year of the programme entails course work involving 11 focus areas, of which one focusses on consulting as process. The learning outcome of this focus area is intended to develop students' capability to explore their consulting profile and develop a personal frame of reference for consulting. During the course work component of the Consulting Psychology doctoral programme, students attend five block weeks of face-to-face training and complete several individual and group-based projects as part of their formative assessment. Another primary focus of the first year is the successful defence of a research proposal, which forms the basis for continued research-based study from the second year until completion of the degree.

### 3.3. Data Protocols

One of the individual tasks students are required to do in their first year of study is to keep a personal diary in which they journal about becoming a consulting psychologist. The students then write self-reflective essays that entail a summative and critical reflection based on periods of five and eight months' journaling. To explore the coping dynamics of students transitioning into a consulting psychology role, self-reflective essays constituted the data protocols for this study. Critical reflexivity refers to the process in which one questions one's positionality and basic assumptions about life, people and oneself, while seeking alternative ways of being and looking at things [61–63]. In this regard, self-reflective practice is akin to identity work as defined by Sveningson and Alvesson [14]. In Nagata's definition of self-reflexivity [64], coping as an adaptational endeavour underlying identity work is furthermore evident. He defined self-reflexivity as having a conversation with the self about one's experiences while one is experiencing them, leading to the ability to regulate internal psychological and unconsciously driven responses. Self-reflective essays were moreover deemed appropriate to the purpose and system psychodynamic orientation of this study, as reflective practice elicits a meta-level of thinking and feeling about the self [64] and makes the "unthought known" observable and conscious [65].

Instructions for the self-reflective essays were based on circular existential and relational questions relevant to experiential learning activities aimed at developing critical reflexivity [61]. The questions were formulated as a guide to facilitate critical self-reflection in line with the objective to engage with and develop a personal frame of reference as a consulting psychologist. The intention with these questions was not to be prescriptive but to offer some direction to the task of self-reflecting. The first question aligns with Cunliffe's existential (who am I?) and relational (who am I in relation to something/someone?) approach [61] and asks the student to reflect on the question: Where am I at this moment regarding my personal frame of reference? The second and third questions refer to the circular influence approach [61], examining how existential and relational learnings influence one's responses and ways of acting, behaving and responding. The second question asks: How does your evolving personal frame of reference affect your understanding of the landscape of consulting psychology? The third is: What is the impact of your evolving personal frame of reference on your personal consulting profile?

Self-reflective essays are shared with lecturers after five months' journaling and again after another three months' journaling. Through purposive, convenience sampling, the transcripts that constituted the data sets for this study included all the self-reflective essays of the current course work students after eight months. By that time, the students had already completed most of their course work-related projects and workshops.

### 3.4. Participants

To ensure the confidentiality and anonymity of participants, they are not identified as individuals but described as a group. The participants included three women and five men. In terms of population group, three were white, one was Indian and four were black, with an age range of 29–58, and average age of 40. Participants were approached face to face during their first block period on campus, and the nature and purpose of the research were explained to them. This was followed up by individual e-mails requesting their participation. Inclusion criteria entailed that participants had to have a master's degree in a specific psychology domain; be registered psychologists with the HPCSA, practising for at least three years; be registered doctoral students in Consulting Psychology; be willing to share their self-reflective experiences and be willing to participate in the research. Four participants were registered in the category of IO psychology, and one each was registered as clinical, educational, counselling and research psychologist. All the participants had gained a minimum of three years' experience since obtaining their professional HPCSA registration. Pseudonyms are used in the findings to indicate the number of the participant, for example, P4 refers to participant number four.

*3.5. Ethical Considerations*

Ethics approval was obtained from the relevant Institutional Senate Ethics Committee (REF#:2017_RPCS_018). Participants consented in writing that data generated from their self-reflective work on being a consulting psychologist could be used for research purposes relevant to the bigger project. The researchers abide by the ethical codes of conduct as prescribed by the UNISA ethics policy, as well as the ethical code of psychologists registered under the HPCSA. The researchers also declare that the research was conducted in compliance with the ethical principles enunciated in the Declaration of Helsinki. As such, ethical principles of anonymity, confidentiality and informed consent were upheld by reporting on the data in a collective, interpretive sense and by using pseudonyms.

*3.6. Data Analysis*

Eight self-reflective essays were loaded as primary data documents in Atlas.ti to ease the management of the data analysis process. Data were analysed through hermeneutic phenomenological analysis according to the analytic stages of naïve reading, structural thematic analysis and comprehensive understanding [66]. The fundamental premise of the hermeneutic circle was applied throughout these stages of analysis. Initially, we considered each individual's experience in relation to the meaning we constructed from the participants' collective experience and vice versa [26]. We also compared findings from each stage with findings in the previous stage in a consistent, circular manner [67]. At this point the themes provided a richness in terms of saturation and no more themes emerged to add to the clear description of coping dynamics from a system psychodynamic perspective.

## 4. Findings

The naïve reading revealed how some students struggled to engage with the task of self-reflection in that they recited textbook definitions of the skills and competencies of a consulting psychologist. Without reflecting on their emotional responses to the personal development journey they were on, they merely noted that they were busy developing the mentioned list of skills and competencies. Others provided less detached accounts of their experience of becoming a consulting psychologist, but their performance anxiety in relation to the task topic of becoming a consulting psychologist was evident in how they engaged in anti-task behaviour. Performance anxiety was coped with in various ways, mostly to avoid the required identity work or to resist the professional identity role transformation they were experiencing. Especially, students who did not engage in self-reflection consistently throughout the year seemed to struggle with performance anxieties and integrating identity tensions. Three themes were constructed in the thematic analysis to describe the students' coping with the identity tensions and demands from a systems psychodynamic perspective. The first theme described how they resisted the primary task of self-reflection and, consequently, resisted identity work through basic assumption (anti-task) behaviour. The second theme described their coping dynamics by applying primary defences against the perceived incongruence of conscious (normative) and unconscious (phenomenological and existential) roles. The third theme described their coping along the more sophisticated defences the students applied to resolve the identity tensions and performance anxiety they experienced. The three themes are conceptualized below in terms of their different sub-themes and related categories. Verbatim data from the participants' narratives are used to illustrate the meaning-making during analysis.

*4.1. Resisting the Primary Task through Basic Assumption Behaviour*

The first sub-theme entails resisting the task of deep self-reflection and identity work through basic assumption (anti-task) behaviour, which manifested in dependency, fight behaviour, pairing and me-ness.

### 4.1.1. Dependency

Students idealised several external objects such as lecturers, the programme and consulting knowledge. In doing so, they expressed their dependency on external structures and authority figures to feel successful in their career or profession. The Consulting Psychology doctoral programme was related to the mother figure and several instances of how the programme or lecturers or the consulting psychology knowledge domain was made the hero or saviour were evident in the data. Such "heroing" usually followed students' expressed feelings of insecurity, low self-confidence and uncertainty or the need to succeed or fantasise about career opportunities and personal change. In this way, some students seemed to cope by relying on the doctoral programme to help them deal with the identity transition they were seeking and working with, while doing the task of self-reflection in the role of becoming a consulting psychologist. Table 1 below summarises pertinent verbatim extracts supporting how students acted out or expressed their dependency in an effort to cope with the identity tensions and insecurities they were experiencing.

**Table 1.** Dependency behaviour as seen in the data cited.

| Condensation | Verbatim Excerpts | Participant |
| --- | --- | --- |
| Expectation that the programme and SP knowledge will bring about change, opportunities and success edge for her | My main objective with doing specifically this programme is to become more skilled … I can see the potential that it holds for the field I am working in, and it makes me very excited. I think using this approach can facilitate change that is required in this field and also provide me with an edge as consultant. I can't help but to wonder how things will change? Will it be a marginal change, or will it open a new and exciting world with new opportunities for me? | P1 |
| Relates PhD and CP to mother and expresses gratitude for being developed through the programme. Hero-ing the programme; finding it a safe space like "mother" | My assumptions were largely drawn from my Mother's experience and interaction with her (she is a consulting psychologist) … There is an obvious soft skill curriculum that comes with attaining the highest form of academic qualification, and in this case, I feel that it, my PhD, will supplement my development nicely. | P3 |
| Dependency is demonstrated through anger at the challenging workload of the programme | I sometimes also feel that the PhD journey that I commenced this year, by registering in the programme, is demanding and stressful with regards to time and workload; however, it has always been my dream to pursue and complete my PhD. It feels as though I do not have enough time to see that my work as a student is always attended to timeously. | P2 |
| Coping with performance anxiety by pairing with the supervisor, fantasy that the supervisor will enable academic success for her | Furthermore, having Professor XX as my research supervisor and the research module co-ordinator has enabled me to have more faith in my own capabilities and strengths. | P2 |
| Finds solace in the power of the programme to help him cope with his limitations | I am thus grateful for this course because it has made me aware of this possible limitation and it thus affords me an opportunity to identify similar feelings of discomfort should they arise in future whilst I am engaged in consulting work—I would then come up with a strategy to either counsel myself to attend to those uncomfortable issues or perhaps ask a suitable colleague to assist me in that regard | P7 |
| Relies on the consulting process to enable his success | However, because I am aware of this possible limitation, I will pay special attention to my interpretations during this stage and put in measures to minimise my biases in order to ensure that a more accurate picture of the client's situation is upheld. Luckily, the consulting process itself (e.g., evaluation phase, stage 4) offers one the opportunity to evaluate their actions in each stage. | P7 |
| Finds the programme supportive of his functioning | However, I must say that after a great exposure through this programme (training as a consulting psychologist) I have come to appreciate that as psychologist, we can help each other through the sharing or exchange of knowledge. | P4 |
| Coping with her low self-confidence by looking to the programme to address her insecurities | I took it upon myself to apply for a consulting psychology programme for professional and academic development and as a challenge to myself to try and succeed in something outside of my scope of practice and the confines of clinical psychology. I have always looked down upon myself and with very low self-esteem. | P8 |
| Idealising the programme as saving her from potential limited way of thinking (performance anxiety—what I know/don't) | Consulting Psychology is different. It has allowed to me adopt a new and different frame of perspective and reference. I am now thinking in a broader and organisational environment. | P8 |

### 4.1.2. Fight

Fight reactions were a way of coping with the need to preserve the self and resist the identity tensions felt during self-reflection. One student attacked the task of self-reflection by denigrating its value in light of preserving what she feels more comfortable with, namely to be a practitioner: "When confronted by theory and philosophy, my question is always what does this look like in practice? If I can't figure that out, or it is not clear to me, it is not useful, no matter how beautiful or elegant it is" (P1). Another student demonstrated fight behaviour by attacking team work as irreconcilable with her leadership style. Again, the student is trying to preserve her known leadership identity. She reveals the identity tension she experiences and demonstrates how she copes with it through fight behaviour: "I naturally take a leadership role (my current role at work requires this) and develop innovative ideas in a team setting; this had to be suppressed at times" (P2).

### 4.1.3. Pairing

Students attempted to pair with an authoritative form or powerful other, by aligning themselves with the programme, with a lecturer or with the class as a team. In expressing her need to be part of the group, P2 copes with the anxiety she experiences in transitioning into the consulting psychologist role: "after the week, I realised that I needed the group. In May, I became more involved with the team, working in groups. Even though it was a bit uncomfortable for me working as a member of a team, I did not just play my role but enjoyed it. Furthermore, the course seemed easier as we shared similar experience and could relate." Several students strongly aligned themselves to the doctoral programme and the consulting psychology knowledge domain, proclaiming a new-found sense of confidence and security in their own abilities. P4, for example, states: "this programme has broadened my horizons. It has help[ed] me to appreciate that there are three different levels of intervention that any qualified and properly trained psychologist must be able to operate in, namely, individual, group and organisational level. The above, is an illustration of the significant impact that this programme is having on me as a professional". Similarly, P8 notes that she feels more confident as a psychologist because "there is no better programme I would recommend to my peers and colleagues than that of the Consulting PhD programme". She continues to align herself to the group in order to cope with the potential loneliness that she experiences in the identity transition process that she embarked on: "My classmates and I have an exceptional relationship where we assist each other not only academically but also personally and professionally." Through pairing, it seems that the students constructively cope with the performance anxieties that emanate from the identity demands they experience in the context of the consulting psychology doctoral programme.

### 4.1.4. Me-Ness

Students experience tension between the identity needs of belonging and of uniqueness. The push and pull between wanting to retain individuality, while wanting to be part of the group, was in particular coped with by emphasising me-ness. Through me-ness, students cope with identity transition, as is seen in a student's denial of personal change: "Looking at my framework of reference when I started the programme, I could not really say that much has changed with regards to my view My frame of reference is still focused on the individual but not excluding the group and organisational factors" (P5). Similarly, P1 notes that she also does not require much personal change and therefore does not see the need for the task of self-reflection: "Reflecting on my personal frame of reference is not something that I usually do, and I am not sure whether I will ever become in the habit of doing so. I like to think that I have a degree of self-awareness". In this way, students resist the task of self-reflection, and in doing so, they are resisting identity work by highlighting their inner strength and adequacy. By resisting the task in this way, students reveal the anxiety they experience when engaging in a task focussed on identity work, a task that uncovers identity tensions consequent to transforming their professional identity. Me-ness illustrates the individual's escape into the inner world that is felt to be safe, comfortable and good [68,69]. Coping with the task of self-reflection (identity work) through self-reliance, independence and distancing from interdependence is evident in the words of P2: "After working for almost fourteen years in management and being offered directorship at the age of thirty-eight years, I began to realise that I needed to trust my capabilities, to assure myself that I have what it takes to 'manage' whatever comes my way" and "The realisation has to some extent shaped a frame of reference for my believing that I do not necessarily need others: I cannot/do not like working as a member of a team, I do not like group work, as I find that I achieve better results, when I complete tasks in my own way and independently."

### 4.2. Primary Defences Defending against Perceived Identity Incongruences

The second theme entails defending against perceived identity incongruences. The perception of identity incongruence was evident in students experiencing tension between their conscious (normative) and unconscious (phenomenological and existential) roles. Coping with the resulting performance anxiety was evident in primary psychodynamic defences such as splitting, projection and projective identification [25]. These have also been referred to as immature defences [70].

#### 4.2.1. Splitting

To deal with the anxiety elicited in the task of self-reflection, students tended to split objects of their role identification into good and bad opposites. One split was evident in juxtaposing the academic role with the role of practitioner, projecting onto the academic role the less worthy and valuable task:

I am not sure that the academic world will ever be home to me. It might become easier for me to understand and adhere to its customs, but I don't think I will ever choose to stay there for longer than I need to. I would rather be in the world outside where I can be doing. I have always performed well academically, and everyone has expected that I would follow an academic career, but that did not interest me. The theoretical and philosophical have never appealed to me, the practical did, and that is where I positioned myself (P1).

Similarly, P6 splits the roles of scientist and practitioner yet in a less obvious manner. For the greater part of his essay, he has copied definitions of the scientist-practitioner roles and aligned himself to each definition, without demonstrating authentic reflection on how he engages with and integrates these roles. His stance remains distanced and detached, as seen in his paraphrased list of roles: "My consulting value proposition and paradigm as a scientist-practitioner is adapting and designing assessment technologies/instruments for purpose of selection, training and vocational assessment; conducting individual assessment for the purpose of selection and training, career and vocational guidance, and leadership coaching; conducting individual assessment on individual wellness and work adjustment (psychopathology and work adjustment); conducting career counselling, advice and therapy " (P6).

Another student copes by creating a split between the consulting psychologist and other "external forces" that inhibit constructive work in the organisation. In doing so, P5 reveals the performance anxiety he experiences in taking up the consulting psychologist role: "I also realise that there are forces internal and external that would make it more challenging for the Consulting Psychologist to find a balance between the organisation, the group and the individual. The challenge for the Consulting Psychologist is to divorce him/herself from these external forces such as politics". The spilt in professional roles is also evident in P4's hero ing of the consulting psychologist role as one that "is equipped with diverse knowledge and experience to help others meaningfully" as opposed to "I have always restricted my dealings with issues and clients strictly to methodologies within industrial psychologist stream."

In working with the tension between her need for belonging and her need for independence, P2 copes by splitting the self from the group: "I found myself concerned with whether the members of the team were up to date with expectations and going to an extent of contacting them through emails and some with a telephonic discussion. I consider this a degree of improvement as at least I thought about the team and did something about that thought". Splits were also evident in the different professional categories of psychology: "I have always recognised myself as a clinician who works only with the individual to target the presented pathology/illness for a desired behaviour, treatment management and curative modalities ... Consulting Psychology is different" (P8).

Splitting seemed to reveal students' perception of identity incongruence in what they normatively expected themselves to do in the role, as opposed to their inherent and obscured insecurities and fears of incompetence. Their conscious and unconscious self-expectations left them with anxious feelings, which they resolved by creating outside forces that were feared or not acceptable. The way participants cope through splitting becomes more evident when integrated with how they invariably use projection as an unconscious defensive coping strategy.

### 4.2.2. Projection

Engaging in the self-reflective exercise created discomfort because it entailed having to face and work with both positive and negative aspects of the self. Students defended themselves against this discomfort by attributing negative parts of the self to others. Narcissism and self-indulgence were projected onto the psychology profession as a whole and onto the task of reflecting on personal paradigms: "My impression is that psychologists have become so obsessed with their own paradigms and frames of reference, and in the process became so self-focused, that they are forgetting why they are here. There is a whole country in dire need of psychological services, but psychologists are pre-occupied with their paradigms, like Nero fiddling while Rome is burning" (P1). Later this student also projects egotism onto the corporate world: "The corporate world needs to learn how to effectively engage with communities, even if it [is] just for the benefit of their own triple bottom line—let's not kid ourselves, the corporate world rarely does something out of pure altruism" (P1).

Students cope with their performance anxiety by projecting inefficiency onto colleagues: "I often have a feeling that it will be better if I can do my work without interruptions from colleagues" (P2). They cope with their own fear of engaging in the self-reflection task (identity work) by projecting the fear of change onto organisations and "other people": "Some organizations may initially be apprehensive of change. The apprehension is normal ... they do not want change because change requires them to work and use energy, time and effort" (P8).

Positive attributes were also projected onto the consulting psychology profession and the consulting psychology doctorate. In a way, the students were idealising consulting psychology as an all-knowing, super-capable and competent force, which they wanted to attain: "although my frame of reference has been limited to models within industrial psychology, this programme has broadened my horizons" (P4). Similarly, P3 projects his ideal self onto the programme: "I feel that my journey through the PhD programme ... will assist in refining my work as a scientific-practitioner."

### 4.2.3. Projective Identification

Students' projective identification was manifested predominantly in identifying with the sense in the class that their careers and professional competence would be adequate if they had *enough* knowledge. The students' projective identification started with them having introjected feelings of inadequacy, low self-efficacy and a sense of knowing too little and having too little experience. P3 comments "Starting off, my point of reference with regard to consulting psychology was nothing" and P2 reflects "I reflected on what I associated the term, capability, with. My impression is that there had been a rooted feeling of doubt, not only self-doubt, but silently kept childhood statements that echoed: 'you will not be able to be on the same level with your class/group'."

The introjected inadequacy led to the fantasy that the doctoral programme would empower them to become successful consulting psychologists (their expressed normative role). Students' valence for thinking they did not know enough was further evident in expressions of rigid self-expectation: "Some workshops really started providing me with more insights in what a Consulting Psychologist is supposed to know .... Consequently, as a Consulting Psychologist I need to become more aware of how the environment, the culture and the political spectrum dictates the functioning of the organisation, secondly I need to be able to understand ..." (P5). Similarly, P8 says: "Consulting Psychology is comprehensive and quite theoretical .... . Consulting psychologists need stringent and empirically tested interventions and methods to be able to target these dynamics."

*4.3. Applying Sophisticated Defences to Cope with Identity Tension and Performance Anxiety*

In psychodynamic theory, rationalisation and intellectualisation are regarded as mature defence mechanisms [45,71]. Through these mechanisms, the students suppressed emotions and proposed rational arguments to explain or describe their experiences.

4.3.1. Intellectualisation

Examples of intellectualisation, where students suppress their emotional experiences, abound. One student, P4, engages with identity work (self-reflection) by speaking about her experience of becoming a consulting psychologist in a detached (theoretical) and absolute (over-generalising) manner: "A consulting psychologist must always be aware of the values and past experiences that he or she brings into the system or relationship with the new client. I have learnt that this helps for one to remain accurate and objective on matters while dealing with the client". Student P8 addresses her fear of incompetence by also providing a rational over-generalisation of what one *should* do when engaging in identity work: "It is important to also be aware of one's biases, personality and behavioural dynamics regarding the process as a psychologist and not allow these to hinder the consulting process". In a critical tone, using absolute terms, P1 defends against engaging personally with the identity transformation: "As a consultant, my work must be practical and useful, yet imbedded in science, and I must be aware of my impact and others' impact on me. Doing this programme is a time for me to grow and develop and acquire new skills."

Intellectualising their personal development in becoming a consulting psychologist, students avoid facing the anxiety that stems from such very personal identity work. In this way, they suppress feelings of discomfort experienced in relation to their evolving and transforming professional identity.

4.3.2. Rationalisation

The belief that knowledge imparted to students by the lecturers and the programme will enable success and competence was found in various essays. This belief shows how students idealise the programme and knowledge as an intellectual defence against the fear of incompetence. In presenting an essay that is fully paraphrased from competency lists and definitions pertaining to consulting psychology, P6 rationalises his professional development and does not engage in self-reflection at all. Also finding solace in knowledge and science, P3 rationalises that the knowledge he will gain as a result of the programme will make him a more effective psychologist: "This potential is where positive psychology can be enlisted as a new frame of reference to carry on the implementation of growth and actualising steps ... So far, conceptually, it would appear that my personal knowledge base, at this stage of my development, would constitute Psychodynamics, Dynamic Systems Theory and Positive Psychology." In his explanation of how he looks at a client context, P5 notes: "As an assessment practitioner, the data tells you something and 'it is what it is'." Similarly, P7 rationalises his consulting skills in relation to a process model of consulting psychology; he seems to find comfort in the belief that the theoretical model will provide him with the ability to ensure success:

In the third stage, intervention phase, care should be given to ensure that the interventions to be implemented are a joint venture (joint action plan) between myself and the client and not necessarily my own prescriptions. The last stage, evaluation phase, is of importance for me because it allows me to be reflective and critical of myself and the consulting process in each stage, so as to address possible challenges and ensure that the consulting process brings about desired outcomes.

In these ways, students do not take ownership of their identity transformation, because of fearing their incompetence. They rather rationalise about how the programme will impart the knowledge they need, or how theoretical models will show them the way to be effective in consulting.

## 5. Discussion

This study aimed to develop an understanding of the coping dynamics that consulting psychology doctoral students employ when transitioning their professional role identity. In working with their professional identity, students show fear of incompetence as well as anxiety in relation to preserving the self. Such fear and anxiety are defined as performance and survival anxiety [30] and are typically also found in other system psychodynamic studies working with identity construction [26,45,72]. To deal with their performance and survival anxieties, the three themes constructed in the findings describe the students' coping dynamics from a system psychodynamic stance in terms of basic assumption or anti-task behaviour, primary and sophisticated psychodynamic defences.

Students firstly engaged in basic assumption or anti-task behaviour such as dependency, fight, pairing and me-ness as a way of coping with their performance and survival anxieties. In doing so, they attempted to regain a sense of competence and self-efficacy while transitioning their professional role identity. In idealising consulting psychology, as well as the doctoral programme and its lecturers, students demonstrated a dependency mentality to resolve the insecurity and self-doubt they felt in having to take up the role of consulting psychologist. Similarly, pairing with these same objects, which they deemed powerful and authoritative, helped them to cope with feeling incompetent. Students therefore sought out the consulting psychology knowledge domain and the doctoral programme with its lecturers, as a container of their performance anxiety. To preserve the self and deal with survival anxieties in taking up the consulting psychology role, students reverted to an anti-task mentality of fight and me-ness. By attacking the task of self-reflection (i.e., the task of identity work) and resisting teamwork, students tried to preserve a sense of self that they knew and felt comfortable with. In doing so, they split the self from the task or from the team and projected onto the task/team the feelings of discomfort and suspicion related to their ability to be a consulting psychologist. By emphasising me-ness, students also attempted to preserve the self in relation to the identity tension they experienced between wanting to belong to the consulting psychology fraternity and retaining their unique psychologist identity.

Anxiety that resulted from perceived identity incongruences in their normative and existential or phenomenological role parts secondly surfaced in how they used projection, splitting and projective identification to deal with the ensuing fears of not being good enough. Taking up a new role entails taking up a role that is given or normative, which refers to the rational and measurable work related to the task [26,73]. At the same time, it includes taking up an informal role, reflected in the personal, frequently unconscious needs and aspirations of the individual [50]. The informal or unconscious role is also referred to as the existential or phenomenological role parts [26]. Students demonstrated these primary psychodynamic defences specifically in relation to coping with the incongruences they experienced in their formal or conscious task and the informal/unconscious task to manage the self in relation to the other. Through splitting the domains and roles of academic/scientist and practitioner, students coped with the tension of not feeling good enough in their existential/phenomenological role. Their self-doubt was recognised in the high expectations they introjected in relation to the consulting psychology role and the concurrent projective identification of not being good enough. Students continued to cope with their performance anxiety by projecting negative attributes such as narcissism and inefficiency onto others in an effort to preserve their sense of self.

Thirdly, they rationalised and intellectualised their work identity and professional role transition, in an attempt not to deal with the unwanted emotions that resulted from their identity work. Further expression of their normative role was also found in students' intellectualisation and *rationalisation* of consulting psychology as an all-powerful knowledge domain. Through these two sophisticated defences, students talked about their consulting psychology role in a way that emphasised absolute knowledge and competence, ultimately demonstrating incongruence with their felt (introjected) sense of incompetence. In this way, reasoning and rational thought were used to avoid dealing with difficult emotions [37,74].

The findings show how consulting psychology doctoral students engage with and express identity work when they reflexively engage and formulate self-reflective thoughts about their experience of taking up a new professional role identity. It is hypothesised that active identity work (such as in self-reflective activities) is valuable for students to deal with their performance and survival anxiety, because in doing so, they uncover their unconscious coping dynamic, a dynamic that is a normal part of facilitating their adjustment to transitioning into a new professional identity. This finding is in support of reflective spaces or identity workspaces, which according to system psychodynamic theory provide a safe space to surface and consciously work with below-the-surface anxieties that result from identity transition [54,55]

Employing certain defence mechanisms to cope with performance anxiety has traditionally been described as maladaptive or pathological [35]. In the present study, the defensive coping dynamic that was evident in the students' identity work is rather proposed to be a natural (normal) and evolving process of transitioning into a new professional role [54]. Some psychodynamic perspectives on coping emphasise hierarchies of defensive coping, depending on the level of reality distortion—from psychotic and immature to mature [70,75]. These theories show that coping *evolves* as a person's cognitive functions mature [76]. This study further supports the notion of defences as behavioural phenomena, rather than referring to defence mechanisms as measurable constructs [77]. It is therefore conjectured that defensive coping is an important, dynamic and interrelated behavioural phenomenon that could potentially be conducive to adjustment, because it is a natural part of identity work. Defensive coping is an important part of understanding the coping phenomenon holistically, and it is in working with all the parts of the coping dynamic that the whole adjustment process can be facilitated. Defensive coping is conducive to adjustment, as it progresses the identity work relevant to professional role transition and brings to the surface the otherwise unconscious coping dynamic. Recently, even quantitative studies exploring the correlation between coping and defences have confirmed that specific adaptive strategies can only be effectively employed when unconscious processes have been attended to [78]. Active identity work is a way to develop consciousness of the self when taking up a new professional role.

Therefore, rather than speaking of either adaptive or immature coping strategies and either psychotic or mature defence mechanisms, coping is a dynamic phenomenon that includes defensive coping. Studies show that people constantly re-evaluate and reconstruct the self in the work context [72]. Unearthing the unconscious dynamics of defensive coping through conscious identity work may develop the resilience required to establish a professional identity in which the self is both interdependent and unique. A study of the systems psychodynamic role identity of academic supervisors [26] similarly demonstrates how focusing solely on conscious behavioural adaptation can limit valuable insight into unconscious adaptive identity work. We argue that the human coping phenomenon is studied only in part if the covert and unconscious social dimensions of coping, such as splitting, projections and projective identification, are ignored. The system psychodynamic stance, with its focus on the unconscious and irrational forces in human behaviour, moves beyond the mainstream cognitive coping theory that focusses on rational, conscious coping, by taking a depth perspective on this phenomenon. To focus solely on conscious behavioural adaptation may thus limit valuable insight into unconscious adaptive identity work.

The findings also have implications for the work of faculty in general and educators in particular. It is evident that students should be provided with more conscious and structured support. Firstly, it would serve to sensitise (creating awareness) faculty and educators to the anxiety-provoking realities of identity transition and identity formation. Secondly, it would create safe, contained spaces for conscious reflection nestled in the different components of the consulting psychology doctoral programme. Finally, identity work, in the form of self-reflective activities, should become the norm (rather than the exception), as a way of developing consciousness of the self. In doing this, coping would be appreciated as a dynamic phenomenon, and in the process, resilience would be nurtured as required in transitioning a new professional identity.

A system psychodynamic perspective to coping would be incomplete if the discussion does not allow for constructing interpretations and hypotheses of the individual as reflective of the larger system and the dilemmas and challenges faced by it [71]. This study therefore also has implications for understanding how the psychology profession in South Africa may be coping with its unresolved identity tensions as a collective. The students' identity work reflects the performance and survival anxieties evident in the psychology profession, as their defensive coping mirrors similar dynamics in the profession as a collective. From the background to this article, the splitting of professional categories, fight behaviour, pairing with the legal system and intellectualisation and rationalisation through the formulation and reformulation of regulations demonstrate how the psychology profession is coping with its identity tensions and are suggestive of an identity transition. Like the students' experiences, the psychology profession will have to find ways to "normalise" this tension (both a challenge and an opportunity) in search of an identity that is relevant and dynamic, given the ever-changing socio-political and economic landscape and changing societal needs.

This study is limited in the extent to which the range of the coping dynamic has been discussed, purely because of the limited scope of taking a specific approach. As such, the authors acknowledge that the findings reflect a specific stance that does not necessarily demonstrate the holistic coping dynamic we are advocating, because it does not deal with overt, cognitive coping strategies. As researchers we acknowledge the limitation associated with small sample sizes and qualitative analysis, and therefore, we do not claim that our findings constitute an absolute or generalisable truth. In true hermeneutic phenomenological fashion, we present the findings as a perspective that may add value to scholarly understanding in working with identity conflicts and related coping dynamics. The value of the psychodynamic approach to coping was celebrated in the findings and highlighted defensive coping as an essential and natural part of the whole coping dynamic. We support research proponents of in-depth, qualitative inquiry into the study of defensive coping [77]. Continuous interpretive inquiry from a psychodynamic stance, to build theory abductively [77], is recommended to enhance our understanding of defensive coping as part of adjustment rather than pure evidence of maladaptive coping. In this vein, we recommend research exploring students' identity work as it evolves in their self-reflective work throughout the year, exploring the aspect of maturation in the phenomenon of defensive coping.

## 6. Conclusions

In general, consulting psychology doctoral students perform well in terms of coping with their professional identity development, as they have completed their first year of doctoral studies successfully and are all well on track towards completing the degree. An exploration of below-the-surface dynamics reveals an interesting defensive coping dynamic that contributes to a more holistic understanding of coping with identity transition in the psychology profession. Consciously, the students engaged in self-reflection about taking up the role of consulting psychologist in a rational and intelligent manner. Unconsciously, while transitioning a professional role identity, they experienced performance and survival anxiety, which became conscious in the process of self-reflection or identity work and in exploring their coping dynamics from a systems psychodynamic perspective. The students also mirrored the identity work and defensive coping of the larger psychology profession. The identity work of the individual professional cannot be detached from the identity work of the collective system but rather provides potential insight into the collective on its adaptive functioning.

System psychodynamic coping is a dynamic process phenomenon that should not be conceptually limited to the discussion of defence mechanisms. Coping with identity tensions and demands includes defensive coping, which is a natural phenomenon and something that is engaged with in everyday life as our professional careers develop. Conscious and active identity work reveals the related unconscious dynamics. The task of self-reflection not only propels the students to do identity work, but in the process, they become aware of their defensive coping and continue to develop reflexivity and resilience in transitioning to a new professional identity. Proponents of system psychodynamics view wellness as a relational and systemic concept [68]. A certain level of congruence between the internal and external reality of the self is needed for wellness to be developed and sustained. This wrestling with identity tension is therefore not only important, but necessary as well. In becoming aware of their identity-related anxieties and defensive coping, professionals, such as the students in this study, can feed this awareness back into the profession and collectively start to model the courage to acknowledge insecurities, take back their projections and repair splits. Psychological well-being and coping result from healthy intrapersonal (within a person) and interpersonal (between people) relations [54].

Developing consulting psychology competence is valuable in facilitating behavioural change in organisations [79,80]. The doctorate in consulting psychology is particularly important in the South African context to bridge professional divides in the psychology profession and draw from the multidisciplinary pool of skills and competence to continue to address the country's mental health needs effectively on all levels.

**Author Contributions:** Conceptualisation, A.B.; data curation, A.B. and A.-P.F.; formal analysis, A.B.; investigation, A.B. and A.-P.F.; methodology, A.B. and A.-P.F.; project administration, A.B. and A.-P.F.; validation, A.-P.F.; writing—original draft preparation, A.B.; writing—review and editing, A.B. and A.-P.F. All authors have read and agreed to the published version of the manuscript.

**Funding:** This research received no external funding.

**Acknowledgments:** The authors would like to acknowledge their Ph.D. students for the selfless sharing of their phenomenological experiences and their permission to use their personal experiences for research purposes.

**Conflicts of Interest:** The authors declare no conflict of interest.

## References

1. Laher, S. Scope of Practice: Boon or Bane? *Psytalk.* Available online: http://psytalk.psyssa.com/scope-practice-boon-bane/ (accessed on 12 November 2019).
2. Pretorius, G. Reflections on the scope of practice in the South African profession of psychology: A moral plea for relevance and a future vision. *South Afr. J. Psychol.* **2012**, *42*, 509–521. [CrossRef]
3. Van Zyl, L.E.; Nel, E.; Stander, M.W.; Rothmann, S. Conceptualising the professional identity of industrial or organisational psychologists within the South African context. *SA J. Ind. Psychol.* **2016**, *42*. [CrossRef]
4. Flax, M. Are Definitions of Practice Hindering the Psychology Profession? Available online: https://www.sacap.edu.za/blog/counselling/mental-health-care-in-south-africa/ (accessed on 4 March 2020).
5. HPCSA. Update on the Review of the Regulations Relating to the Scope of the Profession of Psychology. Available online: https://www.psyssa.com/update-on-the-review-of-the-regulations-relating-to-the-scope-of-the-profession-of-psychology/ (accessed on 15 October 2019).
6. Department of Health. *Government Notice R1169: Notice not to Proceed with the Proposed Regulations Defining the Scope of the Profession of Psychology*; Government Gazette No.42702: Pretoria, South Africa, 2019.
7. Stets, J.E.; Serpe, R.T. Identity Theory. In *Handbook of Social Psychology*, 2nd ed.; DeLamater, J., Ward, A., Eds.; Springer: New York, NY, USA, 2013; pp. 31–60.
8. Pratt, M.G.; Rockmann, K.W.; Kaufmann, J.B. Constructing professional identity: The role of work and identity learning cycles in the customization of identity among medical residents. *Acad. Manag. J.* **2006**, *49*, 235–262. [CrossRef]
9. Breakwell, G.M. *Coping with Threatened Identities*, 2nd ed.; Psychology Press: London, UK, 2015; pp. 1–222.
10. Alvesson, M. Self-doubters, strugglers, storytellers, surfers and others: Images of self-identities in organization studies. *Hum. Relat.* **2010**, *63*, 193–217. [CrossRef]
11. Brown, A.D. Identities and identity work in organizations. *Int. J. Manag. Rev.* **2014**, *17*, 20–40. [CrossRef]
12. Kreiner, G.E.; Hollensbe, E.C.; Sheep, M.L. Where is the "Me" among the "We"? Identity work and the search for optimal balance. *Acad. Manag. J.* **2006**, *49*, 1031–1057. [CrossRef]

13. Snow, D.A.; Anderson, L. Identity work among the homeless: The verbal construction and avowal of personal identities. *Am. J. Sociol.* **1987**, *92*, 1336–1371. [CrossRef]
14. Sveningsson, S.; Alvesson, M. Managing managerial identities: Organizational fragmentation, discourse and identity struggle. *Hum. Relat.* **2003**, *56*, 1163–1193. [CrossRef]
15. Pals, J.L. Narrative identity processing of difficult life experiences: Pathways of personality development and positive self-transformation in adulthood. *J. Pers.* **2006**, *74*, 1079–1110. [CrossRef]
16. Ibarra, H.; Barbulescu, R. Identity as narrative: Prevalence, effectiveness, and consequences of narrative identity work in macro work role transitions. *Acad. Manag. Rev.* **2010**, *35*, 135–154.
17. Kirpal, S. Researching work identities in a European context. *Career Dev. Int.* **2004**, *9*, 199–221. [CrossRef]
18. Saayman, T.; Crafford, A. Negotiating work identity. *SA J. Ind. Psychol.* **2011**, *37*. [CrossRef]
19. Brunning, H. *Psychoanalytic Essays on Power and Vulnerability*; Karnac: London, UK, 2014; pp. 12–98.
20. Stapley, L. *Individuals, Groups and Organizations Beneath the Surface: An Introduction*; Karnac: London, UK, 2006; pp. 12–74.
21. Bion, W.R. *Experiences in Groups*; Tavistock: London, UK, 1961; pp. 1–200.
22. Cilliers, F. The experienced impact of systems psychodynamic leadership coaching amongst professional in a financial service organisation. *SA J. Eco. Manag. Sci.* **2018**, *21*. [CrossRef]
23. Long, S. Socioanalytic methodology. In *Socioanalytic Methods: Discovering the Hidden in Organanisations and Social Systeims*; Long, S., Ed.; Karnac: London, UK, 2013.
24. Krantz, J. Dilemmas of organizational change: A systems psychodynamic perspective. In *The Systems Psychodynamics of Organizations: Integrating the Group Relations Approach, Psychoanalytic, and Open Systems Perspectives*; Could, L., Stapley, L.F., Stein, M., Eds.; Routledge: London, UK, 2001; pp. 133–156.
25. Barabasz, A. Psychodynamic perspective of organizational change. *Management* **2016**, *20*, 155–166. [CrossRef]
26. Cilliers, F. The systems psychodynamic role identity of academic research supervisors. *S. Afr. J. High. Educ.* **2017**, *31*, 29–49.
27. Clarke, S.; Hahn, H.; Hoggett, P. *Object Relations and Social Relations: The Implications of the Relational Turn in Psychoanalysis*; Karnac: London, UK, 2008; pp. 12–74.
28. Blackman, J.S. *101 Defences: How the Mind Shield Itself*; Brunner-Routledge: New York, NY, USA, 2004; pp. 12–32.
29. De Board, R. *The Psychoanalysis of Organisations: A Psychoanalytic Approach to Behaviour in Groups and Organisations*; Routledge: London, UK, 2014; pp. 11–58.
30. Steyn, M.; Cilliers, F. The systems psychodynamic experiences of organisational transformation amongst support staff. *SA J. Ind. Psychol.* **2016**, *42*. [CrossRef]
31. Lipgar, R.M.; Pines, M. *Building on Bion: Branches: Contemporary Developments and Applications of Bion's Contributions to Theory and Practice*; Jessica Kingsley: London, UK, 2003.
32. Gabriel, Y.; Carr, A. Organisations, management and psychoanalysis: An overview. *J. Manag. Psych.* **2002**, *17*, 348–365.
33. Cramer, P. Coping and defense mechanisms: What's the difference? *J. Pers.* **1998**, *66*, 919–946. [CrossRef]
34. Vaillant, G.E. Involuntary coping mechanisms: A psychodynamic perspective. *Dialogues Clin. Neurosci.* **2011**, *13*, 366–370.
35. Aldwin, C.M.; Brustrom, J. Theories of coping with chronic stress. In *Coping with Chronic Stress*; Gottlieb, B., Ed.; Springer: New York, NY, USA, 1997; pp. 75–103.
36. Petriglieri, G.; Stein, M. The unwanted self: Projective identification in leader's identity work. *Org. Stud.* **2012**, *33*, 1217–1235. [CrossRef]
37. Balls, M. Rationalisation and Intellectualisation. *Altern. Lab. Anim.* **2015**, *43*, 49–50. [CrossRef] [PubMed]
38. Vansina, L.S.; Vansina-Cobbaert, M. *Psychodynamics for Consultants and Managers*; Wiley-Blackwell: Hoboken, NJ, USA, 2008; pp. 14–27.
39. Czander, W.M. *The Psychodynamics of Work and Organisations*; Guildford Press: New York, NY, USA, 1993; pp. 1–401.
40. Lazarus, R.S. Coping theory and research: Past, present, and future. *Psychosom. Med.* **1993**, *55*, 234–247. [CrossRef] [PubMed]
41. Du Plessis, M.; Martins, N. Developing a measurement instrument for coping with occupational stress in academia. *SA J. Ind. Psychol.* **2019**, *45*. [CrossRef]

42. Barnard, A.; Clur, L.; Joubert, Y. Returning to work: The cancer survivor's transformational journey of adjustment and coping. *Int. J. Qual. Stud. Health Well-Being* **2016**, *11*, 32488. [CrossRef] [PubMed]

43. Fennell, P.A. *The Chronic Illness Workbook: Strategies and Solutions for Taking back Your Life*; New Harbinger Publications: Oakland, CA, USA, 2001; pp. 12–39.

44. Petriglieri, G. A Psychodynamic perspective on identity as fabrication. In *The Oxford Handbook of Identities in Organizations*; Oxford University Press: Oxford, UK, 2020; pp. 168–185.

45. Cilliers, F.; Harry, N. The systems psychodynamic experiences of first-year master's students in industrial and organisational psychology. *SA J. Ind. Psychol.* **2012**, *38*, 117–126. [CrossRef]

46. Bollas, C. *The Shadow of the Object: Psychoanalysis of the Unthought Known*, 2nd ed.; Columbia University Press: New York, NY, USA, 2017; pp. 13–20.

47. French, R.; Vince, R. Learning, managing, and organizing: The continuing contribution of group relations to management and organization. In *Group Relations Management and Organization*; French, R., Vince, R., Eds.; Oxford University Press: Oxford, UK, 2002; pp. 1–22.

48. Vince, R. Institutional illogics: The unconscious and institutional analysis. *Organ. Stud.* **2018**, *40*, 953–973. [CrossRef]

49. Long, S. The unconscious won't go away—Especially in organisations. *Org. Soc. Dyn.* **2019**, *19*, 218–229. [CrossRef]

50. Sievers, B.; Beumer, U. Organisational role analysis and consultation: The organisation as inner object. In *Coaching in Depth*; Long, S., Newton, J., Sievers, B., Eds.; Routledge: London, UK, 2006; pp. 65–82.

51. Alvesson, M.; Willmott, H. Identity regulation as organizational control: Producing the appropriate individual. *J. Manag. Stud.* **2002**, *39*, 619–644. [CrossRef]

52. Knapp, H.D. Projective identification: Whose projection-whose identity? *Psychoanal. Psychol.* **1989**, *6*, 47–58. [CrossRef]

53. Krantz, J. Work culture analysis and reflective space. In *Socioanalytic Methods: Discovering the Hidden in Organanisations and Social Systems*; Long, S., Ed.; Karnac: London, UK, 2013; pp. 13–22.

54. Petriglieri, G.; Petriglieri, J.L. Identity workspaces: The case of business schools. *Acad. Manag. Learn. Educ.* **2010**, *9*, 44–60.

55. Petriglieri, G.; Petrilieri, J.L.; Wood, J.D. Fast tracks and inner journeys: Crafting portable selves for contemporary careers. *Admin. Sci. Q.* **2018**, *63*, 479–525. [CrossRef]

56. Crowther, S.; Ironside, P.; Spence, D.; Smythe, L. Crafting stories in hermeneutic phenomenology research: A methodological device. *Qual. Health Res.* **2016**, *27*, 826–835. [CrossRef] [PubMed]

57. Davidsen, A.S. Phenomenological approaches in psychology and health sciences. *Qual. Res. Psychol.* **2013**, *10*, 318–339. [CrossRef] [PubMed]

58. Laverty, S.M. Hermeneutic phenomenology and phenomenology: A comparison of historical and methodological considerations. *Int. J. Qual. Methods* **2003**, *2*, 21–35. [CrossRef]

59. Tan, H.; Wilson, A.; Olver, I. Ricoeur's theory of interpretation: An instrument for data interpretation in hermeneutic phenomenology. *Int. J. Qual. Methods* **2009**, *8*, 1–15. [CrossRef]

60. Kafle, N.P. Hermeneutic phenomenological research method simplified. *Bodhi Interdiscip. J.* **2013**, *5*, 181–200. [CrossRef]

61. Cunliffe, A.L. Republication of "On Becoming a Critically Reflexive Practitioner". *J. Manag. Educ.* **2016**, *40*, 740–746. [CrossRef]

62. Giampapa, F. The politics of researcher identities: Opportunities and challenges in identities research. In *The Routledge Handbook of Language and Identity*; Preece, S., Ed.; Routledge: London, UK, 2016; pp. 289–303.

63. Petriglieri, G. F**k science!? An invitation to humanize organization theory. *Organ. Theory* **2020**, *1*, 1–18. [CrossRef]

64. Nagata, A.L. Promoting self-reflexivity in intercultural education. *J. Intercult. Commun.* **2004**, *8*, 139–167.

65. Zietsma, C.; Toubiana, M. The valuable, the constitutive, and the energetic: Exploring the impact and importance of studying emotions and institutions. *Organ. Stud.* **2018**, *39*, 427–443. [CrossRef]

66. Lindseth, A.; Norberg, A. A phenomenological hermeneutical method for researching lived experience. *Scand. J. Caring Sci.* **2004**, *18*, 145–153. [CrossRef] [PubMed]

67. Churchill, S.D. Explorations in teaching the phenomenological method: Challenging psychology students to "grasp at meaning" in human science research. *Qual. Psychol.* **2018**, *5*, 207–227. [CrossRef]

68. Henning, S.; Cilliers, F. Constructing a systems psychodynamic wellness model. *SA J. Ind. Psychol.* **2012**, *38*. [CrossRef]

69. Turquet, P.M. Leadership—The individual in the group. In *Analysis of Groups*; Gibbard, G.S., Hartman, J.J., Mann, R.D., Eds.; Jossey-Bass: San Francisco, CA, USA, 1974; pp. 349–386.

70. Vaillant, L.M. *Changing Character: Short-Term Anxiety-Regulating Psychotherapy for Restructuring Defences, Affects, and Attachment*; Basic Books: New York, NY, USA, 1997; pp. 15–28.

71. Nagel, C. *Psychodynamic Coaching: Distinctive Features*; Routledge: New York, NY, USA, 2020; pp. 11–62.

72. Mayer, C.-H.; Tonelli, L.; Oosthuizen, R.M.; Surtee, S. 'You have to keep your head on your shoulders': A systems psychodynamic perspective on women leaders. *SA J. Ind. Psychol.* **2018**, *44*. [CrossRef]

73. Green, Z.G.; Molenkamp, R.J. The BART System of Group and Organizational Analysis: Boundary, Authority, Role and Task. 2005. Available online: https://www.semanticscholar.org/paper/The-BART-System-of-Group-and-Organizational-Role-Green-Molenkamp/29cdfab7f00f8b4990e802bdad813a6ac750f018 (accessed on 29 February 2020).

74. Mayer, C.-H.; Flotman, A.-P. Constructing identity. In *Muslim Minorities, Workplace Diversity and Reflexive HRM*; Mahadevan, J., Mayer, C.-H., Eds.; Routledge: New York, NY, USA, 2017; pp. 61–76.

75. Haan, N. The assessment of coping, defense, and stress. In *Handbook of Stress: Theoretical and Clinical Aspects*; Goldberger, L., Breznitz, S., Eds.; Free Press: New York, NY, USA, 1993; pp. 258–273.

76. Radnitz, C.L.; Tiersky, L. Psychodynamic and cognitive theories of coping. In *Coping with Chronic Illness and Disability*; Martz, E., Livneh, H., Wright, B., Eds.; Springer: Boston, MA, USA, 2007; pp. 29–48.

77. Mihalits, D.S.; Codenotti, M. The conceptual tragedy in studying defense mechanisms. *Integr. Psychol. Behav. Sci.* **2020**, *54*, 354–369. [CrossRef] [PubMed]

78. Firoozi, M. Interaction of conscious and unconscious mechanisms in coping with chronic pain. *Anesthesiol. Pain.* **2020**, *10*, 1–14.

79. Falender, C.A.; Shafranske, E.P. Consultation in psychology: A distinct professional practice. In *Consultation in Psychology: A Competency-Based Approach*; Falender, C.A., Shafranske, E.P., Eds.; American Psychological Association: Washington, DC, USA, 2020; pp. 11–35.

80. Lowman, R.L. *An Introduction to Consulting Psychology: Working with Individuals, Groups, and Organizations*; American Psychological Association: Washington, DC, USA, 2016; pp. 11–70.

International Journal of
*Environmental Research and Public Health*

*Article*

# Women in Refugee Camps: Which Coping Resources Help Them to Adapt?

**Orna Braun-Lewensohn [1,*], Sarah Abu-Kaf [1] and Khaled Al-Said [1,2]**

[1]  Conflict Management and Resolution Program, Department of Multidisciplinary Studies, Ben-Gurion University of the Negev, Beersheba 8410501, Israel; aks@bgu.ac.il (S.A.-K.); haled70@gmail.com (K.A.-S.)
[2]  Kay Academic College of Education, Beersheba 84536, Israel
*  Correspondence: ornabl@bgu.ac.il; Tel.: 972-8-6461391

Received: 16 September 2019; Accepted: 17 October 2019; Published: 18 October 2019

**Abstract:** The present study aimed to explore the coping resources and mental health of women who have fled Syria to a neighboring European country. To that end, we examined the roles of sociodemographic factors, situational factors, and personal and community sense of coherence (SOC and ComSOC, respectively) in mental-health outcomes. One hundred and eleven refugee women aged 19–70 filled out self-reported questionnaires during August 2018 in a refugee camp in Greece. The questionnaires asked the participants for demographic information (i.e., age, level of education level, and time spent in the camp) and also addressed the situational factors of having received aid from various organizations, appraisal of danger during the war in Syria, and exposure to war experiences, as well as the coping resources of SOC and ComSOC. The results show that time spent in the camp, appraisal of danger, SOC, and ComSOC all play significant roles in predicting the variance of various mental-health outcomes. Together, those factors predict 56% of anxiety, 53% of depression, and 58% of somatization. SOC was also found to mediate the relationships between time spent in the camp and outcome variables, as well as the relationships between the appraisal of danger and the outcome variables. This indicates that SOC is crucial for good adaptation. These results will be discussed in light of the salutogenic theory.

**Keywords:** women; refugees; coping; mental health

---

## 1. Introduction

Since 2011, more than 511,000 Syrians have lost their lives to armed conflict and more than 12 million others have been forced from their homes due to the civil war and the penetration of ISIS forces into Syria. To date, more than 5.6 million of those in need have sought refuge outside Syria, mostly in neighboring countries [1]. In fleeing for their lives, refugees face many other stressors related to their war experiences. They seek to resettle in a new country while having to learn a new language. They also face poverty and a lack of resources, which place them at additional risks of violence, discrimination, and social isolation [2].

Based on the salutogenic model [3,4], the present study sought to explore the coping resources of and common mental-health problems (including anxiety, depression, and somatization) among refugee women who have fled Syria to Greece. Specifically, we aimed to examine the roles of educational levels and the time spent in the refugee camp in these women's adaptation to life in a new country. Based on these sociodemographic factors, we compared the women in terms of several dimensions, namely, a personal sense of coherence (SOC), a community sense of coherence (ComSOC), exposure to the armed conflict, appraisal of danger in the war zone, and whether they had received aid from any of a variety of organizations. In addition, we also wanted to further understand the variables that could explain mental health and adaptation among refugees.

### 1.1. Refugee Women

In times of conflict, women are characterized as powerless victims. During the civil war in Syria, women have faced forms of structural violence from the Syrian regime. Inequalities are emphasized and this affects the ways in which Syrian women experience the process of becoming a refugee. Overall, Syrian women and refugee Syrian women, in particular, are affected by systems of power that marginalize them and their voices [5].

As refugees, these women have been exposed to multiple forms of insecurity and violence. Moreover, a lack of adequate housing adds an additional layer of insecurity and vulnerability to their lives. Some studies have reported that these women face gender-biased violence [6,7]. The lack of suitable housing and access to sanitary facilities also affect the women's physical and mental health and well-being [6]. Indeed, some studies have indicated that women refugees are an especially vulnerable population, with high rates of depression and other mental-health problems [7].

### 1.2. Experiences of War

Direct exposure to the war in Syria have led the participants in this study to flee their homes and become refugees. Exposure to war events refers to the individuals' experience of bombs falling and damaging their neighborhoods and surroundings, as well as harm caused to their acquaintances as a result of the war [8]. This type of exposure to violence is likely to increase the risk of psychological problems such as anxiety, depression, and somatization, especially during the first stage of migration [9]. However, results of studies regarding the cumulative exposure to violent political events are inconclusive [10]. While some research on refugees has shown associations between exposure (i.e., the number of events and their intensity) and various psychological problems [11], other studies that have examined war experiences (e.g., one's community being attacked by rockets/bombs, the experience of someone an individual knows being hurt as result of the war, the experience of having a relative hurt as a result of the war, having been hurt as result of the war, and having had one's home damaged as result of the war) have indicated that the number of events is not the most significant predictor of post-traumatic stress or other internalizing or externalizing psychological problems [12]. In other work, coping resources such as SOC have been shown to mediate the relationship between exposure to war events and stress-related reactions [13]. Thus, it seems important to evaluate the role of these factors in the context of women refugees, in order to understand their adjustment to life in a new country after having experienced and fled from war.

### 1.3. Appraisal of Danger

The primary appraisal is the evaluation of the original threat in order to estimate the current threat, and the secondary appraisal is the assessment of the resources one has in order to deal with the stressor [14]. The evaluation of whether a situation represents a threat or a challenge determines the level of arousal and which of the coping resources one has in his/her repertoire that will be drawn upon to deal with the situation [15]. Studies that have examined this variable in the context of war and terror have shown that women are more vulnerable and report more feelings of danger as compared to men [16]. Research has shown that these feelings seem to be an independent predictor of a variety of mental-health symptoms. That is, the stronger the feelings of danger, the higher the threat appraisal and the more intense the mental-health symptoms [8].

### 1.4. Receiving Aid

There is a debate as to whether humanitarian aid that focuses mainly on material and social support and is funded by a variety of organizations with political agendas actually benefits the refugees who receive it, or whether it harms or does not affect them. Indeed, most studies in this domain lack empirical evaluations [17]. One study that tried to evaluate whether humanitarian aid mitigates or exacerbates the effects of war on stress reactions was based on two interviews and did not draw a clear conclusion. Additionally, despite the problematic idea of organizations driven by external interests, that work did not suggest relying solely on the refugees' needs and priorities [18]. A recent study showed that receiving aid from any of a variety of organizations did not play a significant role in reducing psychological problems among refugee youth and had only a moderate relationship with their expectations [12]. The present study tries to add additional knowledge to fill this lacuna in the research by examining the role of aid from a variety of organizations in reducing psychological problems among refugee women.

*1.5. The Salutogenic Model and Sense of Coherence (SOC)*

The secondary appraisal facilitates the exploration of the resources available to the individual to deal with a stressful situation. In this study, we examined the coping resource of SOC, which is rooted in the salutogenic model [3] and is an important concept in positive psychology [19]. The salutogenic model looks for functions of positive qualities rather than healing from sickness [19,20]. Thus, the present study focuses on coping and resilience resources rather than risk factors. The main construct of this model, SOC, is an enduring tendency to see the world as more or less comprehensible, manageable, and meaningful [4]. In accordance with the salutogenic theory, a person with a strong SOC will be more likely to evaluate a particular stimulus as neutral [19]. Therefore, an individual with a strong SOC is less likely than one with a weak SOC to perceive stressful situations as threatening and anxiety-provoking. SOC determines the ability of individuals to use resources that are available to them to promote their well-being [21]. Moreover, SOC includes components that consolidate resilience and enhance subjective mental health [19]. Indeed, numerous studies have shown that SOC may be considered a protective factor that helps to moderate and mediate stress experiences (e.g., [22,23]). Furthermore, Evans and Davis [22] showed that the ways in which SOC acts through family, community, and cultural dimensions can aid successful coping and reduce stress among marginalized and minority ethnic groups.

*1.6. Community Sense of Coherence (ComSOC)*

From a socio-ecological perspective [24], community is an important resource for various populations and helps them to adjust to new environments [15]. Membership in social groups can act as a social resource [19,22]. In this study, we used the relatively new concept of ComSOC, which has been developed as a culturally sensitive tool. This concept reveals how collective cultures define their SOC through community rather than individual frames, thereby emphasizing societal values. ComSOC embraces the individual's perception of a community in terms of Antonovsky's three components: Comprehensibility, manageability, and meaningfulness [25]. Communal resources of comprehensibility, manageability, and meaningfulness enable members of the community to express and to realize themselves, to feel satisfaction, to challenge and have communal interests, and also amplify feelings of affiliation and social connectedness [26,27]. In various studies, ComSOC has been found to be stronger among collectivistic minority cultures than among Western majority cultures and while it has been found to be negatively correlated with psychological problems, it has also helped to explain job satisfaction [28,29]. We assumed that, in the context of Syrian Arab culture, community coherence plays an important role in the refugees' adjustment to their new environment.

*1.7. Demographic Factors: Age, Level of Education, and Time Spent in the Refugee Camp*

Age serves as a predictor of mental-health problems, with research indicating that older women report more mental-health problems. Moreover, women from traditional and collectivistic societies, such as Arab societies, who are less educated usually report more mental-health symptoms than more educated women [30]. The immigration experience also plays a significant role, with older women and immigrants who are new residents of a country reporting more mental-health problems than younger women who are citizens of that country [31].

*1.8. Research Questions*

In accordance with the literature described above, the following research questions and hypotheses were formulated: (1) Are there differences between women who have resided in a refugee camp for up to a year and women who have resided in a refugee camp for between 1 and 2 years, in terms of exposure to war events, feelings/appraisal of danger, having received aid from any of a variety of sources, the coping resources of SOC and ComSOC, and/or the mental-health outcomes of anxiety, depression, and somatization? Based on a recent study, we hypothesized that a longer stay in the camp would be associated with higher levels of psychological problems, more feelings of danger, and weaker SOC. However, based on previous research [12], we did not expect that exposure to war experiences or having received aid would vary with the amount of time spent in the refugee camp. (2) Are there significant differences in the mental-health outcomes of women refugees (i.e., anxiety, depression, and somatization) depending on their educational level and their exposure to war events, feelings/appraisal of danger, receiving aid from any of a variety of sources, and/or the coping resources of SOC and ComSOC? We expected women with higher levels of education to report stronger SOC and fewer mental-health problems [30,32]. In addition, since no information on the independent variables of exposure to war events, feelings/appraisal of danger, or receiving aid was found, we hypothesized that level of education would not be associated with any differences in exposure to war experiences, appraisal of danger, or having received aid. (3) We evaluated a model in which different demographic variables (i.e., age, time spent in the refugee camp, and education level), as well as situational factors of exposure to war events, appraisal of danger, having received aid from organizations (or family or community members), and coping resources were entered as predictors of anxiety, depression, and somatization. We expected age and education level [33], time spent in the camp, exposure to war experiences, appraisal of danger, SOC, and ComSOC [8,12,26] to be significant contributors to the various mental-health outcomes. We hypothesized that while levels of education and the coping resources would have positive effects, age, exposure to war, and a relatively high appraisal of danger would have negative effects. In addition to evaluating the entire model, we also examined the roles of SOC and ComSOC in mediating various relationships between the demographic or situational variables and the outcome variables.

## 2. Materials and Methods

*2.1. Participants*

One hundred and eleven refugee women aged 19–70 (M = 41.01, SD = 11.42), who reported having between 0 and 19 children (M = 3.88, SD = 2.71), participated in this study during August 2018. The women were residing in refugee camps in Greece; 2.7% reported having arrived 1 month prior to the administration of the questionnaire, 22.5% had resided in the refugee camp between 1 and 6 months, 30.6% between 6 and 12 months, and 35.1% reported having resided in the refugee camp for more than a year. Most of the women (77.5%) reported that they had not had a relative in the refugee camp prior to their arrival. Most of these women (74.8%) were Sunnis. In terms of level of education, 3.6% had not had any formal education, 7.3% had only graduated elementary school, 51.8% had only graduated high school, 25.5% had a non-academic higher-education diploma, and 11.8% had an academic degree.

*2.2. Procedures*

Data were collected by self-reported questionnaires during August 2018 in a refugee camp in Greece. Prior to the administration of the questionnaires, the study was evaluated and approved by the university department's ethics committee (Department of Conflict Management and Resolution, Ben-Gurion University of the Negev). All ethical standards were maintained. All participants were informed that the researchers were interested in their experiences, participation was voluntary, and anonymity was emphasized. The questionnaires were translated into Arabic by an Arabic-language teacher and then reverse-translated into Hebrew to ensure the accuracy of the translation. A researcher who is a native speaker of Arabic approached the women in person, explained to them the nature and aims of the study, and emphasized the voluntary nature of participation and the anonymity of their responses.

*2.3. Measures*

Demographic characteristics included questions regarding age, number of children, education, ethnicity, when they first entered the refugee camp, and if they had any relatives in the camp prior to their arrival.

*Exposure to war events* was assessed using five yes (1)/no (0) questions that referred to whether the individual's community had been attacked by rockets/bombs, whether someone the individual knows had been hurt as result of the war, whether a relative had been hurt as a result of the war, whether the individual herself had been hurt as result of the war, and whether the individual's home had been damaged as result of the war. The answers to the different questions were added up to calculate an index with a potential range of 0–5 (M = 1.35, SD = 0.19).

*Appraisal of danger* was assessed using an index of four questions, each of which was answered using a 5-point Likert scale (1—*not at all*; 5—*very much*). Questions related to how dangerous the situation in Syria was for the study participant, her family, her friends, and civilians in Syria. The mean of the items was calculated to create an index ranging from 1 to 5 (M = 4.36, SD = 0.50).

The variable *receiving aid* was assessed by six questions, each answered using a 5-point Likert scale (1—*not at all*; 5—*very much*). Questions related to receiving aid from family members, Muslim organizations, aid organizations, European governments, and the United Nations. A mean score was calculated to create an index with a range of 1–5 (M = 2.40, SD = 0.54).

*Sense of coherence* (SOC; [4]) was measured using a series of semantic differential items scored on a 7-point Likert-type scale that had anchoring phrases at each end. High scores indicated a strong SOC. An account of the development of the SOC scale and its psychometric properties, showing it to be reliable and reasonably valid, appears in Antonovsky's writings [4]. In this study, SOC was measured using the short-form scale consisting of 13 items, which was found to be highly correlated to the original long version [4]. The scale includes items such as *"Doing the things you do every day is"* with answers ranging from (1) *"a source of pain and boredom"* to (7) *"a source of deep pleasure and satisfaction."* In the present study, the Cronbach's alpha coefficient for the scale was good ($\alpha$ = 0.87).

*Community Sense of Coherence* (ComSOC; [26]). This is a 16-item seven-point Likert-type scale with anchoring phrases at each end. It translates the major themes of Antonovsky's personal SOC—comprehensibility, manageability, and meaningfulness—into community resources. Items include: *"To what extent do you feel you can influence what's happening in your community?"*; *"Living in your community gives meaning to your life in a way that other communities couldn't"*; and *"Do you feel that things that happen in your community have no meaning for you?"*. The Cronbach's alpha coefficient for this scale in the present study was excellent ($\alpha$ = 0.92).

*Brief Symptom Inventory* [34]. We used the short version of the questionnaire comprised of 18 items, which are rated on a 5-point Likert scale (0—*not at all*; 4—*very much*). The questionnaire examined three areas of psychological and psychiatric problems: somatization, depression, and anxiety. The reliability of the short version of the questionnaire and its three subscales has been reported to be good [35]. Here are examples items from each subscale. Somatization: *"To what extent have you suffered from a feeling of fainting or dizziness?"*. Anxiety: *"To what extent have you suffered from a feeling of stress?"*. Depression: *"To what extent have you suffered from a feeling of depression?"*. In this study, the reliability of the somatization subscale was good ($\alpha$ = 0.87), the reliability of the anxiety subscale was good ($\alpha$ = 0.87), and the reliability of the depression subscale was also good ($\alpha$ = 0.84).

*2.4. Data Analysis*

Statistical analyses were conducted using the statistical software SPSS Version 25, (Routledge, Abingdon, UK). A significance level ($\alpha$) of $p < 0.05$ was chosen. First, the frequencies and percentages of the sample's demographic characteristics were explored. Then, we ran *t*-tests for independent samples to evaluate the effects of time spent in the refugee camp and levels of education on the different study variables. Finally, a hierarchal regression was performed to investigate the extent to which variance in the dependent variables (i.e., levels of anxiety, depression, and somatization) could be explained by the selected independent variables. We also used the Sobel test [36,37] to evaluate whether SOC and ComSOC mediated the relationships between the different demographic or situational variables and the mental-health outcomes.

## 3. Results

### 3.1. Differences Among Women Who Had Been in the Camp for Different Periods of Time

Our first question related to the comparison of women who had resided up to 1 year in the camp with women who resided in the camp between 1 and 2 years, in terms of our study variables. The results of this analysis are presented in Table 1.

**Table 1.** Differences among women who had resided in the camp for up to 1 year and women who had resided in the camp for between 1 and 2 years.

| | Up to 1 Year N = 62 | | Between 1 and 2 Years N = 39 | | *t* |
|---|---|---|---|---|---|
| | *M* | *SD* | *M* | *SD* | |
| Appraisal of danger (1–5) | 4.41 | 0.42 | 4.33 | 0.55 | 0.77 |
| Having received aid (1–5) | 2.42 | 0.56 | 2.35 | 0.55 | 0.64 |
| Exposure to war events (0–5) | 1.34 | 0.14 | 1.35 | 0.27 | −0.37 |
| SOC (1–7) | 2.56 | 0.86 | 2.88 | 0.75 | −1.91 ^ |
| ComSOC (1–7) | 2.90 | 1.06 | 3.17 | 1.01 | −1.25 |
| Anxiety (0–4) | 3.43 | 0.54 | 3.12 | 0.58 | 2.75 ** |
| Depression (0–4) | 3.34 | 0.55 | 2.95 | 0.69 | 3.10 ** |
| Somatization (0–4) | 3.27 | 0.62 | 2.95 | 0.78 | 2.26 * |

Note: ^ $p < 0.06$; * $p < 0.05$; ** $p < 0.01$.

Our analysis revealed some prominent differences, especially in terms of anxiety, depression, and somatization. Contrary to our hypothesis, newcomers reported higher levels of these problems than the veteran residents of the camp. It should be noted that marginal effects were exhibited in personal SOC, with women who had spent more time in the camp reporting stronger SOC. However, it should also be noted that among all of the women, personal SOC and ComSOC were lower than the average of the scale; whereas scores for mental-health outcomes were at the higher ends of those scales.

### 3.2. Differences Among Women According to Their Levels of Education

We then examined differences in the study variables corresponding with the different educational levels of the women. The results of this analysis are presented in Table 2. Contrary to our hypothesis, there were no differences in any of the study variables that corresponded to differences in levels of education. That is, education did not seem to serve as a protective factor in this situation.

### 3.3. Explanation of the Various Mental-Health Outcomes

Our last question related to the explanation of the mental-health outcomes—anxiety, depression, and somatization—in terms of the different demographic, situational, and coping-resource variables. The results of this analysis are presented in Table 3. It seems that time spent in the camp, appraisal of danger, and the coping resources of SOC and ComSOC are significant in predicting the variance of various mental-health outcomes. Together, those factors predicted 56% of the reported anxiety, 53% of the reported depression, and 58% of the reported somatization. In addition, age was also a significant predictor of somatization. It seems that older women report more somatization. However, overall, it seems that time is a healing factor and that as time passes, the mental health of these women improves. Moreover, the way one perceives a situation and personal and collective resources all play fundamental roles in shaping one's mental health.

**Table 2.** Differences in the study variables among women with different levels of education.

| | Up to High School N = 69 | | More Than High School N = 41 | | *t* |
|---|---|---|---|---|---|
| | *M* | *SD* | *M* | *SD* | |
| Appraisal of danger (1–5) | 4.32 | 0.50 | 4.46 | 0.48 | −1.46 |
| Having received aid (1–5) | 2.36 | 0.45 | 2.46 | 0.66 | −0.84 |
| Exposure to war events (0–5) | 1.35 | 0.22 | 1.34 | 0.14 | 0.17 |
| SOC (1–7) | 2.81 | 0.77 | 2.60 | 0.91 | 1.26 |
| ComSOC (1–7) | 3.01 | 0.98 | 3.11 | 1.11 | −0.52 |
| Anxiety (0–4) | 3.29 | 0.52 | 3.44 | 0.61 | −0.52 |
| Depression (0–4) | 3.14 | 0.57 | 3.25 | 0.70 | −0.86 |
| Somatization (0–4) | 3.18 | 0.67 | 3.14 | 0.70 | 0.25 |

**Table 3.** Results of hierarchical multiple regression predicting mental-health outcomes.

| | | Anxiety | | | | | Depression | | | | | Somatization | | | |
|---|---|---|---|---|---|---|---|---|---|---|---|---|---|---|---|
| | $R^2$ | *B* | $\beta$ | *SE* | *t* | $R^2$ | *B* | $\beta$ | *SE* | *t* | $R^2$ | *B* | $\beta$ | *SE* | *t* |
| **Step 1** | 0.09 | | | | | 0.10 | | | | | 0.10 | | | | |
| Age | | 0.00 | 0.09 | 0.01 | 0.91 | | 0.00 | 0.05 | 0.01 | 0.49 | | 0.01 | 0.22 | 0.01 | 2.21 * |
| TSC [1] | | −0.36 | −0.31 | 0.12 | −3.05 ** | | −0.42 | −0.32 | 0.13 | −3.17 ** | | −0.39 | −0.27 | 0.14 | −2.69 ** |
| Education | | −0.02 | −0.02 | 0.12 | −0.17 | | 0.02 | 0.01 | 0.13 | 0.14 | | −0.13 | −0.09 | 0.14 | −0.91 |
| **Step 2** | 0.22 | | | | | 0.23 | | | | | 0.16 | | | | |
| Age | | 0.01 | 0.13 | 0.00 | 1.41 | | 0.00 | 0.08 | 0.01 | 0.82 | | 0.01 | 0.25 | 0.01 | 2.60 * |
| TSC | | −0.36 | −0.30 | 0.11 | −3.33 ** | | −0.40 | −0.31 | 0.12 | −3.42 ** | | −0.38 | −0.27 | 0.13 | −2.79 ** |
| Education | | −0.04 | −0.03 | 0.11 | −0.39 | | 0.00 | 0.00 | 0.11 | −0.02 | | −0.15 | −0.10 | 0.13 | −1.12 |
| Exposure index [3] | | 0.01 | 0.00 | 0.26 | 0.02 | | −0.20 | −0.06 | 0.29 | −0.71 | | −0.10 | −0.03 | 0.33 | −0.31 |
| AoD [2] | | 0.53 | 0.44 | 0.11 | 4.97 *** | | 0.60 | 0.44 | 0.12 | 5.08 *** | | 0.54 | 0.37 | 0.13 | 4.01 *** |
| Receiving aid | | −0.11 | −0.11 | 0.09 | 1.26 | | −0.14 | −0.13 | 0.10 | −1.42 | | −0.13 | −0.11 | 0.11 | −1.13 |
| **Step 3** | 0.25 | | | | | 0.20 | | | | | 0.32 | | | | |
| Age | | 0.00 | 0.08 | 0.00 | 1.03 | | 0.00 | 0.04 | 0.00 | 0.54 | | 0.01 | 0.18 | 0.00 | 2.45 * |
| TSC | | −0.23 | −0.20 | 0.09 | −2.66 ** | | −0.28 | −0.21 | 0.10 | −2.71 ** | | −0.21 | −0.15 | 0.11 | −2.03 * |
| Education | | −0.02 | 0.02 | 0.09 | −0.25 | | −0.01 | −0.01 | 0.10 | −0.08 | | −0.11 | −0.08 | 0.10 | 0.10 |
| Exposure index | | 0.11 | 0.04 | 0.21 | 0.53 | | −0.12 | −0.04 | 0.24 | 0.49 | | 0.05 | 0.01 | 0.25 | 0.20 |
| AoD | | 0.20 | 0.16 | 0.10 | 1.97 ^ | | 0.30 | 0.22 | 0.11 | 2.59 * | | 0.07 | 0.05 | 0.12 | 0.60 |
| Receiving aid | | 0.04 | 0.04 | 0.08 | 0.53 | | 0.03 | 0.03 | 0.09 | 0.31 | | 0.07 | 0.06 | 0.09 | 0.78 |
| SOC | | −0.18 | −0.27 | 0.07 | −2.57 * | | −0.28 | −0.37 | 0.08 | −3.38 ** | | −0.18 | −0.24 | 0.09 | 2.30 * |
| ComSoc | | −0.21 | −0.39 | 0.06 | −3.51 ** | | −0.13 | −0.21 | 0.07 | −1.82 | | −0.33 | −0.50 | 0.07 | −4.59 *** |

Note: ^ $p < 0.06$; *** $p < 0.001$; ** $p < 0.01$; * $p < 0.05$. 1 Time spent in the camp, 2 Appraisal of danger, 3 Exposure to war events.

To evaluate the mediating roles of SOC and ComSOC in the relationships between time spent in the refugee camp or appraisal of danger and the various mental-health outcomes, we ran several Sobel tests. The results indicated that SOC mediated the relationships between time spent in the refugee camp and the appraisal of danger and the outcome variables of anxiety ($z = 2.00$, $p < 0.05$, $z = 2.79$, $p < 0.01$, respectively) and depression ($z = 2.15$, $p < 0.05$, $z = 2.87$, $p < 0.01$, respectively). As for somatization, only the effect of appraisal was mediated by SOC ($z = 1.99$, $p < 0.05$). We also found that ComSOC mediated the role of SOC in the explanation of somatization ($z = 2.15$, $p < 0.05$), underscoring the importance of that variable.

## 4. Discussion

In light of the ongoing civil war in Syria, which has forced millions of Syrians to flee to other countries, this study examined whether and how SOC and ComSOC help Syrian refugee women as they adapt to life in a refugee camp. Rather than examining the topic from a pathogenic point of view, we wanted to understand which coping resources assist these women as they adapt to their new situation.

Overall, our data indicate that these refugee women are a vulnerable population. Their SOC and ComSOC levels were very low objectively and relative to those of other populations of women around the world who belong to marginalized minority groups [38]. In the same vein, it seems that their mental-health symptoms of anxiety, depression, and somatization are at the higher end of the scales and our findings in this area resemble those of other studies carried out in similar contexts [6,7]. These results are not surprising considering the long civil war from which these women fled. Additionally, these results can be explained by the fact that their new place of residence and their current lives are characterized by insecurity and their futures are uncertain.

Our first research question related to differences between women who had resided in the refugee camp for at least a year (but no more than 2 years) and women who had arrived more recently. Contrary to our hypothesis, our results point to positive adaptation and healing; as time passes, the levels of anxiety, depression, and somatization among these women decrease. Additionally, it seems that the personal resource of SOC becomes stronger over time spent in the camp. This result is in line with those of studies from other places around the world that have shown that when one is torn from one's home, the first period is a major disturbance, leading to a weakening of various coping-resource systems, but that as time passes, those resources can be recovered [39].

As for the role of education in this setting, contrary to our hypothesis, we found no significant effects of being more or less educated. It seems that in such an extreme context in which women's lives are in danger, higher education does not provide protection and does not significantly aid women as they adapt to life as refugees.

Our last and most important question related to the role of demographics, situational factors, and personal or communal coping resources in reducing mental-health symptoms, to aid these women's adaptation to their new environment. In line with our hypothesis, our results show that time spent in the refugee camp and appraisal of danger play significant roles in explaining various mental-health symptoms. In contrast to adolescent Syrian refugees, among whom spending more time in refugee camps has a negative affect [12], for grown women, time spent in the camp has a healing effect. The longer the women had been in the camp, the better mental health they reported. It could be that contrary to adolescent refugees, adult women who have had some time to understand their new environment and deeply comprehend the situation from which they fled can assign new meaning to and better comprehend their potential futures in their new environment despite the difficulties inherent in their situation. Additionally, as previous studies have also indicated (e.g., [8]), our study shows that the way a woman perceives a situation of war and the meaning she assigns to that situation play significant roles in predicting her mental health. Thus, the greater danger she feels, the more negative mental-health symptoms she will report.

Another interesting finding relates to the contribution of age to somatization. In this study, older women reported higher levels of somatization. This finding is in line with our hypothesis and previous studies that have found that the tendency for individuals to present their distress through somatic complaints is common in countries with collectivistic cultures, such as Arab countries [40–42]. It seems that the older women were more affected by traditional/collectivistic cultural values and tended to report more somatic symptoms.

Although time spent in the refugee camp and appraisal of danger played significant roles in the explanation of mental-health outcomes, as we hypothesized, it is noteworthy that once the personal resource of SOC was entered into the equation, the importance of the amount of time spent in the refugee camp and appraisal of danger decreased dramatically. These results indicate that SOC and ComSOC have the most important roles in explaining and predicting mental-health outcomes. SOC and ComSOC cancel out or significantly weaken the effects of the above-mentioned variables; stronger SOC leads to better mental health and stronger ComSOC leads to fewer anxiety or somatization symptoms. This study continues a line of studies rooted in positive psychology that have tried to look at factors that promote mental health rather than risk factors that lead to pathogenic outcomes. Thus, it seems that when individuals succeed in finding ways to comprehend and manage their situations, they will enjoy better mental health. Additionally, a community that one can trust and on which one can rely serves as a significant protective factor that promotes adaptation to life in a refugee camp.

A small note regarding the non-significant factors: In this study, the situational variables of exposure to war experiences and having received aid were not found to have any significant effects. These results are congruent with those of a line of studies that have yielded similar results, indicating that variables other than these play significant roles in such situations [8,12].

This study had several limitations that should be acknowledged. First, the data were collected via self-report questionnaires, which may be affected by social-desirability issues [43]. Second, the extent to which women's experiences of mental-health difficulties converge with external observations, such as clinical reports, remains to be investigated. Third, in the absence of a base rate for the women's mental-health outcomes prior to the study period, we cannot state with certainty whether or not the observed outcomes are due solely to the impact of exposure to war and the refugee experience. In addition, our research employed a cross-sectional design. All of the variables were measured at the same point in time, so we cannot exclude the possibility that women with higher levels of anxiety, depression, and/or somatization may tend to report low levels of SOC and ComSOC and high levels of appraisal of danger. Future longitudinal studies should shed more light on the nature and the direction of these effects. Finally, a potential degree of sample bias cannot be ruled out as our relatively small sample was not a representative sample of Syrian refugee women.

## 5. Conclusions

To summarize, the present study examined the roles of SOC and ComSOC in reducing various mental-health outcomes among women who were forced from their homes in Syria. The study participants had resided in a refugee camp in Greece for periods of time ranging from a few weeks to two years. Our study shows that those who had resided in the camps for longer periods of time were better adjusted and exhibited fewer mental-health symptoms and stronger SOC. Moreover, our results also show that SOC and ComSOC play the most important roles in explaining anxiety, depression, and somatization, and also mediate the effects of the amount of time spent in the refugee camp and appraisal of danger on those outcomes.

These results have some practical implications. First, it is very important to strengthen the SOC and ComSOC of refugee women, to enable them to better adapt when confronted with a variety of stressful situations. It is also important that women be integrated into societal processes, in order for them to feel in control of their lives and to strengthen their senses of manageability and comprehensibility. Another way to gain control and increase feelings of manageability is to create routine in the daily life of the inhabitants of the refugee camp. When these women feel that they can influence decisions regarding their lives, they will gain a sense of meaningfulness, which is an important aspect of SOC and which will, in turn, benefit their mental health.

**Author Contributions:** All of the authors were involved in designing the study. In addition, K.A.-S. collected the data and O.B.-L. and S.A.-K. prepared the manuscript. All of the authors approved the final manuscript.

**Funding:** This research received no external funding.

**Conflicts of Interest:** The authors declare no conflict of interest.

## References

1. Human Rights Watch. Syria: Events of 2018. Available online: https://www.hrw.org/world-report/2019/country-chapters/syria (accessed on 26 August 2019).
2. Hassan, G.; Ventevogel, P.; Jefee-Bahloul, H.; Barkil-Oteo, A.; Kirmayer, L.J. Mental health and psychosocial wellbeing of Syrians affected by armed conflict. *Epidemiol. Psychiatr. Sci.* **2016**, *25*, 129–141. [CrossRef] [PubMed]
3. Antonovsky, A. *Health, Stress, and Coping*; Jossey-Bass: San Francisco, CA, USA, 1979.
4. Antonovsky, A. *Unraveling the Mystery of Health: How People Manage Stress and Stay Well*; Jossey-Bass: San Francisco, CA, USA, 1987.
5. Alhayek, K. Untold stories of Syrian women surviving war. *Syria Stud.* **2015**, *7*, 1–30.
6. Freedman, J. Sexual and gender-based violence against refugee women: A hidden aspect of the refugee "crisis". *Reproduct. Health Matters* **2016**, *24*, 18–26. [CrossRef] [PubMed]

7. Hollander, A.C.; Bruce, D.; Burström, B.; Ekblad, S. Gender-related mental health differences between refugees and non-refugee immigrants—A cross-sectional register-based study. *BMC Public Health* **2011**, *11*, 180. [CrossRef]
8. Braun-Lewensohn, O.; Celestin-Westreich, S.; Celestin, L.P.; Verté, D.; Ponjaert-Kristoffersen, I. Adolescents' mental health outcomes according to different types of exposure to ongoing terror attacks. *J. Youth Adolesc.* **2009**, *38*, 850–862. [CrossRef]
9. Vostanis, P. Meeting the mental health needs of refugees and asylum seekers. *Brit. J. Psychiatr.* **2014**, *204*, 176–177. [CrossRef]
10. Fazel, M.; Reed, R.V.; Panter-Brick, C.; Stein, A. Mental health of displaced and refugee children resettled in high-income countries: Risk and protective factors. *Lancet* **2012**, *379*, 266–282. [CrossRef]
11. Eggerman, M.; Panter-Brick, C. Suffering, hope, and entrapment: Resilience and cultural values in Afghanistan. *Soc. Sci. Med.* **2010**, *71*, 71–83. [CrossRef]
12. Braun-Lewensohn, O.; Al-Said, H. Syrian adolescent refugees: How do they cope during their stay in refugees camps? *Front. Psychol.* **2018**, *9*. [CrossRef]
13. Braun-Lewensohn, O.; Sagy, S.; Roth, G. Brief report: Adolescents under missile attacks: Sense of coherence as a mediator between exposure and stress-related reactions. *J. Adolesc.* **2011**, *34*, 195–197. [CrossRef]
14. Lazarus, R.S.; Folkman, S. *Stress Appraisal and Coping*; Springer: New York, NY, USA, 1984.
15. Frydenberg, E. *Adolescent Coping: Promoting Resilience and Well-Being*; Routledge: London, UK, 2018.
16. Braun-Lewensohn, O. Coping strategies as mediators of the relationships between objective and subjective exposure to ongoing missile attacks and stress reactions. *J. Child Adolesc. Trauma* **2012**, *5*, 315–326. [CrossRef]
17. Cardozo, B.L.; Crawford, C.G.; Eriksson, C.; Zhu, J.; Sabin, M.; Ager, A.; Olff, M. Psychological distress, depression, anxiety, and burnout among international humanitarian aid workers: A longitudinal study. *PLoS ONE* **2012**, *7*, e44948.
18. Almedom, A.M. Factors that mitigate war-induced anxiety and mental distress. *J. Biosoc. Sci.* **2004**, *36*, 445–461. [CrossRef] [PubMed]
19. Mayer, C.H. *The Meaning of Sense of Coherence in Transcultural Management*; Waxmann Verlag: Münster, Germany, 2011; Volume 563.
20. Mittelmark, M.; Sagy, S.; Eriksson, M.; Bauer, G.F.; Pelikan, J.M.; Lindström, B.; Espnes, G.A. The meanings of salutogenesis. In *The Handbook of Salutogenesis*; Springer Nature: Cham, Switzerland, 2017; pp. 7–13.
21. Mittelmark, M.; Sagy, S.; Eriksson, M.; Bauer, G.F.; Pelikan, J.M.; Lindström, B.; Espnes, G.A. The sense of coherence in the salutogenic model of health. In *The Handbook of Salutogenesis*; Springer Nature: Cham Switzerland, 2017; pp. 91–96.
22. Evans, W.P.; Davis, B. Exploring the relationship between sense of coherence and historical trauma among Native American youth. *Amer. Ind. Alask. Nativ. Ment. Health Res.* **2018**, *26*, 1–25.
23. Erim, Y.; Morawa, E.; Atay, H.; Aygün, S.; Gökalp, P.; Senf, W. Sense of coherence and depression in the framework of immigration: Turkish patients in Germany and in Turkey. *Int. Rev. Psychiatr.* **2011**, *23*, 542–549. [CrossRef] [PubMed]
24. Bronfenbrenner, U. Ecological models of human development. *Int. Encycl. Educ.* **1994**, *3*, 37–43.
25. Elfassi, Y.; Braun-Lewensohn, O.; Krumer-Nevo, M.; Sagy, S. Community sense of coherence among adolescents as related to their involvement in risk behaviors. *J. Commun. Psychol.* **2016**, *44*, 22–37. [CrossRef]
26. Braun-Lewensohn, O.; Sagy, S. Salutogenesis and culture. Personal and community sense of coherence in different cultural groups. *Int. Rev. Psychiatr.* **2011**, *23*, 533–541. [CrossRef]
27. Mayer, C.H.; Boness, C. Interventions to promoting sense of coherence and transcultural competences in educational contexts. *Int. Rev. Psychiatr.* **2011**, *23*, 516–524. [CrossRef]
28. Braun-Lewensohn, O. Coping resources and stress reactions among three cultural groups one year after a natural disaster. *Clin. Soc. Work J.* **2014**, *42*, 366–374. [CrossRef]
29. Braun-Lewensohn, O.; Kalagy, T. Between the inside and the outside world: Coping of Ultra-Orthodox individuals with their work environment after academic studies. *Commun. Ment. Health* **2019**. [CrossRef] [PubMed]
30. Hobfoll, S.E.; Canetti-Nisim, D.; Johnson, R.J. Exposure to terrorism, stress-related mental health symptoms, and defensive coping among Jews and Arabs in Israel. *J. Consult. Clin. Psychol.* **2006**, *74*, 207. [CrossRef] [PubMed]

31. Morawa, E.; Erim, Y. Health-related quality of life and sense of coherence among Polish immigrants in Germany and indigenous Poles. *Transcult. Psychiatr.* **2015**, *52*, 376–395. [CrossRef] [PubMed]

32. Volanen, S.M.; Lahelma, E.; Silventoinen, K.; Suominen, S. Factors contributing to sense of coherence among men and women. *Eur. J. Public Health* **2004**, *14*, 322–330. [CrossRef]

33. Ali, N.S.; Azam, I.S.; Ali, B.S.; Tabbusum, G.; Moin, S.S. Frequency and associated factors for anxiety and depression in pregnant women: A hospital-based cross-sectional study. *Sci. World J.* **2012**, *2012*, 653098. [CrossRef]

34. Derogatis, L.R. *SI-18: Brief Symptom Inventory 18—Administration, Scoring, and Procedures Manual*; NCS Pearson: Minneapolis, MN, USA, 2000.

35. Franke, G.H.; Jaeger, S.; Glaesmer, H.; Barkmann, C.; Petrowski, K.; Braehler, E. Psychometric analysis of the Brief Symptom Inventory 18 (BSI-18) in a representative German sample. *BMC Med. Res. Method.* **2017**, *17*, 14. [CrossRef]

36. Baron, R.M.; Kenny, D.A. The moderator–mediator variable distinction in social psychological research: Conceptual, strategic, and statistical considerations. *J. Personal. Soc. Psychol.* **1986**, *51*, 1173. [CrossRef]

37. Preacher, K.J.; Leonardelli, G.J. Calculation for the Sobel Test: An Interactive Calculation for Mediation Tests. Available online: http://quantpsy.org/sobel/sobel.htm (accessed on 27 August 2019).

38. Braun-Lewensohn, O.; Abu-Kaf, S.; Al-Said, H. Analysis of the differential relationship between the perception of one's life and coping resources among three generations of Bedouin women. *Int. J. Environ. Res. Public Health* **2019**, *16*, 804. [CrossRef]

39. Braun-Lewensohn, O.; Sagy, S.; Sabato, H.; Galili, R. Sense of coherence and sense of community as coping resources of religious adolescents before and after the disengagement from the Gaza Strip. *Isr. J. Psychiatr. Rel. Sci.* **2013**, *50*, 110–116.

40. Abu-Kaf, S.; Shahar, G. Depression and somatic symptoms among two ethnic groups in Israel: Testing three theoretical models. *Isr. J. Psychiatr. Rel. Sci.* **2017**, *54*, 32–39.

41. Hamdi, E.; Amin, Y.; Abou-Saleh, M.T. Problems in validating endogenous depression in the Arab culture by contemporary diagnostic criteria. *J. Affect. Dis.* **1997**, *44*, 131–143. [CrossRef]

42. Kleinman, A. Culture and depression. *New Engl. J. Med.* **2004**, *351*, 951–953. [CrossRef] [PubMed]

43. Mundia, L. Social desirability, non-response bias and reliability in a long self-report measure: Illustrations from the MMPI-2 administered to Brunei student teachers. *Ed. Psych.* **2011**, *31*, 207–224. [CrossRef]

International Journal of
*Environmental Research
and Public Health*

*Article*

# "Enclave in Transition": Ways of Coping of Academics from Ultra-Orthodox (Haredim) Minority Group with Challenges of Integration into the Workforce

Tehila Kalagy

Department of Public Policy and Administration, Ben-Gurion University of the Negev, P.O.B. 653 Beer-Sheva, Israel; kalagy@bgu.ac.il

Received: 27 February 2020; Accepted: 30 March 2020; Published: 31 March 2020

**Abstract:** Traditional societies around the world face various challenges with the introduction of "modern" values as a result of various globalization processes occurring worldwide. In the research literature, these groups are generally referred to as a "transitional societies." The focus of the research discourse on "a society in transition" is the social change derived from the undermining of that traditional society and the weakening of its constituent values with the acquisition of higher education and modification of traditional division of roles in the family. In the last two decades, the ultra-Orthodox society in Israel has undergone far-reaching changes that are reflected in the acquisition of higher education and the accelerated entry into the employment market. In light of these changes, this study seeks to examine how the academic ultra-Orthodox deal with this integration into a work place outside the "enclave." Methodologically, the study is based on qualitative content analysis of four focus groups, two for men and two for women, as is customary in ultra-Orthodox society. During the group discussion, participants were asked to describe how they cope with conflicts and their general professional challenges in the workplace. The findings of the study show that both the men and the women, described themselves as adaptable and coped well, despite the social difficulties facing their community and professional challenges in the employment space. The analysis of the major themes relies on the Stress and Coping theories.

**Keywords:** ultra-Orthodox; minorities; workforce; stress; coping; conservatism

---

## 1. Introduction

A conservative community frequently faces two major dilemmas: should it open to the external environment and its effects, and if so, to what extent? Both dilemmas stem from the central ambition of the divergent enclave to preserve its conservative values [1,2]. However, community members, whose existence depends on external society, are debating whether and how they can combine the "outside" and the "inside" without compromising the values of the community. In this context, they are very much concerned with their proper attitude toward acquiring an academic education, which is the salient manifestation of this combination, since acquiring higher education, and then a job, could potentially create a value conflict between the conservative Torah world and the changing modern world [3], thereby undermining stability. During the 20th century, globalization processes were expanded, as well as markets offering a variety of modern cultures and styles. This global media trend has, to some extent, forced conservative societies to open up and encompass a variety of concepts [1,4]. Conversely, traditional conservative societies often feel hostile to trends that offer diversity, change, and materialism that undermine spiritual life and tradition. The same conservative groups have had to deal with mass media, accelerated computing and the internet, penetrating and undermining the boundaries of the "enclave" and the existing traditional social order. Secular modernity combined with state-of-the-art technology, with its lack of values, posits man as an individual, free from any tribal and traditional affinity [3,5–7].

This threatens the leadership's position and triggers a negative reaction to those who have chosen a different way in the community. Negative attitudes toward non-traditional education and employment processes place the individual in ongoing conflict situations and make it difficult to integrate. Moreover, it may affect the legitimacy of the individual in the community to which it belongs.

The current study sought to examine the ways in which the academic ultra-Orthodox deal with the "outside" during the integration into the employment market. The phenomenological analysis of the research materials was done in light of the interaction model for coping with the pressure by Richard Lazarus and Susan Folkman [8]. The two researchers have tried to answer how people deal with stress situations given their personal characteristics and, of course, the characteristics of the particular situation in the context in question.

## 2. Conceptual Framework and Theoretical Background

### 2.1. Transitional Society

A key term used in the research literature in the context of describing traditional societies is "transitional society." This term combines two seemingly contradictory theoretical terms—traditional society and modern society. Traditional society is characterized by strong emotional ties between its members and a strong sense of belonging to the community and family. Formal education is not so important to it and is characterized by conservative religious traits, which are reflected in traditional employment patterns [9]. The focus of the research discourse on "transition society" is the social change derived from the undermining of the traditional structure of the ultra-Orthodox society and the weakening of constituent values such as patriarchal family structure and traditional division of roles in the family.

Societies undergo processes that affect their identity, following the infiltration of new values, such as acquiring higher education and then integration into an out-of-community workplace. In Israel, conservative minority groups are generally considered "transitional societies," as is the ultra-Orthodox society. Proponents of modernity assume that modernization of the individual and society is made possible by processes, one of which is an increase in the level of education and changing employment patterns of men and women. These processes inevitably affect the individual, the family, and society, when combining dominant cultural values, with those of their own culture [10,11].

### 2.2. Ultra-Orthodox Society—An Enclave in Transition

An "enclave in transition" is a phrase that came up in the current study, which expresses in broad terms the changes that are taking place in Israeli ultra-Orthodox society in its current form [12]. The combination combines two common terms in the study of traditional minority groups. The first term is "transitional society" as described above, and the second, "enclave culture," rests on Douglas's theory and expresses the concept of the seclusion of conservative societies [1,2]. In this context, most researchers ask about the identity of the enclave and their ways of dealing with the impact of global changes on conservative societies in general and on ultra-Orthodox society in particular. One of the significant changes in modern society is the need to acquire education for occupational integration.

Collins (1979) [13] defines the social necessity of acquiring "qualifications" in order to integrate into the modern society that is defined as giving the qualifications. In ultra-Orthodox society, acquiring an academic education is an act that, from a value point of view, has not yet received general public consent and is, from a practical point of view, full of difficulties. One of the key issues that the ultra-Orthodox society is debating is the proper attitude toward acquiring a general education—which is one of the characteristics of the "external" society—because it also serves as a source for creating and disseminating modern values and is at odds with the value of community self-preservation [14]. Although the acquisition of higher education by the ultra-Orthodox, and especially by women, enjoys some legitimacy in the community, many of the rabbis and a large part of the ultra-Orthodox public still boycott it.

As mentioned previously, the ultra-Orthodox community in Israel has undergone a significant process in the past two decades in relation to the world of employment and the acquisition of higher education—there has been a significant increase in the number of ultra-Orthodox students studying in academic and professional institutions, as opposed to the past [15]. The escalation in the numbers of ultra-Orthodox learners in academic institutions, and working is due to a number of factors: the penetration of modern norms into the ultra-Orthodox community, worsening economic distress in this community due to cuts in allowances and the significant reduction in contributions from abroad, the culture of abundance, and an increase in living standards. Another factor that is of particular concern to ultra-Orthodox women is the change in the status of the teaching profession, which was until recently the main source of livelihood for them [16].

Integrators face complex barriers during entering the employment market. In the work space itself, they have to deal with issues related to being religious people, such as men and women working together, prayer times, and kosher food. In addition to all of these, there are cultural gaps that make it difficult to socialize with other employees as well as lack of professional knowledge. The men, in particular, have difficulty interacting with technology (computer use, etc.) and often lack English language knowledge [16].

In addition, they have to deal with conflicts in the family–community space—for example, the need to confront the ideal of segregation and seclusion from Western culture and to continue working in adult life within the community according to customary norms.

The purpose of the present essay is to examine how academic ultra-Orthodox deal with stressful situations in an employment space outside the enclave. This may help to formulate ways for the best integration of secluded minorities into the general employment space, and the integrators may be a model for the rest of the community, a model that symbolizes coping capacity in a different environment from their community.

*2.3. Coping with Stress and Conflict Situations*

The dialectical process of the departure of ultra-Orthodox from their known boundaries into the "outside" world of employment is accompanied by practical and emotional conflicts. The hopping between a religious and closed conservative society and a non-religious liberal society poses complex challenges for the ultra-Orthodox and can often present conflict in their workplace. Socially, his employers and co-workers expose him to relationships and emotional and social interactions he did not know before. This new environment encourages him to assimilate the importance of emotional flexibility and adapt to social and personal stressors [12].

Coping is the realization of behaviors that are the product of assessments the individual makes about situations he experiences. These behaviors do not appear in a vacuum but are the result of a process that began even before encountering the circumstances. For example, recognizing that a response (as opposed to a lack of response) is available to the individual and has the power to recreate the feeling as less threatening than it appears at first glance. In another example, if a response (versus a lack of response) does not help to the extent that the individual expected it to help, it is possible for him/her to reassess the level of threat or the effectiveness of his response.

The complex system known as the "coping process" is a continuous cycle of the individual's interaction with stress situations. As we mentioned above, there are two main coping strategies. The first is problem-focused coping—taking action to solve the problem. That is, the individual needs to do something to reduce the stress. The other is emotion-focused coping—managing the emotions that arise in response to the situation [8]. Although stress situations usually produce both types of strategies, the problem-focused strategies are usually the more dominant ones in that one feels able to do something meaningful and constructive in order to effect change. In contrast, the emotion-focused strategies are more dominant when one feels that the stress situation is unchangeable and hence provides no choice other than enduring it.

The interactional model for studying stress situations by Lazarus and Folkman (1984) [8] addresses the question of how individual and specific characteristics affect coping situations. Stress status, according to this model, is a subjective phenomenon. Each person's response depends on how much he or she considers the situation as threatening or challenging and the coping resources available to them in dealing with the experience. Lazarus and Folkman's approach [8] relies on the assumption that coping with stress is an active process associated with cognitive, emotional and behavioral aspects. External coping resources, such as social and family support, are of great importance for successfully coping with stressful situations. That is, positive relationships make a big contribution to developing coping strategies, expanding employee emotional resources, and effectively coping with stress and conflict situations.

Lazarus and Folkman distinguish between two types of coping strategies—problem-solving and emotion-focused coping. The problem-solving strategies obviously focus on solving the problem, while the emotion-focused strategies are centered on one's attempt to change one's own perception. Problem-solving strategies include, for example, planning or addressing religion or social support. Emotion-focused strategies involve, for example, venting emotions, self-blame, or denial [17].

The current study sought to examine the ways in which the academic ultra-Orthodox deal with the "outside" during the integration into the employment market. To address this, three key questions were defined:

(1) What strategies do ultra-Orthodox academics use in the realm of employment outside their community?
(2) Is there a difference in the coping patterns and coping resources of the ultra-Orthodox who remain in the ultra-Orthodox enclave compared to those who work "outside"?
(3) Are there differences in coping patterns between women and men?

The phenomenological analysis of the research materials was done in light of the interaction model for coping with the pressure by Richard Lazarus and Susan Folkman [8]. The two researchers have tried to answer how people deal with stress situations given their personal characteristics and, of course, the characteristics of the particular situation in the context in question.

## 3. Methodology

### 3.1. Research Procedure

The research was conducted for the Israel Democracy Institute and received the approval of the Ethics Committee for the Dispute Management and Conflict Program at Ben Gurion University. The participants were informed that we are interested in understanding their perceptions and opinions on the subject under study. The confidentiality of the work was also emphasized.

Participants were selected using the snowball sample method [18]. The decision to use this technique arose from the difficulty of accessing likely subjects, in particular those belonging to closed religious communities that prefer to avoid exposure. Such subjects are objectively difficult to reach due to their insulation [19]. The choice of this sampling method was also influenced by the suspiciousness characterizing religious communities and "hidden populations" that are difficult to locate and engage [20].

The research leans on the qualitative paradigm that claims to understand the phenomenon being explored in its daily natural environment [21], thus gaining insights into their experiences and meaning [22].

The data analysis was based on the "Grounded Theory" approach [23], which is relevant to research and raises general questions, as in the present study. This approach assumes that people with shared life circumstances also have common social and psychological patterns that grow out of their shared experiences. Accordingly, the researcher attempts to identify these key patterns and describe them in order to explain the phenomenon under study. The analysis according to this approach includes two steps: first, general topic analysis, which looks for key themes and patterns that emerged in the interviews; and second, providing an interpretation of those themes and the hidden and implicit meanings of the more visible ones.

### 3.2. The Focus Groups

In order to examine the central issue regarding the ways of dealing with the ultra-Orthodox, in the realm of employment, the central tools used in this study are focus groups. These groups were in fact a form of focus groups, whose central purpose was to explore the participants' personal opinions and understand their feelings in depth [24].

The focus groups have a clear advantage of the dynamics between the participants which can reveal fascinating information about them. This is information that individual interviews will not provide. The same dynamics and sharing produce a diverse discourse, different from any other research tool. In addition, within a group, a focus on listening and special diagnostic ability is required. It requires the researcher to develop listening and analysis skills. In women's groups, the group seems to be productive and unique in that it helps to address significant social issues that are sometimes taboo or met with silence [25].

Four focus groups were convened throughout December 2015. The groups were divided by gender, as is accepted in ultra-Orthodox society; 15 women (Table 1) and 11 men between the ages of 20 and 56 participated in the groups. The participants had varied occupational backgrounds (e.g., municipal management, teaching, software engineering, swimming instruction) and also belonged to different streams in Orthodox society (e.g., Lithuanian, Hassidic, Mizrahi, modern). Furthermore, participants in the focus groups reported that they work in a variety of workplaces (16 employees at ultra-Orthodox workplaces, two employees at national religious workplaces, and eight at secular workplaces).

**Table 1.** Participants in focus group for women—personal data.

| Name of Participant | Community Membership | Age | Family Status | Children | Place of Residence | Profession | Place of Employment |
|---|---|---|---|---|---|---|---|
| Ronit | Lithuanian | 23 | Married | ✓ | Jerusalem | Educational Advisor | Ulpana |
| Sarah | Sephardic | 21 | Single | - | Jerusalem | Accountant | Accounting office |
| Gila | Outsider | 34 | Married | ✓ | Jerusalem | Doctoral Student—Biology | University |
| Ruti | Hassidic | 29 | Married | - | Jerusalem | Clinical Psychology | Psychological Services |
| Yaffa | Sephardic | 23 | Married | ✓ | Jerusalem | Accountancy and Information Systems | Ministry of Education |
| Rachel | Lithuanian | 32 | Married | ✓ | Jerusalem | Computer Programmer | Finance Ministry |
| Hodaya | Sephardic | 28 | Married | ✓ | Jerusalem | Computer Sciences | Government Ministry |
| Ora | Outsider | 28 | Married | ✓ | Jerusalem | Accountancy and Information Systems | Accounting Office |
| Shifra | Modern | 24 | Single | ✓ | Jerusalem | Accountancy and Business Management | Accounting Office |
| Shoshana | Hassidic | 22 | Single | ✓ | Jerusalem | Professional Trainer—Gymnastics | Primary School |
| Miriam | Hassidic | 21 | Married | ✓ | Jerusalem | Professional Trainer—Gymnastics | Center for People with Special Needs |
| Dvorah | Sephardic | 43 | Married | ✓ | Jerusalem | Conflict Resolution and Management | Jerusalem Municipality |
| Yehudit | Sephardic | 21 | Married | - | Jerusalem | Lifeguard and Swimming Coach | Private Framework |
| Bayli | Hassidic | 22 | Married | - | Jerusalem | Degree in gymnastics, Wingate | Private Framework |
| Bluma | Lithuanian | 22 | Married | - | Jerusalem | Professional Training—Graphics | Advertising Agency |

The focus groups were recorded with the permission of the participants and then transcribed into text. In the article, all original names have been replaced with pseudonyms. The accumulated findings were interpreted through content analysis of the statements made by the subjects and used to construct the central themes presented below. The group discussions focused on the issue of coping with challenges of integration into the workforce.

## 4. Findings

The research materials relate to three main questions that the research sought to examine—What coping strategies do ultra-Orthodox academics use in the employment space outside their community? Is there a difference in the patterns and coping resources of the ultra-Orthodox who remain in the ultra-Orthodox enclave compared to the ultra-Orthodox who move outside? Are there differences in women's and men's coping patterns?

In general, the research materials indicate two major trends: first, the formulating of unique coping patterns of most participants with conflicts that have arisen in their work situation. The second, the use of problem-focused strategies as opposed to the use of ineffective (emotion-focused) strategies. I will address the three key questions posed by the research in detail.

### 4.1. Coping Strategies in the "Out-of-Community" Employment Space

The first research question, therefore, focused on examining the coping strategies that academic Ultra-Orthodox m take in their encounter with the "outside" working world, one characterized by modern secular values and culture. The different strategies have been grouped into two main strategies—adaptive (i.e., problem-focused) and non-adaptive (i.e., emotion-focused). The work place itself forces them to deal with conflicts associated with belonging to a conservative community with norms different from those outside. There are also work skills that the ultra-Orthodox lack due to their lack of experience.

#### 4.1.1. Instrumental Social Support

The coping strategy that stood out among the research participants was the appeal for instrumental social support. Joining other ultra-Orthodox workers in the workplace is a convenient and effective strategy for dealing with the conflicts that arise.

Participants' comments reveal a combined strategy of seeking emotional support and a united format for solving problems. This form of coping stands out in workplaces with a number of ultra-Orthodox employees, who have gained power by instituting "ultra-orthodox" norms in the workplace, which help the integration of the ultra-Orthodox without harming the output of other workers.

*We do not come to an orientation day. They tried to persuade us in the past and tried to figure out why not. It includes accommodation, and we decided as a team that we're not even giving it a chance, and they really respect that. And even if some of them curse, they really refrain to do so when we're around. [ … ] For example, there is a toast and they want us to come, so we come and everything is kosher by our standards, but we are observers and do not interfere as long as it is during work hours. But once there are fun days, etc., we don't participate at all.*

Ruth speaks in terms of "they" and "us" as she describes her coping in the workplace. There is no doubt that social support from like-minded women serves as a pillar to help her personally cope with the difficulties but also provides a tool for dealing with employers when she and her friends present different demands.

Because it is a closed society, an ultra-Orthodox person is usually not exposed to his peers from the general population. As the ultra-Orthodox press (mostly printed, and also online) operates a strict filter of news and events in general society, ultra-Orthodox people are rarely exposed to behaviors and situations that are not appropriate for their community. We therefore asked how they emotionally cope with exposure to complex social situations. This question mainly answered by women who work in the fields of care and education in non-non ultra-Orthodox frameworks, as Michal replied:

*All in all … there are different views, so I'm not too involved. And I come from a home where my mother works and in the past, studied at the Hebrew University, where it's really heresy, right against religion, and she always told us, "There's nothing to do. We are in a world with lots of people and sectors and need to know how to cope." I don't intend to come and fight, I just keep to myself, the main thing is that as for moral deterioration, everything is okay, and parties and stuff I don't interfere, and in opinions, too, I'm not fighting. I just keep to myself.*

### 4.1.2. Dealing with Identity Dissonance

Zvika expanded on the difficulty of dealing with the complexity of maneuvering between the different identities—the private world versus the workplace.

*It's not that you have to do something in practice, it's just learning to live with that complexity. You need broad shoulders to learn to live with the conflict in this part, on this issue, of friends and identification. You have to learn to deal with the conflict, and may be harder for some and then they will have to make a change, maybe with the values or something.*

Zvika's way of coping reflects another coping strategy that stood out in the results of the study—positive framing, i.e., the reading of the situation in positive terms. A positive interpretation helps manage the feelings of stress and not necessarily deal with the stress itself.

Most focus group participants demonstrated a positive worldview and described their workplace as positive, both at the interpersonal level of employers and employees and at the level of conditions and wages. They expressed high satisfaction with these variables, and the positive framing of the work was very prominent in their strategy of coping with the "outside."

### 4.1.3. Ways to Deal with the "Home vs Outside" Conflict

One of the challenges of a person who integrates into a workplace outside of his community is being in a space with a different value scale. This is especially the case for the ultra-Orthodox, a cultural minority group in Israeli society, since encountering different values of behavioral norms can create difficulty and even contradiction between the two worlds. Participants in the focus groups were asked, among other things, how they cope with a work space that has a different value scale from their own, in view of the fact that they come from a closed society that has advocated religious segregation and conservatism. The findings of the study indicate that the planning strategy is widely used, the most prominent of the strategies used by ultra-Orthodox academics. One of the difficulties that accompanies the ultra-Orthodox is the conflict between home and work. In this context, questions such as, Do I, and to what extent, share information with household members about my work? Will my integration in a non- ultra-Orthodox workplace negatively impact my personal status in the community? etc. Rafi describes his way of dealing with these aspects:

*I made a total separation between the two worlds. At work, even though I fit in there, I never associated with the employees, I never invited them to any event of mine, even if it might have been possible in some way. I made a total separation. Definitely, if you're already raising the matter of matchmaking for example, they ask what the father does [...] I've never given exact details "teaches computers, deals with computers" ... it is referred to as "dealing with". And even when I'm at work and all, I don't try to bridge the two worlds, nor bring my world into work. On the contrary, maybe it is right or maybe wrong to do it. I try just to be ... And that's it. Not trying to connect the two worlds and even trying to separate the two worlds.*

The new situation in Rafi's life puts him in a work–home conflict in terms of values. His remarks reflect on his dilemma of how to combine the conservative values of home, with new values of his non-ultra-Orthodox workplace. Rafi takes the planning strategy, which means creating an action strategy idea that will take precedence over the action that should solve the problem. Rafi sees the difficulty in bridging the two worlds in which he operates and therefore "makes a total separation" between them. This strategy of action helps him move between the two poles and deal with the cultural and practical gaps between home and work. Rafi thus describes a kind of separation between the "part of the Torah," namely his personal and family values, and that of dealing with the obligations of the workplace. The coping strategy is present in all focus groups as a common and unique coping strategy that creates a separation between the "outside" and the "inside." The distinction between the private world and the surrounding world is thus a way of coping employed by many of the research participants in their encounter with the "outside" values.

The other strategy that has been very prominent among the participants is the active coping strategy, the one that takes practical steps. This strategy removes or bypasses the pressure factor and instead improves its results. Such direct actions characterize the focused approach to solving the problem. In each conflict raised for discussion, participants said they strive to work around the barrier or improve the situation.

*4.2. Coping Methods of Employees outside the Enclave vs. Workers Inside the Enclave*

4.2.1. Employment outside the Community

The research materials show that there are real differences between those employed within the ultra-Orthodox community and those employed outside. The essential difference between the two groups of workers is the environment they need to deal with. While intra-community workers are not required to deal with heterogeneous and different societies, those who have been involved in work outside the community often face conflicts—both value conflicts and practical conflicts.

The voice of the participants in the focus groups reveals the professional challenge that exists in any workplace, but the struggle of coping with the value conflict in non-ultra-Orthodox workplaces is not felt at all by those employed in intra-community workplaces. The findings show that employees who choose to work in an ultra-Orthodox workplace do so deliberately because the key criteria for choosing this workplace rather than another are more important to them than the terms of wages, as Rachel demonstrates:

> Over the years I have been looking for work. I see that it's hard for me to be in a place that's not ultra-Orthodox. It might be a kind of conflict. Even when I am offered a double salary or things like that, once it's in a mixed environment, I stay where I am, I forget about it.

The thought of integrating in a place outside the community causes pressure among the integrators. There are those who have chosen to cope and integrate in a mixed place. However, there are those who plan ahead to integrate only in an intra-community work place. Planning brings up ideas for action and thinking about the steps to be taken to best address the problem. This activity inherently belongs to focused problem-solving strategies, but its distinctive feature is that it precedes the action of problem solving and actually belongs to the pre-action phase of the second assessment of the situation.

4.2.2. Spiritual Coping

One of the main challenges the integrators face is in the spiritual aspect, meaning religious feelings and connection to the workplace. Those who integrate into community-based employment do not report significant spiritual difficulties, in contrast to those who have joined non-ultra-Orthodox places of employment. This inner conflict is expressed in a single word repeated in all focus groups—"toughening," or as Shira expresses:

> *I became tougher and more exposed. Obviously you don't stay the same as in the seminar. I attended some group and one of my seminar teachers that was there had a student that "declined" spiritually while doing office work. She was righteous and yet "fell", and then that teacher started a group on how to deal with office jobs, she prepared us and it helped. It's a constant struggle. I make a strong separation [ … ] there is a spiritual effect and that is a fact! Unless you come, do a quiet job and go without talking to anyone, it makes an impact and I'm not really a loner, but there is counterweight—on Shabbat I read books of morals and spiritual things that will strengthen me.*

The "toughness" is actually a coping strategy. Its use indicates that ultra-Orthodox people develop a defense mechanism, a kind of spiritual armor, when leaving for work outside. Taking this strategy helps them protect themselves from external influences.

### 4.2.3. Traditional Clothing in a Secular Environment

Shira's portrayal concerns spiritual aspects that can influence the level of religiosity of the integrators. Another aspect of religious norms is the traditional outer wear. The ultra-Orthodox population in Israel usually dress according to certain codes. Among men, this code is more pronounced. They wear a white shirt and a dark suit, and a brimmed hat. In this secular space this dress code is very unusual and prominent. It bothers the newcomers at first until they adapt to the place and are being observed less. There are also few who share that they come to work without their suit and hat, in order to stand out less.

Moreover, some feel that their usefulness at work is evident from Amos, who is a lawyer by profession. In his remarks, he expressed serious concern that his unique attire would endorse the stereotype that exists for the ultra-Orthodox, thereby damaging the client he represents:

> *Before representation in law, there is always this fear of appearing as an ultra-Orthodox person. This ... how did Oded tell me? If I show up with my beard, then I'm actually harming the customer. But when representing the client there is work needed to impress the judge and to gain his sympathy, despite the ultra-Orthodox appearance that I feel does not invite good treatment. Same thing just when I come with my suit and hat ... I make sure to go with a hat and suit regularly.*

### 4.2.4. Coexistence in the Employment Space

Most of the focus group participants reported that they felt that they were able to co-operate with common work norms on both sides. Compared to a closed ultra-Orthodox space that does not need coexistence, since the norms are monotonous and widely accepted. As Yaakov describes,

> *The firm is mixed. It is secular, it also has quite a few ultra-Orthodox, but it's mixed. And thank God, we manage [ ... ] to co-exist very well.*

The success of coexistence can be dependent on the fact that the integrators use adaptive strategies and seek practical solutions to all the conflicts that arise at work. From the second point of view, the participants say that employers address all the difficulties of the ultra-Orthodox workers and try to solve any conflict that arises in a practical way.

In sum, most participants take active coping and planning as coping strategies to deal with conflict situations in their workplace. Most of them stated that, on a personal level, they are paving the way for success and dealing with stress situations on a daily basis. Socially, most of them share the feeling that they are able to reach coexistence and mutual understanding with both their employers and their non ultra-Orthodox colleagues in the workplace.

### 4.2.5. Employment in a National-Religious Space

In spite of the dichotomy that we initially aimed for in the study, i.e., religious versus secular, another challenge was raised and that is working with people from the national-religious sector. As has emerged from all focus groups, the challenge of working with national-religious is the most difficult challenge, as it presents the most complex conflict situations to ultra-Orthodox workers, on the basis of the supposedly shared religious identity, as Shoshi describes,

> *It was precisely there [in the secular workplace] that it was much easier for me to maintain my boundaries, because I think the differences were very clear, unlike today in the [accountants'] office where II work with a lot of religious people, and what is the difference between me and them?—"I'm religious too." And sometimes I feel that it is wrong to present yourself as more religious than someone*

*else because he is also strict about halacha, so you cannot tell him "it is wrong." It is much easier to work there [in a secular workplace] than in a religious place, because there [in the secular workplace] my boundaries were stronger.*

The participants' statements reveal the wide gap in Israel between the ultra-Orthodox and the national–religious society and teach that sociological differences are also important in the employment world. In fact, the participants' statements indicate that conflicts on a religious cultural background arise not only between the ultra-Orthodox minority group and the secular majority group but also between the ultra-Orthodox minority group and another religious minority group. Indeed, ultra-Orthodox research participants also point out the difficulty of socializing with other workers who are national–religious in their workplace.

### 4.3. Comparing Women's and Men's Coping Strategies

The third question the study seeks to answer relates to the way in which men and women deal with employment. The main differences between men and women were found in their coping strategies—effective versus ineffective. Women use strategies of emotional support, instrumental social support, distraction, and emotion venting more than men do.

In the women's focus groups, they talked more about the difficulties there and sought emotional support and a framework for expressing their feelings. In the men's focus groups, the conversation focused more on the ideological level and the settlement of the contradictions between the world of employment and the ultra-Orthodox world. Hence, the ultra-Orthodox women feel less need than the ultra-Orthodox men to take into account the reactions of the environment to their departure for work, perhaps because they have been integrating into the general economy for years. The men, on the other hand, were already dealing with the environmental reaction regarding the ideological question of Torah study versus going to work. We note that in ultra-Orthodox society, men are commanded to study Torah and most totally aspire to it, compared to women who are required to support them. This "role reversal" may well explain the different ways of coping between women and men.

As mentioned, the family is an influencing factor on the integrators. Various studies have found that traditional families that embrace modern values have a major impact on the individual in coping with the adoption or rejection of those values, as well as in how she/he copes leaving the community for education or employment [12].

In the focus groups, men described the reality in which they must combine the ultra-Orthodox community and the family with the workplace, but their descriptions were characterized by striving to find practical solutions. Such as the solution of Moshe given below. Moshe describes the awkward conflict of exposing the children to the reality that Dad works in a non-religious place instead of studying Torah like the fathers of his children's friends:

*Since I work where I work, I know there is life at home and at work and I don't mix them up. When at home, I try not speak loudly on the phone so as not to change the atmosphere. Why? Because the kids are in a very specific framework and want to see that their dad is the same as their friends' dad, and I don't think that they should be exposed to this at their age. When they grow up they can do whatever they want.*

The women, on the other hand, used discussion in the focus groups as part of their strategy to get emotional social support in their trying coping efforts. At the end of the discussions, the women thanked us for the invitation to attend (despite the difficulty of attending them in the evening). They said the discussions game them food for thought about the process they were experiencing. They also offered to have such support groups, each in her own community, and reasoned that their identification with their peers' difficulties helps them—both because they allow them to express their feelings aloud and because they help them find practical solutions to difficulties in the workplace.

A significant difference was found between ultra-Orthodox women working as a single individual in a foreign system and ultra-Orthodox women working together to form a cohesive group. In the public service, such organizing is particularly evident in the designated sections for ultra-Orthodox women, a kind of enclave. A similar arrangement also exists in the Jerusalem municipality, in one of the divisions, where it was not intended in advance to separate the ultra-Orthodox workers from the other workers, but in practice the ultra-Orthodox women were grouped together under a division manager who happens to be ultra-Orthodox herself, and also serves as their informal representative. The same division manager participated in one of our discussion groups, and the strategy she presented in her workplace is active coping, mostly receiving instrumental social support. She explained that her power to dictate norms in her department is based on the incorporation of the ultra-Orthodox employees, who together are empowered when facing the City Council.

Dina (division manager) said,

*These conflicts occur and I deal with them, my employees deal with them, I guide them how to deal with certain situations [ … ] there are some kind of fun days with a program that I see are not appropriate for me. I inform the Division Manager that this is not right for me and we are not coming, and I back up my employees when they also decide not to come. I've created some kind of consensus when it comes to sports department, where there needs to be greater sensitivity. If there is a man in the environment who expresses himself in an improper or rude way, I am not embarrassed. It may not be for everyone. I get up and explain to him that we are married ultra-Orthodox women and we have children [ … ] and do not talk in this way [ … ] and they have learned to respect this over the years.*

### 4.3.1. Lack of Professional Knowledge

One of the gaps in professional knowledge in the workplace is the knowledge of English. Moshe argues that the lack of knowledge of the English language eventually translates into a lack of professional knowledge, and in his words, "occupational disability."

The lack of knowledge of the English language links the cultural gaps between the ultra-Orthodox society and the entire Israeli society, and this issue is more felt and spoken of by the men who are already in the work market or those who want to join it. This is undoubtedly one of the biggest barriers to entry into the labor force, especially among men, since they do not learn English in their study settings. Some have used a strategy of repression or reconciliation with the matter, and others have reported that they are trying to learn in parallel with working. The ultra-Orthodox men encounter the problem of lack of professional knowledge in their encounters with technology, computer use, etc. With this difficulty, they deal relatively easily, study in the workplace, and report a speedy closure of the gaps.

### 4.3.2. Level of Religiosity

Women report a higher level of religiosity than men. Women also attach importance to the Halacha and hold views that are at odds with what is customary in their workplace. The feeling is that women are constantly trying to settle their conflicts with the "outside." The men, on the other hand, feel relatively relaxed even though they are aware of the problems that arise when they work "outside" the community. There is also a difference between young women and older women in their religious-spiritual adaptation to the workplace. This difference is noticeable in the words of Rachel, the oldest participant in the group, who is already a grandmother:

*I was much more hysterical. I went through some complex situations. I am 25 years among these people. The people have changed, the dynamics changed and higher education has also allowed me to look at things differently. When I left the seminar I was a 19-year-old and came to the sports department and was scared of everything and came home pale and told my husband what I was going through. That's no longer today because I have a bachelor's degree in educational administration, we have experienced some insights, and we have acquired a master's degree in conflict management and settlement, which in itself has given us insights into accepting the other without sacrificing our own principles. My sons and husband are "Talmidei Chachamim" (very learned), my kids study at holy yeshivas, and what shouldn't come into our home doesn't come in. But you really need some resilience. Do not be alarmed, live it without being hysterical and give it time.*

## 5. Discussion

The present research is of importance in the field of social policy and service planning for ultra-Orthodox society in particular and Israeli society in general. It is characterized by a particular observation of the researched community and the formulation of appropriate policy in accordance with the results of the qualitative research based on the voices of the participants in the various focus groups.

The empirical research materials combined with the theoretical aspects form the basis for the conceptualization strategies of members of the ultra-Orthodox community in their encounter with the "outside." In other words, the participants' comments allow for an overall view of ultra-Orthodox society in the context of changing processes in the employment space. It seems to us that a new model of working ultra-Orthodox people is emerging, one that adopts the integration strategy and at the same time practices different degrees of differentiation. In the working ultra-Orthodox community there is integration in the general society and there is a desire for them to be partners in the Israeli economy, and at the same time they draw an emotional and practical boundary line between themselves and the sector outside, and in this case, employers and their colleagues in the workplace.

Various studies that examine minority employment areas report primarily on a form of integration that does not recognize individual liberties and particular heritage. The present study is connected to other studies [26] that seek to assimilate the cultural complexity of integration and enable them to sustain their way of life in the workplace as well. Wasserman and Frankel argue that this is how we can ensure better integration of women along with maintaining their autonomy.

In fact, even when acquiring higher education, the enclave members are dismantling and reassembling modern values. In fact, the ultra-Orthodox integrators use problem-solving strategies as opposed to less effective ones. Using these strategies is a dual process of integrating into the majority group by adopting modern traits (higher education and employment) and strictly adhering to the traditional values of the minority group culture from which they come.

The ultra-Orthodox men and the ultra-Orthodox women adopt the integrated strategy differently. The academic ultra-Orthodox women, from the various streams, advocate conservatism and adherence to the values of the ultra-Orthodox community, i.e., Torah study for men and their children, and a desire for gender segregation. Women also rank their religiosity in a place higher than that of men, although they have been in the labor and education market longer than men and also define themselves as more open to different environments. Despite the acceptance of "outside" values by the women, they show stronger loyalty to the basic conservatism upon which they were educated. There are two distinct groups: integrating in mixed workplaces and maintaining their social religious values, and advocating segregated integration, i.e., claiming a homogeneous "enclave" in the work place that avoids social integration.

The use of adaptive strategies by most research participants is broader and converges to formulate the unique coping patterns of most participants with conflicts that have arisen in the employment space. The intention is to adopt a transition strategy that combines conservatism with the use of modern features and adaptation to Western work patterns. In fact, ultra-Orthodox academics are becoming the mediators between the enclave and the "foreign" population.

To sum up, it seems that, paradoxically, out of conservatism and the need for coping with those in a secular space, an innovative model of academic conservatives joining modern and advanced workplaces is emerging. The same model is worthy of development and assistance as presented in the solutions below.

## 6. Conclusions

On the basis of these findings, it seems that three main directions of action should be recommended. First, a psycho-educational program should be developed, during their academic studies, toward the integration of the ultra-orthodox academics into the employment space. The program will help strengthen the inner world of those who integrate and reduce tensions and fears of personal identity. In addition, the program will pay special attention to building effective coping strategies when joining the world of work. Furthermore, the program will accompany that group in the employment space in the early stages of entering the workplace, monitor their dilemmas and difficulties, and provide answers by way of the resources and coping strategies found to be effective in research. (Adaptive strategies include acceptance strategy, active strategy, and instrumental support.)

Creating a work support service system, in the early years of work, will help solve the professional difficulties brought up by participants during their work. This recommendation is related to the phrase "occupational disability," conceived by one of the research participants to illustrate the difficulty of coping with one of the main barriers to ultra-Orthodox society in integration into academia and employment: lack of English proficiency. This gap is connected to a wider lack of knowledge, especially amongst ultra-Orthodox men. Therefore, it is recommended to incorporate those who intend to go to work in pre-employment preparation courses for obtaining tools and skills for optimal integration, and to assist them by providing professional knowledge as a tool for work, with emphasis on English studies and working with computers.

The third idea is aimed at developing, in cooperation with employers, employment training when the ultra-Orthodox employee is already working. Since employers in the economy have discovered the potential of the ultra-Orthodox workforce, both in terms of high output and in terms of the moral values, it is advisable to continue developing employment training programs for academic ultra-Orthodox graduates, led by employers and government encouragement. These training programs will improve the quality of work of the ultra-Orthodox workforce in reducing conflicts and tension in the workplace and will of course greatly contribute to the Israeli economy.

*Study Limitations*

At the end of the present study, I can point to three methodological limitations that should be addressed in future studies. First, we propose to conduct a more comprehensive study that will also include the population of employers and will reflect the integration process from their perspective. The second refers to the examination of ultra-Orthodox society as one unit regarding their integration into the workforce. In our opinion, it would be desirable to examine this phenomenon among each ultra-Orthodox stream separately in order to better understand the way these societies struggle with the important trend discussed in this paper. The third and last recommendation is a comparison of the ultra-Orthodox community to another conservative religious group outside of Israeli society. Such a comparison is likely to bring a wider perspective to processes of change among traditional religious women who seek an education and intend to enter the workforce.

**Funding:** This research received no external funding.

**Conflicts of Interest:** The authors declare no conflict of interest.

## References

1. Sivan, E.; Almond, G.A.; Appleby, S.R. *Modern Religious Fanaticism: Judaism, Christianity, Islam, Hinduism*; Chaim Hertzog Center for the Research of the Middle East and Diplomacy: Tel Aviv, Israel, 2004. (in Hebrew)
2. Douglas, M. *Cultural Bias*; Royal Anthropological Institute of Great Britain and Ireland: London, UK, 1978.
3. Barzilai-Nahon, K.; Barzilai, G. Cultured Technology: The Internet and Religious Fundamentalism. *Inf. Soc.* **2005**, *21*, 25–40. [CrossRef]
4. Stadler, N. Fundamentalism. In *Modern Judaism: An Oxford Guide, Nicholas de Lange and Miri Freud-Kandel*; Oxford University Press: Oxford, UK, 2005; pp. 216–227.
5. Rashi, T. The Kosher Cell Phone in ultra-Orthodox Society: A Technological Ghetto within the Global Village. In *Digital Religion: Understanding Religious Practice in New Media Worlds*; Campbell, H., Ed.; Routledge: Philadelphia, PA, USA, 2013; pp. 173–181.

6. Campbell, H. *When Religion Meets New Media*; Routledge: London, UK, 2010.
7. Cahaner, L. Space, society and community: The spatial structure of the ultra-Orthodox community in Israel in an era of change. *J. Law Soc. A* **2018**, 259–298. (in Hebrew).
8. Lazarus, R.S.; Folkman, S. *Stress, Appraisal and Coping*; Springer: New York, NY, USA, 1984.
9. Herzog, H. The Status of Women in Israel: A Fifty Years Perspective. In *Israel Culture, Religion and Society 1948–1998*; Cohen, S.A., Shein, M., Eds.; Jewish Publications: Capetown, South Africa, 2000; pp. 53–74.
10. Fuchs, H. Education and employment of Arab youth. In *Taub Jerusalem: Center for Social–Policy Research in Israel*; Weiss, A., Ed.; Report on the Status of the State Society; Economy and Policies: Jerusalem, Israel, 2018; pp. 221–264. (in Hebrew)
11. Harel-Shalev, A.; Kook, R.; Yuval, F. *Gender Relations in Bedouin Communities in Israel: Local Government as a Site of Ambivalent Modernity*; Place and Culture: Gender, Glasgow, UK, 2018.
12. Kalagy, T.; Braun-Lewensohn, O. *'Outside Versus Inside': The Integration of Ultra-Orthodox Academics in the Israeli Workforce*; Research Document for The Israel Democracy Institute: Jerusalem, Israel, 2017. (in Hebrew)
13. Collins, R. *The Credential Society*; Academic Press: New York, NY, USA, 1979.
14. Brown-Hoizman, I. "I Shall Work": Ultra-Orthodox Women Shouldering the Burden of Breadwinning: Its Justifications and Consequences. *Democr. Cult.* **2012**, *14*, 45–92.
15. Malach, G.; Cahaner, L. *The Yearbook of Ultra-Orthodox Society in Israel 2018*; The Israel Democracy Institute: Jerusalem, Israel, 2018.
16. Kalagy, T.; Braun-Lewensohn, O. Agency of preservation or change: Ultra-Orthodox educated women in the field of employment. *Community Work Fam.* **2018**, *22*, 229–250. [CrossRef]
17. Carver, C.S.; Scheier, M.; Weintraub, J.K. Assessing coping strategies: A theoretically based approach. *J. Personal. Soc. Psychol.* **1989**, *56*, 267–283. [CrossRef]
18. Vogt, W.P. Ceiling Effect. In *Entry, Dictionary of Statistics and Methodology: A Nontechnical Guide for the Social Sciences, 3d ed.*; SAGE: Thousand Oaks, CA, USA, 2005; p. 40.
19. Atkinson, R.; Flint, J. Accessing hidden and hard-to-reach populations: Snowball research strategies. *Soc. Res. Update* **2001**, *33*, 14.
20. Watters, J.; Biernacki, P. Targeted sampling: Options for the study of hidden populations. *Soc. Probl.* **1989**, *26*, 416–430. [CrossRef]
21. Denzin, N.; Lincoln, Y. The Discipline and Practice of Qualitative Research. In *Handbook of Qualitative Research*; Denzin, N.K., Lincoln, Y.S., Eds.; SAGE: Thousand Oaks, CA, USA, 2000; pp. 1–32.
22. Bogdan, R.; Biklen, S.K. *Qualitative Research for Education: An Introduction to Theories and Methods*; Allyn and Bacon, Inc.: Boston, MA, USA, 1998.
23. Corbin, J.; Strauss, A.L. *Basics of Qualitative Research*; SAGE: Thousand Oaks, CA, USA, 2008.
24. Berg, L.B.; Lune, H. *Qualitative Research Methods for the Social Sciences*; Pearson: Essex, UK, 2011.
25. Kook, R.; Harel-Shalev, A.; Yuval, F. Focus Groups and the Collective Construction of Meaning: Listening to Minority Women in Israel. *Women's Stud. Int. Forum* **2019**, *72*, 87–94. [CrossRef]
26. Wasserman, V.; Frenkel, M. The politics of (in) visibility displays: Ultra-orthodox women manoeuvring within and between visibility regimes. *Hum. Relat.* **2019**. [CrossRef]

International Journal of
*Environmental Research
and Public Health*

*Article*

# The Contribution of Long-Term Mindfulness Training on Personal and Professional Coping for Teachers Living in a Conflict Zone: A Qualitative Perspective

**Tal Litvak-Hirsch** [1,*] **and Alon Lazar** [2]

[1]   Conflict Management & Resolution Program, Department of Multidisciplinary Studies, Ben-Gurion University of the Negev, Beer Sheva 8410501, Israel

[2]   Program for Education, Society and Culture, Ono Academic College, Kiryat Ono 5545173, Israel; alon.la@ono.ac.il

*   Correspondence: litvakhi@bgu.ac.il

Received: 4 May 2020; Accepted: 4 June 2020; Published: 8 June 2020

**Abstract:** It has been suggested that mindfulness training can provide teachers with coping mechanisms and influence their perceptions of self and others. However, how does mindfulness help teachers cope in a stressful security situation both as Israeli citizens who live in a war zone and as teachers who are responsible for their students' lives? Fifteen female teachers, who lived and worked in the western Negev and who had completed two-years of mindfulness training, were interviewed. Interviewees reported that their coping skills had been heightened as result of being able to put aside intrusive thoughts and feelings that used to paralyze them and to focus on active coping, centered on what they needed to do promptly. Most also noted a more accepting attitude of themselves, without self-criticism or blame for what they should have or should not have done when facing the stressful situation. In relation to their students, they were more accepting of the behaviors and emotions expressed by their students and reported being more compassionate. The results will be discussed through the prism proposed by Lazarus and Folkman (1991). Educational implications of the outcomes of mindfulness training for those living in areas under the shadow of war will be suggested.

**Keywords:** coping; mindfulness; teachers; terror

## 1. Introduction

The past few decades have witnessed a growing interest in research on both the negative and the positive outcomes of ongoing exposure to terror and war, particularly with regard to the coping mechanisms of both the population at large and professionals living in such areas [1,2].

Israeli society, whose members experience such situations, struggles to remain vibrant and flourishing while threatened by war and terror. In recent years, the situation has become particularly acute for those living within a 40-km range from the Gaza Strip, namely, in Israel's western Negev. These citizens face rocket bombardments and flaming balloons on an ongoing basis. The current study focuses on the reactions of teachers who live and work in this area and have undergone mindfulness training, aiming to assess their coping skills as individuals and professionals. In the literature review, we begin with stress and coping theory. We then discuss mindfulness, and finally, focus on ways of coping and mindfulness among teachers.

## 1.1. Stress and Coping

In the transactional model of coping [3], stress is conceptualized as a situation in which a person feels that s/he is forced to garner considerable resources in order to face external and/or internal demands.

The model also suggests that when attempting to cope with a stressful situation, two cognitive appraisals, primary and secondary, are employed. Primary appraisal is directed by the question, "What do I have at stake in this encounter?", and is accompanied by emotions such as fear, worry, anger, and shame. The question, "What can I do or what are my options for coping?" marks secondary appraisal. During these appraisals, problem-focused and emotion-focused coping are available to the individual. Problem-focused coping means that the individual supposes that the situation can be dealt with effectively, while emotion focused coping directs the individual to assign new meanings to the stressful situation and restrain and handle negative emotions [3].

## 1.2. Mindfulness and Coping

A key element of mindfulness training is the cultivation of skills for dealing with challenging, uncertain, and stressful situations, by bringing the individual to focus their attention to a certain purpose, as it takes place in the present moment, and acting non-judgmentally toward the experience [4]. This is made possible as a result of the non-judgmental curiosity encouraged by mindfulness, which promotes attention to the stream of consciousness without emotional reactivity, mental rigidity, or rejection [5].

In accord with the transactional model of coping [3], it has been suggested that mindfulness enhances clarity and accuracy in the assessment of both the stressor (primary appraisal) and the available resources (secondary appraisal), resulting in more effective coping responses [6]. Mindfulness often forms a connection with compassion.

Compassion denotes focusing one's attention on the other, aiming to assist, and it is based upon positive feelings [7]. Findings suggest that short-term compassion training increased positive affect toward a suffering other [8].

Mindfulness is regarded as one of the keys to self-compassion, since mindfulness aids people to become aware that they are struggling and encourages them to act with kindness toward themselves [9]. Furthermore, when individuals treat themselves with self-compassion, they have more to give to others, and the loving connected presence that they feel for themselves will resonate on others [10]. Self-compassion may be a valuable coping resource for people experiencing negative life events [11]. It relates most strongly to positive cognitive restructuring and involves thinking about stressful situations in ways that enhance coping [12]. This suggests that practicing mindfulness can encourage compassion toward the self as well as toward others.

The effects of mindfulness training have been extensively discussed in areas such as health and education, noting its positive outcomes with regard to stress reduction [13], enhanced self-compassion, and a shared sense of greater self-awareness and self-acceptance, all leading to an improved capacity for engaging with the present in ways that reduce critical self-judgment [14].

## 1.3. Teachers and Mindfulness Training

The teaching profession is noted for its' complex nature [15]. Teachers are required to meet a wide range of demands and responsibilities that require skillful social and emotional conduct such as providing emotionally responsive support to students, cultivating a nurturing classroom environment, modeling exemplary emotion regulation, coaching students through conflict situations with sensitivity, successfully (yet respectfully) managing the challenging behaviors of disruptive students, and handling the growing demands imposed by standardized testing. Studies assessing the effects of mindfulness-based interventions for teachers have found consistent improvements in emotion regulation and mindfulness [16], lowering of anxiety [17], more positive handling of job stress, and the tendency to evaluate challenging students in a more positive affective light [18]. In addition, mindfulness training for teachers was found to promote their compassion, self-compassion, and care [19], and to lead to an improvement in their relations with the students and a better classroom climate and management [20].

There is enough evidence attesting to the positive outcomes of participation in mindfulness training when it comes to teachers. However, the existing literature relies upon American samples of teachers, who live in tranquil environments [16,20], rather than their counterparts in conflict zones, who are exposed to terror on a daily basis, as is the case with many teachers in Israel, especially those residing in the western Negev.

*1.4. Stress and Coping by Israelis Residing in Western Negev, Israel*

Beginning in 2008, residents of the western Negev, Israel, have lived under the threat of terror attacks directed to them from the bordering Gaza Strip in the form of missiles, balloon and kite-borne bombs, and more recently bomb carrying drones.

There is a plethora of studies aiming to identify the reactions of Jewish-Israelis living in the area. A review in [21] indicated that area residents exhibit fairly high levels of Post Traumatic Stress Disorder (PTSD) and depression during temporary breaks in the conflict and these levels rise significantly during periods of escalation. In contrast, very little research has studied the coping modes of these residents. One study found that Israeli women living in proximity to the Gaza Strip have been found to use both problem-focused and emotion-focused coping as well as optimism, humor, and denial [1]. In another study, the relations between PTSD and Post Traumatic Growth were found to be mediated by problem-focused coping among the area inhabitants [22].

*1.5. Mindfulness Training for Teachers in the Western Negev, Israel*

To aid teachers in the western Negev, mindfulness training programs have been offered by Sapir College and the Shaar Hanegev Psychological Clinic since 2015. For the first three years, the programs lasted 12 weeks, similar to programs offered to teachers in the USA [23]. These initial stages did not undergo evaluation. In 2018, the program was extended to include a two-year training period in order to provide participants with more tools and techniques. During the first year of the program, between September and June 2018, training was held every second week for three hours. During the intermissions between each meeting, participants were asked to practice and reflect upon the skills they had learned.

For the first two months, the focus was on Buddhist principles and the concepts of compassion and awareness. In the following months, participants practiced various techniques of meditation. These included breathing meditation, in which teachers learned to focus their awareness on their breathing experience; open awareness meditation aiming to bring participants to an awareness and focus upon the here and now, and current feelings, thoughts, and body sensations; and compassion training. Here, each participant was asked to think first of another person, one they felt very close to, and how they approached that person in a kindly manner. Participants were then asked to think about how these warm and kind feelings could be extended and directed toward others including strangers, and in the final stage of the training, how to direct such feelings to those one has difficult relations with.

During the months of July and August, when meetings were not held, participants were asked to continue daily meditation individually, at their homes. In the second year of training, between September and June 2019, each session was held for three hours each second week. These meetings, like the first year of the program, aimed at assisting participants to become more confident in their meditation practices, and to explore further Buddhist principles. In addition, the teachers became familiar with modes of teaching their students the basic principles of mindfulness, mainly, by focusing on the here and now, and discussed their practice with their instructors.

The teachers joined the program on a voluntary basis and received credits from the Israeli Ministry of Education for their attendance, as the course is recognized for its contribution to advanced professional training.

The present study aims to provide preliminary answers to an under-discussed topic in the literature: What is the contribution of a long-term mindfulness program to teachers' coping as individuals and as teachers living in a conflict zone?

## 2. Materials and Methods

### 2.1. Participants and Procedures

According to the most recent report of the Israeli Ministry of Education, female teachers comprise 85% of the teaching profession in Israel [24]. There are no available public records recording their exact numbers in the western Negev.

All teachers living in the western Negev and eligible to participate in teachers continuing education programs offered by the Israeli Ministry of Education were given the chance to participate in this program. Past research has indicated that Israeli teachers prefer short continuing educational programs that are closely related to their teaching profession [25]. In this respect, the participants in the program discussed in this study are likely to differ from typical participants in the programs offered by the Israeli Ministry of Education. Fifteen female teachers took part in the program and were informed at its beginning that they could leave at the end of the first year. All completed the program and agreed to take part in the research after they were approached by their instructors, who informed them of the study. The teachers received no form of reimbursement for their participation in the study.

The teachers ranged in age between 45 and 57 years old. Of them, four were kindergarten teachers, five were elementary school teachers, and six were high school teachers. Except for two (one divorced, one widowed), all were married, had children, and lived in cities, towns, moshavim, and kibbutzim, at a distance of a few kilometers to 30 km from the Gaza Strip. After receiving the institutional review board IRB approval from the Ben Gurion University Board (2019-06), the teachers were approached by the first author and her graduate students, and were all interviewed in depth, using semi-structured interviews lasting 1–2 h. Interviews were taped and transcribed verbatim.

The interviews were conducted between the end of June 2019 and August 2019, two weeks to a month and a half after the end of the program. Interviews were chosen to assess the participants' experiences following the program, as qualitative methods such as interviews in mindfulness training assessment have been recommended for their ability to more fully grasp the complexities of a subjective nature as experienced by participants in such programs, that are otherwise not detected by quantitative research [26]. Thus, interviews make it possible to underpin what participants feel, what they have experienced, and how they make sense of the practices they have acquired [27].

The interview guide was designed to explore coping practices employed by the participants and the contribution to these of mindfulness.

Each interview began with the introductory demographic question "Please tell me about yourself", and this provided information about the participants' age, family status and number of children, number of years living in the area, number of years as a teacher, and so forth. The teachers were then asked to explain why they decided to join the program, given that it was time consuming and required considerable dedication. The question aimed to identify the profile of the participants in terms of age and teaching experience, and their psychological, emotional, and/or practical expectations regarding the outcomes of the training, considering the fact that the program differs significantly from other programs offered by the Ministry of Education. The third question aimed at learning about the personal experiences of the participants as residents of an area struck by ongoing terror. They were asked to discuss their modes of coping, and emotions and behaviors when encountering difficulties at the personal, family, and/or professional levels. Finally, the teachers were asked to discuss whether, in their opinions, they had detected any impact of their participation in the program on their modes of coping as individuals and/or professionals. In combination, these questions aimed to understand how professionals living in an area struck by terror understand and employ the principles and insights they acquired during their mindfulness training.

*2.2. Analysis*

The thematic approach [28] was applied. This approach suggests looking at each interview as a holistic unit and then tracing the main themes that emerge from the material according to the research questions. After each of the researchers read the interviews and identified the themes for each question, the themes identified independently were compared to assess inter-rater agreement, reaching 0.94 (k = 0.94) and the discrepancies were discussed until reaching agreement [29]. The themes accepted by the research team are presented in the following.

*2.3. Research Ethics*

The teachers were interviewed only after receiving the IRB approval from the Ben Gurion University board (code number 2019-04), and they were informed of the aims of the research, and that they could stop their participation in the interview without the need to explain their decision. All agreed to take part in the study. In order to ensure their anonymity, responses are presented using pseudonyms, with no mention of any identifying information such as age, place of residence, number of children, and so forth.

## 3. Results

*3.1. Motivation to Join the Training*

All of the teachers had heard about mindfulness before their participation in the program, yet none had practiced it, and they had joined the program in order to acquire new skills needed to assist themselves and their students in coping with life in a conflict zone. As Shira suggests, "I have taught science for many years, and I feel confident about my knowledge of the subject. I feel the need to expand my knowledge of skills and abilities required to help my students and myself in stressful situations, as is the case with rockets bombarding our schools and homes".

*3.2. Contribution of Mindfulness Training for the Teacher's Coping Skills*

The teachers discussed four life domains that they felt had improved following their participation in the program. First, many of the teachers noted an enhanced acceptance of themselves, free of self-criticism or blame. Sigal explained, "As a result of my mindfulness training, I learned to accept myself, to be less critical of myself. I accept myself and my fears; I am not angry with myself anymore; I do what I can, and it is good enough".

Second, improved skills needed to cope with stressful events related to their family life were indicated. All noted that when facing a potential heated argument with a family member, they took a moment to relax before acting, and this was attributed to the impact of their participation in meditation sessions. Rachel pointed out, "Following the program, I reminded myself to take a breather, to stop for a moment, to be in the here and now, to avoid an impulsive reaction. I manage to view the stressful situation from a distance". Miri added: "My husband was an army officer for many years, and as a result, I used to run the home my way, as he was at home only for a few days each month. During his stay, he demanded that things only be done his way. That was a source of ongoing tension. Following the program, I've learnt to let go, to breathe and distance myself from the situation, to allow him to make his role at home clear. We became more relaxed and easygoing. He has recently become a civilian, and this transition, and the fact that he is now at home all the time and all that entails, happened smoothly". However, the teachers noted that these changes did not occur immediately, but rather were a result of their ongoing attendance in the program, as best exemplified by Ronit, "The effects of mindfulness cannot be achieved in one smooth breath; they requires effort, time and practice".

The third life domain discussed was coping with life in a conflict zone. When relating to the stressful security situation, the interviewees reported that, following their participation in the program, their ability to face their daily stressful reality had improved. Rachel noted that "we have lived in this war zone for years; there are ups and down, with stressful periods and those that are more tranquil.

You find your way to manage this situation; yet it is always a challenge. Participation in the program gave a set of skills I was not aware of, and for me, they do the job".

A skill mentioned as relating to stressful situations was strengthening the awareness of the situation and of the self as part of the situation. Yaara explained, "During the training, I expanded my awareness, first of my breathing. That helped me focus on what was going on inside me in the here and now, looking at the stressful situation from a little distance instead of running into it with full emotional reactions. That was very helpful when the sirens sounded, and you had to focus on action. Regarding the stressful security situation, the interviewees reported that as a result of the intervention, their coping had been aided by the acquired ability to put aside the intrusive thoughts and feelings that served to paralyze them—or at least to 'soften their volume,' as Yael explained—and to focus on active coping in the here and now such as running to the shelter, helping the students, and more. Another example of the contribution of mindfulness to active coping was given by Irit: "As a result of the training, I learned to understand that my thoughts and feelings are not reality; they are the creations of my mind. So when there is an alarm, I am aware of my fears and my intrusive thoughts but they do not paralyze me. I can breathe and make myself relax; then I can be active and do everything I have to, run to the shelter and help my students or my own children at home. That is a very big positive change for me".

Many of the teachers were very open and claimed that "the fear never goes away", even with the help of mindfulness skills. However, the volume of the anxiety decreased and the focus on the here and now served as an anchor for active coping.

Finally, the teachers noted the impact of mindfulness training on their role as teachers as well as its influence on their students, in several important aspects of daily conduct in class.

All of the teachers emphasized that following their mindfulness training, their ability to take a step back and not react emotionally to infractions by their students, has become a major force in their professional conduct. Sivan said, "With the help of the training, I have learned to observe my students and their needs and to be more focused on helping them, especially when it comes to the more challenging ones". Dikla stated, "There is one student in class. Time and again she has tested my boundaries, and at first, the whole class followed her. I told myself, take a deep breath, and be nice to her, although she deserves no such reaction. In the end, I was the one who triumphed!".

Some of the teachers noted that they aimed to teach their students some of the principles and skills they had acquired during the program, thus implementing one of goals of the second year of training.

Yael explained, "There is this student who tends to act with no concern for the consequences. One day after class, I asked him to practice a breathing technique with me that I had learnt at the training program. He refused at first, but finally agreed. Nowadays, I see him take a breather and relax, before acting out". Shosh pointed out, "We speak the language of mindfulness in class, emphasizing the need to be attuned to the needs of each student, to stretch a helping hand to those struggling. From a class with raging and arguing students, they have become a calm and cohesive group". Another aspect noted was acceptance of members of minority groups by the students. Aliza said: "I have an assistant teacher, Nadav, who recently came out of the closet. Prior to my training, the kids used homosexual as an insult. Following the training, I gathered the children with Nadav present, and explained that there was nothing wrong with homosexuals and emphasized the need to accept one another's lifestyle. The kids now refrain from calling each other homosexuals." Another example was described by Aviva: "An Ethiopian student joined my class. The school headmaster explained to me that she believed that my pupils, following their familiarity with mindfulness, would accept that student more easily than any other class. She was right, as the other students accepted the Ethiopian girl warmly and helped her integrate quickly".

The last aspect discussed was an improved ability to help students deal with the stressful security situation, yet noting the need to receive more professional guidance. Naama maintained, "In the second year of the training, I taught them [the students] some breathing skills and we spoke a lot about here and now and on focusing. However, I am not a professional mindfulness guide; there is very little that I can do. To my mind, incorporating mindfulness within the school curriculum, would benefit both students and staff". The teachers stressed the contribution of the long training, which enabled them to experience mindfulness personally as well as pass it on to their students in some ways; nevertheless they felt that they were not professional enough to teach it actively as part of their class curriculum.

In summary, all the teachers suggested that long term mindfulness training could help teachers who live in conflict zones cope with stressful situations, both personally and professionally, and enhance their coping skills and their self-compassion.

## 4. Discussion

Coping with terror by civilians, especially in the Israeli context, is a subject of ongoing interest to scholars [1,22] who assess problem-focused and emotion-focused coping [3] in the context of living under the threat of terror. The current effort aimed to add to this literature by looking into the effects of mindfulness training, considered to have positive effects among teachers [16,21], and especially among Israeli female teachers living in a conflict zone. In general, the current results corroborate previous suggestion that mindfulness training enhances its practitioners' ability to assess both the stressors they face and their available resources to handle these stressors in a more serene manner [6]. This means that mindfulness training helps to improve the employment of primary and secondary appraisals.

Following their participation in the training, several changes were noted by the interviewees. First, they indicated that mindfulness training helped them to regulate their fear reactions when under the threat of terror, both at home and at school. In this respect, employment of problem focused coping, instead of emotion focused coping, was attributed to participation in the program. Second, when conducting their relations with their family members, again, problem focused coping was the dominant mode of conduct noted, following the program. This was also the case at school, along with being able to express more compassion toward their students as well as their peers. This finding echoes those previously discussed noting the positive impact of mindfulness training on these aspects of conduct [14,15].

This is most likely the result of a change related to self-compassion following the program. It was argued that self-compassion indicates the awareness that each one is struggling to achieve, and thus acting with kindness to oneself leads to acting kindly to others, thus becoming more compassionate toward them [17]. Similar to the reports in the literature [19], the teachers pointed to a more positive classroom climate and management. It is worth noting that the teachers pointed out that the tools they have acquired to help them deal personally with stressful situations should be expanded further in order to assist them in teaching mindfulness to students. In this respect, it seems that these teachers acknowledge that the current training they have received has it limits, and they need further instruction on how to implement the guidelines offered by mindfulness training within their classes.

This study is the first of its kind, as it aimed to assess the impact of long-term mindfulness training among teachers living and working in an area, time and again struck by terror, noting its positive impact both at the professional and family levels. These findings suggest long-term mindfulness training should be part of the professional tool kit available to teachers living and working in areas dealing with the dangers of ongoing terror, and potentially also be accessible to their students.

However, several research limitations require future studies in order to expand these initial findings. First, coping was assessed through self-report only, with no control group comparison. Thus, in order to further explore the insights participants discussed, such an evaluation is needed. Second, participants had long-standing careers as teachers, and as such, their motivation was to acquire tools to manage the stressful situations they experienced as teachers in a conflict zone. This directs attention to the need to assess how novice teachers in conflict zones, who must also deal with the stressors emerging from their new career, discuss the impact of their participation in mindfulness training. A gender difference should also be researched, looking at the contribution of mindfulness training to males in comparison to females. Finally, a comparison between short- and long-term mindfulness training programs in Israel as well as in other areas where teachers must handle the threat of terror and war as part of their work should be examined.

## 5. Conclusions

The emerging picture that the current study suggests is that female teachers working and living in a conflict zone do benefit from long-term mindfulness training in the professional domain as well as when it comes to their individual and family conduct. This implies that in order to understand more fully the impacts of such training, attention should also be directed at populations of teachers in societies under the conditions of ongoing conflict in other parts of the world as well as to post-conflict societies. It is also important to find out how the students of teachers, like those studied here, reflect upon the outcomes following the conduct of their teachers as well as their family members.

**Author Contributions:** Conceptualization, T.L.-H. and A.L.; methodology, T.L.-H. and A.L.; validation, T.L.-H. and A.L. formal analysis, T.L.-H. and A.L.; investigation, T.L.-H. and A.L.; resources, T.L.-H. and A.L.; visualization, T.L.-H. and A.L.; writing—original draft preparation, T.L.-H. and A.L.; writing—review and editing, T.L.-H. and A.L.; supervision, T.L.-H. and A.L.; project administration, T.L.-H. and A.L. All authors have read and agree to the published version of the manuscript.

**Funding:** This research received no external funding.

**Conflicts of Interest:** The authors declare no conflict of interest.

## References

1.  Litvak-Hirsch, T.; Lazar, A. Experiencing processes of growth: Coping and PTG among mothers who were exposed to rocket attacks. *Traumatology* **2012**, *18*, 50–60. [CrossRef]
2.  Maguen, S.; Papa, A.; Litz, B.T. Coping with the threat of terrorism: A review. *Anxiety Stress Cop.* **2008**, *21*, 15–35. [CrossRef] [PubMed]
3.  Lazarus, R.S.; Folkman, S. The concept of coping. In *Stress and Coping*; Monat, A., Lazarus, R., Eds.; Columbia University Press: New York, NY, USA, 1991; pp. 191–226.
4.  Kabat-Zinn, J. *Full Catastrophe Living: Using the Wisdom of Your Body and Mind to Face Stress, Pain, and Illness*; Delacorte Press: New York, NY, USA, 1990.
5.  Cullen, M.; Brito, C.G. *Mindfulness-Based Emotional Balance: Navigating Life's Full Catastrophe with Greater Ease and Resilience*; New Harbinger Press: Oakland, CA, USA, 2014.
6.  Walsh, R.; Shapiro, S.L. The meeting of meditative disciplines and Western psychology. *Am. Psychol.* **2006**, *61*, 1–13. [CrossRef] [PubMed]
7.  Singer, T.; Klimecki, O.M. Empathy and compassion. *Curr. Biol.* **2014**, *24*, R875–R878. [CrossRef] [PubMed]
8.  Klimecki, O.M.; Leiberg, S.; Ricard, M.; Singer, T. Differential pattern of functional brain plasticity after compassion and empathy training. *Soc. Cognit. Affect. Neurosci.* **2014**, *9*, 873–879. [CrossRef] [PubMed]
9.  Neff, K.D. *Self-Compassion: The Proven Power of Being Kind to Yourself*; Harper Collins Publishers Inc.: New York, NY, USA, 2015.
10. Neff, K.D. The development and validation of a scale to measure self-compassion. *Self Identity* **2003**, *2*, 223–250. [CrossRef]
11. Allen, A.B.; Leary, M.R. Self-Compassion, stress, and coping. *Soc. Personal. Psychol. Compass* **2010**, *4*, 107–118. [CrossRef] [PubMed]
12. Rudaz, M.; Twohig, M.P.; Ong, C.W.; Levin, M.E. Mindfulness and acceptance-based trainings for fostering self-care and reducing stress in mental health professionals: A systematic review. *J. Context. Behav. Sci.* **2017**, *6*, 380–390. [CrossRef]
13. Perridge, D.; Hefferon, K.; Lomas, T.; Ivtzan, I. "I feel I can live every minute if I choose to": Participants' experience of a positive mindfulness programme. *Qual. Res. Psychol.* **2017**, *14*, 482–504. [CrossRef]
14. Jennings, P.A. Promoting teachers' social and emotional competencies to support performance and reduce burnout. In *Breaking the Mold of Preservice and Inservice Teacher Education: Innovative and Successful Practices for the Twenty-First Century*; Cohan, A., Honigsfeld, A., Eds.; Rowman & Littlefield Education: Lanham, MD, USA, 2011; pp. 133–144.

15. Emerson, L.M.; Leyland, A.; Hudson, K.; Rowse, G.; Hanley, P.; Hugh-Jones, S. Teaching mindfulness to teachers: A systematic review and narrative synthesis. *Mindfulness* **2017**, *8*, 1136–1149. [CrossRef] [PubMed]

16. Hwang, Y.S.; Bartlett, B.; Greben, M.; Hand, K. A systematic review of mindfulness interventions for in-service teachers: A tool to enhance teacher wellbeing and performance. *Teach. Teach. Educ.* **2017**, *64*, 26–42. [CrossRef]

17. Taylor, C.; Harrison, J.; Haimovitz, K.; Oberle, E.; Thomson, K.; Schonert-Reichl, K.A.; Roeser, R.W. Examining ways that a mindfulness-based intervention reduces stress in public school teachers: A mixed-methods study. *Mindfulness* **2016**, *7*, 115–129. [CrossRef]

18. Lavelle Heineberg, B.D. Promoting caring: Mindfulness and compassion-based contemplative training for educators and students. In *Handbook of Mindfulness in Education: Integrating Theory and Research into Practice*; Schonert-Reichl, K.A., Roeser, R.W., Eds.; Springer: New York, NY, USA, 2016; pp. 285–294.

19. Sharp, J.E.; Jennings, P.A. Strengthening teacher presence through mindfulness: What educators say about the cultivating awareness and resilience in education (CARE) program. *Mindfulness* **2016**, *7*, 209–218. [CrossRef]

20. Oberle, E.; Schonert-Reichl, K.A. Stress contagion in the classroom? The link between classroom teacher burnout and morning cortisol in elementary school students. *Soc. Sci. Med.* **2016**, *159*, 30–37. [CrossRef] [PubMed]

21. Greene, T.; Itzhaky, L.; Bronstein, I.; Solomon, Z. Psychopathology, risk, and resilience under exposure to continuous traumatic stress: A systematic review of studies among adults living in southern Israel. *Traumatology* **2018**, *2018 24*, 83–103. [CrossRef]

22. Regev, I.; Nuttman-Shwartz, O. Coping styles and aggregate coping styles: Responses of older adults to a continuous traumatic situation. *J. Loss Trauma* **2019**, *24*, 159–176. [CrossRef]

23. Weare, K. *Promoting Mental, Emotional and Social Health. A Whole School Approach*; Routledge: London, UK, 2013.

24. The Israeli Educational System in International Perspective. Education at a Glance. 2019. Available online: http://meyda.education.gov.il/files/MinhalCalcala/hoveret_lasar_EAG2019_10092019.pdf (accessed on 5 April 2020).

25. Avdor, S. Teachers in-service training is only the beginning of journey: Personal and organizational contexts of teachers' development. *Dapim* **2015**, *59*, 231–264. (In Hebrew)

26. Sauer, S.; Walach, H.; Schmidt, S.; Hinterberger, T.; Lynch, S.; Büssing, A.; Kohls, N. Assessment of mindfulness: Review on state of the art. *Mindful* **2013**, *4*, 3–17. [CrossRef]

27. Moss, D.; Waugh, M.; Barnes, R. A tool for life? Mindfulness as self-help or safe uncertainty. *Int. J. Qual. Stud. Health Well-being* **2008**, *3*, 132–142. [CrossRef]

28. Lieblich, A.; Tuval-Mashiach, R.; Zilber, T. *Narrative Research: Reading, Analysis, and Interpretation*; Sage: Thousand Oaks, CA, USA, 1998.

29. Huynh, T.; Hatton-Bowers, H.; Smith, M.H. A critical methodological review of mixed methods designs used in mindfulness research. *Mindful* **2019**, *10*, 786–798. [CrossRef]

International Journal of
*Environmental Research
and Public Health*

*Article*

# CB-Art Interventions Implemented with Mental Health Professionals Working in a Shared War Reality: Transforming Negative Images and Enhancing Coping Resources

**Dorit Segal-Engelchin *, Netta Achdut, Efrat Huss and Orly Sarid**

Spitzer Department of Social Work, Ben-Gurion University of the Negev, POB 653, Beer-Sheva 84105, Israel; nettaach@bgu.ac.il (N.A.); ehuss@bgu.ac.il (E.H.); orlysa@bgu.ac.il (O.S.)
* Correspondence: dorsegal@bgu.ac.il; Tel.: +972-50-558-3764

Received: 1 March 2020; Accepted: 24 March 2020; Published: 28 March 2020

**Abstract:** Research on mental health professionals (MHPs) exposed to a shared war reality indicates that they are subject to emotional distress, symptoms of posttraumatic stress disorder, and vicarious trauma. This article focuses on a CB-ART (cognitive behavioral and art-based) intervention implemented during the 2014 Gaza conflict with 51 MHPs who shared war-related experiences with their clients. The intervention included drawing pictures related to three topics: (1) war-related stressors, (2) coping resources, and (3) integration of the stressful image and the resources drawing. The major aims of the study were (1) to examine whether significant changes occurred in MHP distress levels after the intervention; (2) to explore the narratives of the three drawing and their compositional characteristics; and (3) to determine which of selected formats of the integrated drawing and compositional transformations of the stressful image are associated with greater distress reduction. Results indicate that MHP distress levels significantly decreased after the intervention. This stress-reducing effect was also reflected in differences between the compositional elements of the 'stress drawing' and the 'integrated drawing,' which includes elements of resources. Reduced distress accompanied compositional transformations of the stressful image. MHPs can further use the easily implemented intervention described here as a coping tool in other stressful situations.

**Keywords:** mental health professionals; shared war realty; distress; art-based intervention; war-related stressors; coping resources

## 1. Introduction

Since the onset of the Al-Aqsa Intifida in September of 2000, Israeli society has been witnessing continual terrorist attacks by Hamas and other terrorist organisations, including suicide bombings, drive-by shootings, knife and gun attacks, and missile attacks in urban settings launched from the Gaza strip. Hamas's ongoing threat against Israeli civilians has led to several military operations. The current study was conducted during Operation "Protective Edge", also known as the 2014 Gaza conflict. This operation was launched in the summer of 2014 in response to the substantial increase in Hamas's rocket attacks against Israeli communities, firing on an almost daily basis [1].

During Operation "Protective Edge", which lasted 50 days, more than 4500 rockets were launched towards Israel from Gaza. This operation and the period immediately preceding it represented an intense period of rocket and mortar fire against Israel's civilian population. Although the range of these rockets covered more than 70% of Israel's civilian population, those residing in communities near the Gaza Strip were most affected, having only 15 seconds to seek shelter. During this time, six civilians and 67 soldiers in Israel were killed, more than 1600 civilians were harmed, and an estimated 10,000 civilians evacuated their homes. In the Gaza Strip, approximately 2,125 Palestinians were killed [1].

Mental health professionals (MHPs) are among the first responders to address the needs of traumatized people following exposure to large-scale disasters, including terrorist attacks and wars. In the southern region of Israel, which has been subject to missile attacks from Gaza since 2001, MHPs encounter a double exposure to war-related trauma as community members and as professionals providing service to terror victims [2]. This situation in which MHPs are coping with the same traumatic event as their clients is referred to as "shared trauma," "shared tragedy," or "shared traumatic reality" [3–5]. Shared traumatic situations typically occur in communal disasters, such as natural disasters and war [6]. MHPs working in shared traumatic situations face multiple levels of vulnerability to traumatization, including direct, secondary, and vicarious traumatization [3].

The negative consequences of shared reality situations have been well documented. These consequences may include emotional distress during a traumatic event [6], as well as immediately after a traumatic event and up to a year later [5]. Cohen et al. [7], who interviewed therapists working with traumatized children following the shared traumatic reality of the Second Lebanon War, found high levels of anxiety, stress, and symptoms of posttraumatic stress disorder (PTSD) among the therapists. Similar findings emerged from Finklestein et al.'s [8] study of MHPs working in areas affected by repeated rocket attacks from the Gaza Strip, indicating that MHPs were at risk for both PTSD and vicarious trauma (VT) symptoms. Those who lived in the more affected area were at even greater risk for developing PTSD and VT symptoms. Additional studies have shown similar associations among level of exposure to terror attacks, PTSD symptoms, and emotional distress [6,9], providing further support for an incremental dose effect [6]. However, other studies have not found an association between exposure levels and emotional distress [10,11]. Increased levels of PTSD symptoms also have been reported among physicians and nurses exposed to a shared war-related reality in Israel [10,12,13] and in Gaza [14,15].

Work under shared reality conditions exposes MHPs to the blurring of boundaries between professional and personal lives [4,16,17], including boundaries between work and family loyalties [5,18]. Research also points to the blurring of boundaries between MHPs and their clients [5], manifested in their difficulty separating their personal experience from that of their clients [19].

Several studies on the effects of working in a shared war reality have reported a decrease in perceived professional competence among MHPs [5] and a sense of being deskilled [3]. However, other studies reveal a strong perception of professional competence [19] and high levels of professional confidence [9] among these MHPs.

Work in a shared traumatic reality also has been associated with positive consequences. These consequences may include a sense of growth, both personal and professional [4,10,16,19], and a sense of resilience [19]. Post-traumatic growth also has been reported among nurses working in a shared war–related reality in Israel [10] and in Gaza [15]. Positive consequences also can include heightened intimacy in the therapeutic relationship [4,20], a strong therapeutic alliance [19], a high level of work satisfaction, and a sense of agency and helpfulness [7].

The overall picture that emerges from studies of MHPs exposed to a shared traumatic reality stresses the importance of interventions designed to alleviate their emotional distress, particularly among those operating in areas highly exposed to armed conflicts and terror attacks. Interventions for MHPs in traumatic events mainly consist of group support [21,22], individual or group supervision [7], and debriefing sessions [7,9]. Research findings, however, have called into question the effectiveness of debriefing methods in alleviating symptoms of stress among MHPs and disaster workers [9,23]. One possible explanation for the ineffectiveness of debriefing methods is that MHPs view participation and sharing as integral to the organizational culture of mental health services, rather than as a unique intervention tailored to alleviate their war-related stress [9]. Another plausible explanation is that emotional turmoil and thoughts related to traumatic experiences do not lend themselves to easy verbalization [24]. The inadequacy of conventional verbal methods in the context of disasters points to the need to search for alternative methods of self-care for MHPs in shared war situations. To address this need, Huss, Sarid, and Cwikel [25] developed an art-based intervention model for stress reduction and self-care for social workers operating in a war zone during the Iron Cast Operation (2008). They based this intervention model on the use of a single drawing in a single group session. In the first stage of the intervention, social workers were asked to draw one image of their war experience as social workers. They were then instructed to identify the sources of their stress and their stress reactions within the artwork and to change their artwork by adding sources of coping and resilience. Allowing the social workers to change their artwork helped them to gain a sense of control over diffuse sources of anxiety. Huss and Sarid [26] have found that transformation of stressful visual images through drawing and in the imagination is linked to decreased levels of work-related stress among health care professionals. These findings demonstrate the efficacy of transforming a stressful image without extensive verbalization for stress reduction.

This article focuses on a CB-ART (cognitive behavioral and art-based) intervention for distress reduction that was developed based on these earlier findings. This intervention was implemented with MHPs who shared war-related experiences and distress with their clients during Operation "Protective Edge".

*The Current Study*

The conceptual framework used in this study was based on the Stress and Coping Model (SCM) [27] and on the "art as therapy" orientation, which highlights the healing qualitaties of art making [28–31].

The aims of the current study were (1) to examine whether significant changes occurred in distress levels among MHPs at the end of the intervention; (2) to explore the narratives of the three drawing (e.g., stress, resources, and integrated drawings) and their compositional characteristics; (3) to determine which compositional elements of the stressful image that were transformed within the 'integrated drawing' were associated with greater distress reduction; and (4) to determine which of the selected formats for the 'integrated drawing' (e.g., a new sheet of paper, 'stress drawing,' or 'resources drawing') were associated with greater distress reduction.

## 2. Methods

Three CB-ART workshops were implemented with MHPs in southern Israel during the 2014 Gaza conflict at the social work department in Ben-Gurion University of the Negev. MHPs were recruited through advertisements on the university website, emails sent after the war began to health and social service agencies in the community, and through a snowball sampling technique. All participants were employed in health or social services agencies and were both working and living in the war zone.

Before the workshop began, the authors described the objective and procedure of the study and emphasized that participation in the workshop did not require participation in the study. All MHPs who participated in the workshops chose to take part in the study. They were asked to note their level of distress at the beginning and end of the workshop. At the end of the workshop, they also were asked to note the compositional elements that they used in the 'stress drawing' and to describe the transformations that they had made in the compositional elements of the stressful image within the 'integrated drawing.' To enable pre–post comparisons on an individual level, MHPs were asked to provide the last four digits of their national ID number on both the questionnaire and the three drawings. They were told that this information was needed for statistical purposes and would not be used to identify them.

The research was approved by the departmental ethics committee at Ben-Gurion University of the Negev. All participants signed consent forms agreeing to have their drawings and questionnaire used in research.

*2.1. CB-Art Intervention Description*

The two-hour workshop started with a short introductory lecture on stress responses to disasters and the debilitating effect of a negative distressing image, symptom, or memory on negative mood states. We explained how drawings can be analyzed by both the narrative attached to them and their compositional elements, such as shape, size, colors, and placement of the images on the paper [32,33].

In the first phase of the intervention, participants were asked to draw their current emotions and thoughts relating to the war situation (referred to here as the *stress drawing*). They were then asked to write a short description of their artwork on the back of their drawing. Participants presented their drawings within the group setting, describing what they had drawn, and shared their narrative. The group then discussed the compositional characteristics of each drawing. The aim of this phase was to enable MHPs to identify the sources of their war-related stress as well as the compositional elements that characterized their stressful image.

After drawing about their current condition, participants were asked to draw a new picture that reflected on their personal and social resources that could enable them to better cope with stressful situations (referred to here as the *resources drawing*). Again, they were asked to write a short description of their artwork on the back of it. Drawings were shared within the group setting and their compositional elements noted. The aim of this phase was to enable MHPs to identify coping resources at their disposal as well as the compositional elements that characterized their resources drawing.

In the last phase of the intervention, participants were asked to draw a picture that integrated the stressful image and the resources drawing (referred to here as the *integrated drawing*). Participants had the option to draw a new picture or to add elements to either the stress drawing or the resources drawing. The integrated drawings were shared within the group setting and their compositional elements discussed and compared to those of the stress and resources drawing. The purpose of the integrated drawing was to enable participants to learn how to "build bridges" between their resources and their distress image, symptom, or memory. The drawings in all three phases described above were created on A-4 paper with oil pastels.

*2.2. Sample*

To ensure the homogeneity of the sample, participants in the three workshops were compared by their demographic variables. Chi-square tests revealed no significant differences in gender, marital status, country of birth, education level, religion, degree of religious observance, or perceived financial situation. For this reason, we pooled the three groups into one. Table 1 presents the demographic characteristics of the pooled sample. The average age was 37 years, and most were female (86%) and Israeli born (82% ), were married or lived with a partner (63%), and had children (52%). Almost all participants were Jewish (96%), and most defined themselves as secular (72%). About half of the participants (52% ) viewed their financial situation as fair.

Participants included: 45 (88.2%) social workers, 5 (9.8%) psychologists and one psychiatrist (1.9%). All participants drew the three drawings included in the CB-ART intervention described above. With regard to the selected format for the integrated drawing, 21 participants chose to draw a new picture; 3 participants chose to add elements to the stress drawing; and 27 participants chose to add elements to the resources drawing.

**Table 1.** Demographic characteristics of study participants ($N = 51$).

| Characteristics | $n$ | % or Mean (SD) |
| --- | --- | --- |
| Age | | 37.5 (12.5) |
| 23–35 | 26 | 52.0 |
| 36–54 | 19 | 38.0 |
| 55 and above | 5 | 10.0 |
| Gender | | |
| Male | 7 | 14.0 |
| Female | 43 | 86.0 |
| Marital status | | |
| Single | 16 | 32.0 |
| Married | 23 | 46.0 |

**Table 1.** *Cont.*

| Characteristics | *n* | % or Mean (SD) |
|---|---|---|
| Cohabiting | 9 | 18.0 |
| Divorced/Separated | 2 | 4.0 |
| Has children | | |
| Yes | 26 | 52.0 |
| No | 24 | 48.0 |
| Country of birth | | |
| Israel | 41 | 82.0 |
| Other (US, Europe, FSU) | 9 | 18.0 |
| Religion | | |
| Jewish | 49 | 96.0 |
| Muslim | 2 | 4.0 |
| Degree of religiosity | | |
| Secular | 36 | 72.0 |
| Traditional or religious | 14 | 28.0 |
| Education | | |
| Academic-B.A. degree | 36 | 72.0 |
| Academic-M.A. degree | 14 | 28.0 |
| Perceived financial situation | | |
| Bad | 4 | 8.0 |
| Fair | 26 | 52.0 |
| Good or very good | 20 | 40.0 |

## 2.3. Measures

### 2.3.1. Distress Level

Participant distress level was measured using the Subjective Units of Distress Scale (SUDS) [34]. Respondents were asked to assess their level of distress on an 11-point scale ranging from 0 (absence of distress) to 10 (extreme level of distress). The SUDS has been used in previous studies that evaluated the efficacy of art-based interventions in reducing stress [26,35].

### 2.3.2. Compositional Elements

We used a compositional element scale, based on the compositional analysis of image transformation [26,36], to examine the compositional elements of the stressful image and its transformed elements within the integrated drawing. This scale covered five compositional elements: object, color, placement, size, and background. Participants were asked to fill in this scale for both the stress drawing and the integrative drawing.

### 2.3.3. Statistical Analyses

To investigate whether significant changes occurred in participant distress levels following the intervention, in a first step, we employed a paired sample *t*-test to compare pre–post SUDS scores. In a second step, we used descriptive statistics to analyze the compositional elements of the three drawings. Statistical tests for examining differences in these characteristics were not carried out because of the constraint of small cells. In a third step, we used independent sample *t*-tests to examine whether transformations in the compositional elements of the stress drawing within the integrated drawing were related to a greater reduction in SUDS scores. For this purpose, we computed a variable based on the difference between the SUDS score at (t2) and the SUDS score at (t1) (referred to here as *SUDS-difference score*). We then examined three types of transformations within the integrated drawing: (1) object transformation (e.g., addition versus change in or omission of objects), (2) color transformation (e.g., number and types of colors used), and (3) size transformation (e.g., reduction versus non-reduction of the stressful image size).

An independent samples *t*-test was also used to examine whether the selected format for the integrated drawing (e.g., a new sheet of paper, addition of elements to the stress drawing, or addition of elements to the resources drawing) was related to a greater reduction in the SUDs scores. Because only three participants drew the integrated drawing on the resources drawing, we created two categories: the first comprised those who added elements from the resources drawing to the stress drawing, whereas the second comprised participants who either drew a new picture or added elements from the stress drawing to their resources drawing.

## 3. Results

### 3.1. Pre–Post SUDS Scores

Table 2 presents the results of the paired sample *t*-tests conducted to examine differences in MHP SUDS scores between the beginning and end of the intervention. As Table 2 shows, the scores significantly decreased after the intervention, suggesting efficacy of the process. The mean difference was 1.51 (on a 0–10 scale).

**Table 2.** Pre–post Subjective Units of Distress Scale (SUDs) scores ($N = 51$).

|  | SUDS Score Mean (SD) | Paired Sample *t*-Test |
|---|---|---|
| SUDS score at t1 | 5.97 (2.22) |  |
| SUDS score at t2 | 4.46 (1.98) |  |
| Mean difference | 1.51 (1.39) | 7.41 ** |

** $p < 0.01$.

### 3.2. Compositional Characteristics of the Mhps' Drawings and Their Explanatory Narratives

The second aim of the study was to explore the compositional characteristics of the three drawings and their explanatory narratives. Here we present the descriptive statistics of the compositional characteristics of the three drawings, followed by two illustrative examples of drawings and explanatory narratives.

As Table 3 shows, there were substantial differences among the three drawings in the compositional elements. Almost half of the stress drawings (49%) had no background, compared to only a little more than a third of the resources drawings (37.3%), and less than a third of the integrative drawings (29.4%). A quarter of the stress drawings (25.5%) were composed of a single object, and in 37.3% of the drawings, the stressful image was placed at the center. In contrast, less than 10% of the resources drawings (7.8%) and of the integrative drawings (9.8%) were composed of a single object, and a considerably smaller percentage of these drawings had their image placed at the center of the drawing (19.6% and 13.7%, respectively). More than half of the stress drawings (54.9%) and of the resources drawings (54.9%) consisted of medium and large objects, compared to 45.1% of the integrated drawings. Additionally, in almost half of the stressed drawings (47.1%), black emerged as the dominant or only color, compared to only 3.9% of the resources drawings and 21.6% of the integrated drawings. A similar pattern was observed with regard to the dominant use of grey within the stress drawings (11.8%), compared to its dominant use within the resources drawings (2.0%) and the integrated drawings (5.9%). In contrast, in about a third of the resources drawings and the integrated drawings, green emerged as the dominant or only color, compared to only 7.8% of the stress drawings.

**Table 3.** Descriptive statistics of the compositional characteristics of the three drawings ($N = 51$).

|  | Drawings | | |
|---|---|---|---|
|  | Stress Drawing (%) | Resources Drawing (%) | Integrated Drawing (%) |
| Background |  |  |  |
| No background | 49.0 | 37.3 | 29.4 |

**Table 3.** *Cont.*

| | Drawings | | |
|---|---|---|---|
| | **Stress Drawing (%)** | **Resources Drawing (%)** | **Integrated Drawing (%)** |
| Had background | 51.0 | 62.7 | 76.6 |
| Number of objects | | | |
| One | 25.5 | 7.8 | 9.8 |
| Several | 74.5 | 92.2 | 90.2 |
| Object placement | | | |
| Center of the drawing | 37.3 | 19.6 | 13.7 |
| All over the drawing | 62.7 | 80.4 | 86.3 |
| Object size | | | |
| Large-medium | 54.9 | 54.9 | 45.1 |
| Small or mixed | 45.1 | 45.1 | 54.9 |
| Colors | | | |
| Black | | | |
| None or minor | 52.9 | 96.1 | 78.4 |
| Dominant or only | 47.1 | 3.9 | 21.6 |
| Green | | | |
| None or minor | 92.2 | 68.6 | 66.7 |
| Dominant or only | 7.8 | 31.4 | 33.3 |
| Grey | | | |
| None or minor | 88.2 | 98.0 | 94.1 |
| Dominant or only | 11.8 | 2.0 | 5.9 |

## Illustrative Examples of the Drawings

In this section, we present two illustrative examples of these drawings and explanatory narratives: one by an MHP who selected a new sheet of paper as the format for her integrative drawing, and the other by an MHP who selected the stress drawing as the format for his integrative drawing.

With regard to the stress drawing in Example 1 (drawing at far left in Figure 1), the MHP explained her drawing as follows: "I drew the emotional turmoil that I have been experiencing because of my concern over the safety of my family, including my children and grandchildren, family friends who are in active duty military service during the war, and my clients and staff members". These three groups of people are displayed in the three vertexes of the triangle that appears in the drawing, with the clients and staff members placed at the upper vertex, as noted in the words added to her drawing. With regard to the resources drawing (drawing at the center in Figure 1), the MHP noted the following: "I drew my home and family members, which I view as a major resource enabling me to better cope in stressful situations". With regard to the integrative drawing (drawing at the right in Figure 1), the MHP stated that "The drawing expresses my lessened feelings of emotional turmoil".

**Figure 1.** Selecting a new sheet of paper as the format for the integrative drawing.

Analysis of the compositional elements of her three drawings reveals substantial differences among them. As can be seen, the stressful image was represented by a single, primarily black, large object, placed at the center of the drawing, with no background. In contrast, the resources and the stress drawings are characterized by the use of several mixed-sized objects and lighter, "optimistic" colors. In terms of the transformed compositional elements of the stressful image within the integrated drawing, the MHP explained that "I had omitted the triangle representing the sources of my emotional turmoil and decreased the size of the stressful image within the integrated drawing". One can also see that she has moved the stressful image from the center of the page. This compositional transformation appears as an additional indicator of her lessened feelings of distress.

The stress drawing in Example 2 (left in Figure 2) initially included only the image of a man holding a huge ball. The MHP described his drawing as follows: "In this drawing the focus is on myself. The huge ball symbolizes the extreme stress and burden that I have been encountering during the war in multiple domains due to the need to address my family and clients' needs simultaneously". With regard to the resources drawing (right, Figure 2), he noted that "I drew the resources that usually help me calm down in stressful times: my home, family, music, and the beach". With regard to the integrative drawing (right, Figure 2), the MHP stated that "I have added my family and home to the stress drawing because they help me to cope with any problem or stressful situation that I encounter".

**Figure 2.** Selecting the stress drawing as the format for the integrative drawing.

Analysis of the compositional elements of the drawings in Example 2 also reveals substantial differences between the drawings. Whereas the initial stressful image was represented by a single large object with no background, the resources drawing is characterized by the use of several mixed-sized objects, scattered all over the paper. With regard to the integrated drawing, the addition of objects to the stress drawing enabled the MHP to alter the proportion of the stressful image within the drawing, as well as to situate it within his everyday social context. These compositional transformations reflected the decrease in his feelings of distress.

The two examples of drawings and themes emerging from the explanatory narratives presented above, which were evident in the explanatory narratives of the vast majority of participants, agree with the overall picture emerging from the descriptive statistics of the compositional characteristics of the three drawings displayed in Table 3.

*3.3. SUDS-Difference Scores by the Selected Format of the Integrated Drawing and Compositional Transformations*

In the current study, we also sought to determine which of selected formats and the transformed compositional elements of the stressful image within the integrated drawing are associated with greater SUDS reduction. Table 4 presents the SUDS-difference scores by the selected format for the integrated drawing and transformed compositional elements. As shown in Table 4, we found statistically significant differences with regard to size transformation within the integrated drawing and to the selected format of the integrated drawing. Participants who reduced the initial size of the stressful image within the integrated drawing had a greater SUDS-difference score (−1.87 [1.22]) than those who maintained the initial size of the stressful image (−1.13 [1.48]). Additionally, participants who drew the integrated drawing on either the resources drawing or on a new sheet of paper had a higher SUDS-difference score (−1.73 [0.15]) than participants who drew the integrated drawing on the stress drawing (−0.92 [0.86]).

**Table 4.** SUDS-difference scores by the selected format of the integrated drawing and compositional transformations: stress-drawing versus integrated drawing (*N* = 51).

| | SUDS-Difference Mean (SD) | Independent Sample *t*-Test |
|---|---|---|
| Selected format for the integrative drawing | | 1.828 * |
| Stress drawing | −0.92 (0.86) | |
| Resources drawing or a new drawing | −1.73 (0.15) | |
| Shape transformation | | 0.150 |
| Addition of shapes | −1.50 (1.40) | |
| Change or omission of shapes | −1.60 (1.51) | |
| Color transformation | | −0.78 |
| Yes | −1.59 (1.36) | |
| No | −1.20 (1.54) | |
| Size transformation | | −1.786 * |
| Reduction of initial size | −1.87 (1.22) | |
| Maintenance of initial size | −1.13 (1.48) | |

* *p* < 0.05.

## 4. Discussion

This study focused on a CB-ART intervention implemented with MHPs who shared war-related experiences and distress with their clients during the 2014 Gaza conflict. Results indicate that MHPs' levels of distress significantly decreased after the intervention, suggesting its efficacy. Further evidence of the efficacy of a CB-ART intervention in reducing disaster-related distress is derived from Segal-Engelchin et al.'s [35] study of Nepalese students living in Israel during the 2015 Nepal earthquake, who were indirectly exposed to the disaster that struck their country. A plausible explanation for the decline in MHPs' levels of distress at the completion of the intervention may be that drawing and identifying the war-related stressors as well as personal and social resources increased their sense of control in the war situation. Transforming the compositional elements of the stress drawing within the integrated drawing may have also enhanced MHPs' sense of control. Previous studies on art-based interventions suggest that the active management of a stressful image leads to an enhanced sense of control [25,26]. Additionally, it may be that modifying the compositional elements of the stress image within the integrated drawing allowed the MHPs to modify its emotional content into a more enabling meaning [32]. This possibility is reflected in the explanatory narrative ascribed by the MHP in Example 1 to her integrative drawing, in which the image of war-related stress was modified, indicating that the drawing mirrored the lessened feelings of emotional turmoil. Another possible explanation lies in the integrated drawing, in which the stressful image and the coping resources were simultaneously displayed. This depiction may have enabled the MHPs to view war-related stressors and coping resources at their disposal as two interrelated entities of their war experience. This, in turn, may have led them to perceive the war-related stressors as less threatening and more manageable. The use of arts as a tool that enhances manageability has been reported by Huss and Samson [37] in their study of a group of recovering cancer patients.

One aim of the study was to explore participants' narratives of the three drawings (e.g., stress, resources, and integrated drawings) and their compositional characteristics. Their narratives of the stressful image, as demonstrated in the above two examples, reflected their emotional turmoil, feelings of extreme distress, and concern about the safety of their loved ones and their clients. A further prominent theme in their narratives, which was also evident in the narratives of the other participants, was the emotional burden that stemmed from assisting their clients while coping with their own anxieties and caring for their family members. These narratives corroborate findings of previous studies indicating high levels of anxiety and emotional distress among MHPs in shared trauma situations [6–9] as well as the blurring of boundaries between work and family loyalties [5,18,25].

Two major resources that enabled better coping with stressful situations emerged from the narratives of the resources drawing. One was the family and home environment, and the other was related to social-leisure activities, such as listening to music and going to the beach. Previous studies indeed indicate that family is an essential source of support for MHPs in shared war realities [5]. Their narratives of the integrated drawing reflect lessened feelings of emotional turmoil as well as a perceived ability to better cope with the shared war reality.

The different narratives ascribed to the three drawings are also expressed in the different compositional characteristics of these drawings. Our quantitative findings revealed substantial differences among the three drawings in the compositional elements, as illustrated in the two examples given. The stress drawings were generally characterized by a single, predominantly black, large–medium-sized object, placed at the center of the drawing, with no background. In contrast, both the resources and integrative drawings were typically characterized by the use of several mixed-sized objects and lighter optimistic colors, scattered all over the drawing. The use of single large objects and intense black lines to depict war-related stressors have been reported previously [25], and comparable compositional characteristics of stressful images also have been found in a study of children with cancer [38]. The use of the color black is associated with stress and depression in the diagnostic art therapy literature [39–42]. While diagnostic analyses of color is based on universal measures, other art therapy directions point to the cultural significance of specific colors [43]. On an Israeli cultural level, the color black is also associated with negative experiences such as mourning, a situation in which people wear black clothes, and negative moods, which are described as "black".

An additional objective of the current study was to determine which of the transformed compositional elements of the stressful image within the integrated drawing was associated with greater SUDS score reduction. The results indicate that size transformation of the stressful image within the integrated drawing was the only transformed compositional element significantly associated with greater SUDS score reduction. This finding can be interpreted in two ways. MHPs who experienced greater distress reduction after identification of their war-related stressors and coping resources in the framework of the first two phases of the intervention tended to decrease the size of the stressful image within the integrative drawings. It could also be, however, that the greater distress reduction resulted from the size transformation of the stressful image rather than being the cause of the size transformation. It is possible that the size modification of the stressful image and its proportion relative to the objects symbolizing various coping resources enabled MHPs to feel an increased sense of control and competence, resulting in their greater SUDS score reduction. Further research is needed to determine the direction of causality between SUDS score reduction and size transformation of the stressful image.

Of interest, we found an association of selecting the resources drawing or a new sheet of paper for the integrated drawing with significantly greater SUDS score reduction. This finding suggests that in the process of transforming the stressful image, attention should also be given to the context where the stressful image is placed, in addition to the transformed compositional elements. The placement of the stressful image in a new context that encompasses MHPs' personal and social resources may lessen its threatening features, resulting in an enhanced sense of agency. It is also possible that the selection of a new context for the stressful image is an indication of reduced levels of distress. Further investigation may shed light on the causal relationship between the format selected for the integrative drawing and SUDS score reduction.

Several limitations of the current study should be acknowledged. The first is related to the cross-sectional design that does not allow for determination of the long-term impact of the CB-ART intervention. Follow-up studies are needed to investigate the long-term effectiveness of this intervention.

The second limitation is related to the lack of a control group. This limitation is inherent to this type of quick-response intervention in times of disaster, where the evaluative research must be conducted rapidly and with regard for the well-being of all the people involved. A third limitation is the relatively small sample size, which precluded rigorous statistical tests to examine the differences among the various compositional elements of the three drawings. Future studies using larger samples may enrich our understanding of the role that compositional elements play in shaping participants' perception of stressful images, resource images, and integrated images. Further research is also needed to examine the contribution of the different components of the intervention (e.g., each of the three different drawings and the group setting) to stress reduction.

## 5. Conclusions

The current study marks the first attempt to examine the effectiveness of a CB-ART intervention implemented with MHPs under actual wartime conditions. In this study, the concepts presented in the Stress and Coping Model [SCM, 27] served as a framework to examine the ways that a CB-ART intervention, based on drawing, can help MHPs to express their stress as well as to acknoewledge their coping strategies, and integrate these two elements in the context of war. Within the drawing process, the identification of the compositional elements of the three drawings and the creation of manipulations within the integrated drawing enabled participants to actively appraise which compositional transformations decreased their distress and enhanced their adjustment and coping [44,45]. Our findings suggest that participants' conscious cognitive processing of the compositional transformations altered their interpretation of the stressful images, which in turn decreased their distress levels.

The study findings make an initial contribution to understanding the ways that stressful images and resource images are integrated on the paper using compositional transformations, as well as the ways in which this process assists in reducing MHP distress levels when they are operating in a shared war reality. The findings add to the art therapy literature on positive psychology and on the healing qualitaties of art making as in " art as therapy" orientations [28–31].

On a practical level, this study offers an easily implemented tool for distress reduction among MHPs in shared trauma situations. The CB-ART intervention provided MHPs an opportunity to depict their war-related stressors as well as their coping resources on the page and to discuss both images and access new thoughts and understanding regarding ways to manage stress in extremely stressful situations. As such, the CB-ART intervention not only may have enriched their coping resources but also have become a coping resource in itself, which they can use in traumatic situations in the future as a self-care strategy.

**Author Contributions:** Conceptualization, D.S.-E. and O.S.; Formal analysis, N.A. and O.S.; Investigation, D.S.-E., E.H. and O.S.; Methodology, D.S.-E. and O.S.; Writing—original draft, D.S.-E.; Writing—review & editing, O.S. All authors have read and agreed to the published version of the manuscript.

**Funding:** This research received no external funding.

**Conflicts of Interest:** The authors declare no conflict of interest.

## References

1.  Israel Ministry of Foreign Affairs. The 2014 Gaza Conflict: Factual and Legal Aspects Jerusalem: Ministry of Foreign Affairs. 2015. Available online: https://mfa.gov.il/MFA/ForeignPolicy/IsraelGaza2014/Pages/2014-Gaza-Conflict-Factual-and-Legal-Aspects.aspx (accessed on 30 January 2020).
2.  Baum, N. Professionals' double exposure in the shared traumatic reality of wartime: Contributions to professional growth and stress. *Br. J. Soc. Work* **2014**, *44*, 2113–2134. [CrossRef]
3.  Saakvitne, K. Shared trauma: The therapists' increased vulnerability. *Psychoanal. Dialogues* **2002**, *12*, 443–450. [CrossRef]
4.  Tosone, C.; Nuttman-Shwartz, O.; Stephens, T. Shared trauma: When the professional is personal. *Clin. Soc. Work J.* **2012**, *40*, 231–239. [CrossRef]

5. Dekel, R.; Baum, N. Intervention in a shared traumatic reality: A new challenge for social workers. *Br. J. Soc. Work* **2010**, *40*, 1927–1944. [CrossRef]
6. Freedman, S.A.; Mashiach, R.T. Shared trauma reality in war: Mental health therapists' experience. *PLoS ONE* **2018**, *13*, e0191949.
7. Cohen, E.; Roer-Strier, D.; Menachem, M.; Fingher-Amitai, S.; Israeli, N. "Common-Fate": Therapists' benefits and perils in conducting child therapy following the shared traumatic reality of war. *Clin. Soc. Work J.* **2014**, *43*, 77–88. [CrossRef]
8. Finklestein, M.; Stein, E.; Greene, T.; Bronstein, I.; Solomon, Z. Posttraumatic stress disorder and vicarious trauma in mental health professionals. *Health Soc. Work* **2015**, *40*, e25–e31. [CrossRef]
9. Dekel, R.; Hantman, S.; Ginzburg, K.; Solomon, Z. The cost of caring? Social workers in hospitals confront ongoing terrorism. *Br. J. Soc. Work* **2007**, *37*, 1247–1261. [CrossRef]
10. Lev-Wiesel, R.; Goldblat, H.; Eisikovits, Z.; Admi, H. Growth in the shadow of war: The case of social workers and nurses working in a shared war reality. *Br. J. Soc. Work* **2009**, *39*, 1154–1174. [CrossRef]
11. Pruginin, I.; Segal-Engelchin, D.; Isralowitz, R.; Reznik, A. Shared war reality effects on the professional quality of life of mental health professionals. *Isr. J. Health Policy Res.* **2016**, *5*, 17. [CrossRef]
12. Ben-Ezra, M.; Palgi, Y.; Wolf, J.J.; Shrira, A. Psychiatric symptoms and psychosocial functioning among hospital personnel during the Gaza War: A repeated cross-sectional study. *Psychiatry Res.* **2011**, *189*, 392–395. [CrossRef] [PubMed]
13. Ben-Ezra, M.; Palgi, Y.; Wolf, J.J.; Shrira, A.; Hamama-Raz, Y. Somatization and psychiatric symptoms among hospital nurses exposed to war stressors. *Isr J. Psychiatry Relat Sci.* **2013**, *50*, 182–187. [PubMed]
14. Abu-El-Noor, N.I.; Aljeesh, Y.I.; Radwan, A.S.; Abu-El-Noor, M.K.; Qddura, I.A.I.; Khadoura, K.J.; Alnawajha, S.K. Post-Traumatic Stress Disorder among health care providers following the Israeli attacks against Gaza Strip in 2014: A call for immediate policy actions. *Arch. Psychiatr. Nurs.* **2016**, *30*, 185–191. [CrossRef] [PubMed]
15. Shamia, N.A.; Thabet, A.A.M.; Vostanis, P. Exposure to war traumatic experiences, Post-Traumatic Stress Disorder and post-traumatic growth among nurses in Gaza. *J. Psychiatr. Ment. Health Nurs.* **2015**, *22*, 749–755. [CrossRef] [PubMed]
16. Shamai, M.; Ron, P. Helping direct and indirect victims of national terror: Experiences of Israeli social workers. *Qual Health Res.* **2009**, *19*, 42–54. [CrossRef] [PubMed]
17. Somer, E.; Buchbinder, E.; Peled-Avram, M.; Ben-Yizhack, Y. The stress and coping of Israeli emergency room social workers following terrorist attacks. *Qual. Health Res.* **2004**, *14*, 1077–1093. [CrossRef] [PubMed]
18. Dekel, R.; Nuttman-Shwartz, O.; Lavi, T. Shared traumatic reality and boundary theory: How mental health professionals cope with the home/work conflict during continuous security threats. *J. Couple Relatsh. Ther.* **2016**, *15*, 121–134. [CrossRef]
19. Lavi, T.; Nuttman-Shwartz, O.; Dekel, R. Therapeutic intervention in a continuous shared traumatic reality: An example from the Israeli–Palestinian conflict. *Br. J. Soc. Work* **2015**, *47*, 919–935. [CrossRef]
20. Tosone, C. Therapeutic intimacy: A post-9/11 perspective. *Smith Coll. Stud. Soc. Work* **2006**, *76*, 89–98. [CrossRef]
21. Mitchell, J.T.; Everly, G.S., Jr. Critical incident stress management and critical incident stress debriefings: Evolutions, effects and outcomes. In *Psychological Debriefing: Theory, Practice and Evidence*; Raphael, B., Wilson, J.P., Eds.; Cambridge University Press: Cambridge, UK, 2000; pp. 71–90.
22. Wilson, J.; Sigman, M. Theoretical perspectives of traumatic stress. In *Psychological Debriefing: Theory, Practice and Evidence*; Raphael, B., Wilson, J.P., Eds.; Cambridge University Press: Cambridge, UK, 2000; pp. 58–68.
23. Kenardy, J.A.; Webster, R.A.; Lewin, T.J.; Carr, V.J.; Hazell, P.L.; Carter, G.L. Stress debriefing and patterns of recovery following a natural disaster. *J. Trauma. Stress* **1996**, *9*, 37–49. [CrossRef]
24. Huss, E.; Hafford-Letchfield, T. Using art to illuminate social workers' stress. *J. Soc. Work* **2019**, *19*, 751–768. [CrossRef]
25. Huss, E.; Sarid, O.; Cwikel, J. Using art as a self-regulating tool in a war situation: A model for social workers. *Health Soc. Work* **2010**, *35*, 201–209. [CrossRef] [PubMed]
26. Huss, E.; Sarid, O. Visually transforming artwork and guided imagery as a way to reduce work related stress: A quantitative pilot study. *Arts Psychother.* **2014**, *41*, 409–412. [CrossRef]
27. Lazarus, R.S.; Folkman, S. *Stress, Appraisal, and Coping*; Springer: New York, NY, USA, 1984.

28. Rosal, M. Cognitive Behavioral Art Therapy. In *Approaches to Art Therapy: Theory and Technique*; Rubin, J., Ed.; Brunner and Routledge: New York, NY, USA, 2001; pp. 210–225.

29. Rubin, J. Introduction. In *Approaches to Art Therapy: Theory and Technique*; Brunner and Routledge: New York, NY, USA, 2016; pp. 1–15.

30. Simon, M.; Graham, A. *Self-Healing Through Visual and Verbal Art Therapy*; Jessika Kingsley: Philadelphia, PA, USA, 2005.

31. Wilkinson, R.A.; Chilton, G. Positive art therapy: Linking positive psychology to art therapy theory, practice, and research. *J. Am. Art Ther. Assoc.* **2013**, *30*, 4–11. [CrossRef]

32. Sarid, O.; Huss, E. Image formation and image transformation. *Arts Psychother.* **2011**, *38*, 252–255. [CrossRef]

33. Sarid, O.; Cwikel, J.; Czamanski-Cohen, J.; Huss, E. Treating women with perinatal mood and anxiety disorders (PMADs) with a hybrid cognitive behavioral and art therapy treatment (CB-ART). *Arch. Women's Ment. Health* **2017**, *20*, 229–231. [CrossRef]

34. Wolpe, J. *The Practice of Behavior Therapy*; Pergamon Press: New York, NY, USA, 1969.

35. Segal-Engelchin, D.; Sarid, O. Brief intervention effectiveness on stress among Nepalese people indirectly exposed to the Nepal earthquake. *Int. J. Ment. Health Addict.* **2016**, *14*, 1–5. [CrossRef]

36. Sarid, O.; Huss, E. Trauma and acute stress disorder: A comparison between cognitive behavioral intervention and art therapy. *Arts Psychother.* **2010**, *37*, 8–12. [CrossRef]

37. Huss, E.; Samson, T. Drawing on the arts to enhance salutogenic coping with health-related stress and loss. *Front. Psychol.* **2018**, *9*, 1612. [CrossRef]

38. Rollins, J.A. Tell me about it: Drawing as a communication tool for children with cancer. *J. Pediatric Oncol. Nurs.* **2005**, *22*, 203–221. [CrossRef]

39. Edwards, M. Jungian Analytic Art Therapy. In *Approaches to Art Therapy: Theory and Technique*; Rubin, J., Ed.; Brunner and Routledge: New York, NY, USA, 2001; pp. 81–94.

40. Goodnow, J. *Children's Drawings*; Harvard University Press: Cambridge, MA, USA, 1997.

41. Khellog, R. *Analyzing Childrens Art*; Mayfeilds Publishing Company: California, CA, USA, 1993.

42. Wadeson, H. The anti-assessment devils advocate. *J. Am. Art Ther. Assoc.* **2002**, *19*, 37–41.

43. Kaplan, F. Now and future ethno-cultural issues. *J. Am. Art Ther. Soc.* **2000**, *19*, 65–79.

44. Ntoumanis, N.; Edmunds, J.; Duda, J.L. Understanding the coping process from a self-determination theory perspective. *Br. J. Health Psychol.* **2009**, *14*, 249–260. [CrossRef] [PubMed]

45. Tedeschi, R.G.; Calhoun, L.G. Posttraumatic growth: Conceptual foundations and empirical evidence. *Psychol. Inq.* **2004**, *15*, 1–18. [CrossRef]

International Journal of
*Environmental Research and Public Health*

*Article*

# Psychological Resilience of Volunteers in a South African Health Care Context: A Salutogenic Approach and Hermeneutic Phenomenological Inquiry

Antoni Barnard [1,*] and Aleksandra Furtak [2]

[1]   Department of Industrial and Organisational Psychology, University of South Africa,
      Pretoria 0003, South Africa
[2]   Department of Human Resource Management, University of South Africa, Pretoria 0003, South Africa;
      hyraam@unisa.ac.za
*   Correspondence: barnaha@unisa.ac.za; Tel.: +27-82-375-2696

Received: 13 March 2020; Accepted: 28 March 2020; Published: 24 April 2020

**Abstract:** Volunteering in non-Western countries, such as South Africa, is subject to poor infrastructure, lack of resources, poverty-stricken conditions and often conducted by volunteers from lower socio-economic spheres of society. Sustaining the well-being of volunteers in this context is essential in ensuring their continued capacity to volunteer. To do so, it is important to understand the psychological resilience of these volunteers and the resistance resources they employ to positively adapt to their challenging work-life circumstances. The aim of this qualitative hermeneutic phenomenological study was to explore volunteers' psychological resilience from a salutogenic perspective. In-depth interviews were conducted with eight volunteers servicing government-run hospitals. Data were analysed through phenomenological hermeneutical analysis. Findings show a characteristic work-life orientation to be at the root of volunteers' resilience. Their work-life orientation is based on a distinct inner drive, an other-directedness and a "calling" work orientation. It is proposed that this work-life orientation enables volunteers in this study context, to cope with and positively adapt to challenging work-life circumstances and continue volunteering. The elements of their work-life orientation are presented as intrapersonal strength resources fundamental to their psychological resilience. It is suggested that organisations invest in developmental interventions that endorse and promote these intrapersonal strengths.

**Keywords:** coping; general resistance resources; positive adaptation; psychological resilience; salutogenic; volunteering

---

## 1. Introduction

Volunteering is an essential and natural part of cultures across the world rendering significant multi-dimensional benefits to individuals, organisations and society [1–4]. Volunteers contribute to economic development and boost socio-economic phenomena such as social cohesion, citizenship, community development and social transformation [3,5–7]. Per definition, volunteering is an unpaid, planned, proactive helping activity where someone's time, effort and energy is given freely for the benefit of other people, groups or organisations [8] to help solve social problems [9]. To continue volunteering over time, such commitment typically incurs personal costs, frequently under very difficult personal and economic circumstances [10,11]. Volunteers often operate in emotionally taxing environments, with limited organisational resources and inadequate training and they suffer stress and burnout [12,13]. Therefore, even though volunteering holds physical and psychological well-being benefits for the individual [2,14,15], these benefits may be compromised by the challenges volunteers face [16]. This may be especially relevant in an African context where volunteerism is typically constrained by poverty [6] inadequate training, poor support and lack of supervision, as well as logistical and financial limitations [17].

Volunteering infrastructure in non-Western countries such as South Africa is fast growing and promises to bridge the challenges that international volunteers face by volunteering in non-Western contexts [18]. For the 2017/2018 financial year, research with 74 leading South African companies showed that 80% of these companies run formal employee volunteer programmes and 46% employed designated full or part-time staff to manage volunteers [19]. Considering the high unemployment rate in the country, it is worthy to note that the reported 610.4 million volunteer hours in 2014, were equivalent to more than 293,000 full-time jobs, valued at R9.8 billion [20]. Volunteering in South Africa has played an important part in addressing key socio-economic and political challenges, yet lack of research and government support hamper its effectiveness [21]. The call for consistent research on volunteering in low income contexts [16] further underscore the value of such research in the South African context.

The study of volunteering in non-Western countries is important, however frequently relate to Western, international volunteers who come to Africa to volunteer [22,23]. Although this is similarly true of South African volunteering and native South African volunteers also stem from higher socioeconomic spheres of society, volunteering in the country is frequently conducted by disadvantaged people [6,24,25] who suffer the same physical and psychological health needs as the people who they care for [17]. It is this type of volunteer that stimulated the interest of the researchers because of the particular resource challenges they experience and have to cope with. There are many volunteers from less privileged backgrounds in the South African context. A study on volunteer characteristics in the country show that Black people volunteer more than double the hours that people from other population groups do and these volunteers report significantly lower levels of education than white and Coloured volunteers [24]. Of the Black volunteers in the study, 61.1% were unemployed and 37.6% of White volunteers were also unemployed [24]. Contrary to European, UK and American studies linking a higher level of education [26,27] and a higher social class and income [28] to volunteering, the 2014 South African volunteering activities survey (VAS) reported no relationship between hours spent volunteering and education and income [20].

Volunteer well-being in the work context is as important as that of paid employees [29,30] and understanding their coping resources and positive adaptation is beneficial to developing and sustaining volunteers [31]. In view of South Africa's socio-political uniqueness, high unemployment and poverty rates, there is a need to conduct research on developing and managing the well-being of volunteers in this country's context. South African volunteers working in high-risk medical care, further highlights a distinctly stressful and psychologically demanding work environment, with high performance demands, yet very limited support [32].

Psychological resilience is defined as a dynamic process of positive adaptation in the face of adversity [33,34]. Stressful work-life circumstances increase the risk for poor mental health, yet many people resile despite the difficulties they must endure [35]. Psychological resilience—the process whereby individuals maintain well-being despite adversity—is, among others, attributed to intrapersonal coping resources or positive psychological strengths that facilitate adaptive coping [35]. Salutogenesis originated as a stress and coping model [36] and is defined as a meta-theoretical paradigm focusing on the resources for health [37]; or a stress resistance resource approach emphasising one's capacity to effectively apply available coping resources [38,39]. Central to salutogenic theory is the sense of coherence (SOC) construct, which is described as a wellness-protecting orientation to life that facilitates coping and positive adaptation in trying circumstances [40,41]. People with a strong SOC view life's challenges as meaningful to engage with and believe that they have the ability to comprehend, manage and respond constructively to challenges [42]. These beliefs reflect the three SOC subcomponents of comprehensibility (cognitive component), manageability (behavioural component) and meaningfulness (motivational component). SOC plays a predominant role in promoting psychological resilience under stressful circumstances [43].

Another core construct in the salutogenic model namely generalized resistance resources (GRR) denotes person, group or environment characteristics that facilitate positive adaptation and coping despite stressful circumstances [36]. Generalized resistance resources play a dual role in positive adaptation. On the one hand they strengthen a person's SOC and on the other they enable the use of specific resistance resources (SRR) in one's immediate environment [44]. The aim of this study was to explore the psychological resilience of eight volunteers in a South African public health context from a salutogenic perspective. This study contributes to the body of knowledge by offering an in-depth understanding of the GRRs that strengthen volunteer resilience.

## 2. Materials and Methods

In this section, the research methodology and the research setting are provided, and the research methods are described in terms of sampling and participants, data collection, data analysis and ethical considerations.

### 2.1. Research Methodology

A qualitative study was conducted following a hermeneutic phenomenological approach and the epistemological notions of social constructionism. In this tradition, knowledge generation is based on the researcher's interpretation of participants' lived experience in a social context [45–48]. Findings of the study present the co-constructed meaning between the researcher and researched [49]. Findings do not claim a single or ultimate truth, but rather a perspectival, socially constructed meaning [46,50]. Such an approach is particularly appropriate to context specific research because meaning is derived from participants located in specific social and cultural contexts [51,52]. The hermeneutic agenda calls for critical interpretation by employing an established meta-theory in making sense of the research phenomenon [47,53]. The meta-theoretical orientation applied in this study pertain to the salutogenic perspective on well-being.

### 2.2. Research Setting

The study was conducted in a faith-based non-profit organisation (NPO) operating in 13 hospitals in the Gauteng and Western Cape provincial health sectors. The hospitals are government run; some situated in developed urban suburbs, and some in townships on the outskirts of a city. Government hospitals in South Africa are characterised by poor service delivery and hygiene, old and poorly maintained infrastructure and medical negligence [54]. The volunteers provide spiritual care and counselling, as well as emotional, social, trauma and physical support to patients and their families. Most of the volunteers come from poor communities and are faced with unemployment and poverty challenges.

Access to the research setting was gained through the management of the NPO, who provided written permission for the study to be conducted. A volunteer coordinator at the NPO was appointed as gatekeeper and assisted to identify and contact volunteers fitting the research inclusion criteria. Eight participants were contacted telephonically and informed about the nature of the study, the researcher was introduced as a psychologist and doctoral student, and they were requested to participate on provision of anonymity, confidentiality and their right to withdraw. All eight agreed to participate and interview logistics were arranged. Before proceeding with the interviews, each participant signed a consent form after perusing a participant information sheet, explaining the nature and purpose of the study as well as their rights as participants.

### 2.3. Researcher Roles and Preconceptions

The study became possible because of the second author's involvement with her faculty's community service project with the said NPO. At the time she was a lecturer in human resource management and a registered industrial and organisational psychologist with the Health Professions Council of South Africa (HPCSA). While the community service project focused on mentorship at the time, the second author's interest in well-being, coping and retention of volunteers working at the NPO evolved into a research project for her PhD. This article forms part of her PhD and was co-conceptualised with her promoter, the first author. The first author holds a doctoral degree and she is a full professor in industrial and organisational psychology and a registered psychologist with the HPCSA in the categories of industrial and organisational as well as counselling. Both authors' research interests focus on employee well-being in the workplace through in-depth qualitative inquiry. The second author conducted the interviews and both authors contributed to the methodology, data analysis and writing of this article.

### 2.4. Sampling and Participants

A convenient, purposive sampling strategy was employed to select information-rich participants [55,56]. Inclusion criteria were based on the definition of a formal volunteer which entails voluntary, non-paid service to others over an extended period through a formal organisation or agency [11]. Eight participants performing volunteer services through the NPO for 12 months and more, were invited and were interviewed in the period between April and July 2016. The eight volunteers, who each serviced one of four Gauteng-based hospitals constituted an adequate sample size for phenomenological research [55]. Table 1 below provides an overview of the participant profiles.

**Table 1.** Participant profiles.

| Participant Acronyms. | Gender | Population Group | Age | Employment | Living Conditions | Hospital Situated in |
|---|---|---|---|---|---|---|
| PR1 | Female | Black | 34 | Part time employment | Rural, low socio- economic upbringing | Developed urban area |
| PR2 | Female | White | 52 | Unemployed | Lives in urban area. Receives financial support | Developed urban area |
| PR3 | Male | Black | 61 | Self-employed | Rural, low socio- economic upbringing | Developed urban area |
| PR4 | Male | White | 58 | Unemployed | Looks after sick mother. Lives in urban area | Developed urban area |
| PR5 | Female | Black | 42 | Unemployed | Low socio-economic living circumstances | Township |
| PR6 | Male | Black | 60 | Unemployed | Low socio-economic living circumstances | Township |
| PR7 | Female | Black | 58 | Unemployed | Low socio-economic living circumstances | Township |
| PR8 | Female | Black | 53 | Unemployed | Low socio-economic living circumstances | Township |

### 2.5. Data Collection

Eight initial in-depth interviews (60–90 min each) and three follow-up interviews (20–30 min each) were conducted. The interviews followed a flexible, thematic approach to elicit rich information by exploring the lifeworld of the participant [57]. After 11 interviews, data saturation was attained based on the depth (richness and thickness) thereof [58]. The in-depth interview allows flexibility to adjust thematically prepared questions during the interview, to facilitate a natural conversation flow and develop a narrative of lived experience, in which the research phenomenon is revealed [59,60]. To understand the antecedents that promote volunteers' resilience and positive adaptation, the theme of the interview questions centred on the lifeworld experiences of volunteering as reflected in Table 2.

Two interviews were conducted on the premises of a district hospital. The other nine were done at the NPO head office, situated on the grounds of two public hospitals. Interviews were digitally recorded and transcribed by a professional transcriber. The software package Atlas.ti was used to store and manage the data.

**Table 2.** Thematic interview questions.

| Interview Questions |
| --- |
| I would really like to know more about you. Can you tell me about your life story? |
| How did it come about that you started volunteering? |
| Can you tell me about your experiences of being a volunteer? |
| What has happened since you started volunteering? |
| What is it like to volunteer? |

*2.6. Data Analysis*

Data were analysed through phenomenological hermeneutical analysis entailing a naïve reading, constructing a structural thematic analysis and developing a comprehensive understanding [61,62]. Throughout the three stages, the metaphorical action of the hermeneutic circle is constantly applied, causing the researcher to move back and forth between the three stages, using each as a critical reflection and verification of the other [62].

The naïve reading entails repetitive reading of the transcriptions to get a sense of it as a gestalt [62] without any thematising [63]. Thereafter, to construct the thematic structural analysis, sections of meaningful text are identified and condensed in everyday language [62]. Condensed text is then reviewed, interpreted and clustered into sub-themes and main themes while continuously reflecting back on the naïve understanding and while constantly considering the research aim [61,62]. The comprehensive understanding is lastly constructed by reflecting on the holistic meaning in relation to the naïve reading, the thematic structural analysis, the research aim, the context of the study, the author's preunderstanding and relevant meta-theoretical literature [62]. The findings reported below focus on the themes constructed in the structural analysis and the discussion that follows reflect the comprehensive understanding.

*2.7. Ethical Considerations*

Ethics approval was obtained from the relevant Institutional Ethics Committee (reference no. 2015_CEMS/IOP_050) and written permission to do the research was provided by the NPO in which the study was conducted. The study was conducted in line with the Ethics Policy of the University of South Africa (UNISA) and the Rules of Conduct for the Profession of Psychology of the HPCSA. Participants signed an informed consent prior to their participation. In reporting the results pseudonyms are used to ensure anonymity. Participant pseudonyms were used according to the abbreviation PR with the number of the participant following, for example, PR6 denotes participant six.

## 3. Findings

Based on the naïve reading, volunteers' resilience seems strengthened by a distinct intrapersonal disposition or orientation to work and life. The structural thematic analysis conceptualises and synthesises this disposition or work-life orientation at the hand of three themes. The three themes describe how volunteers' resilience is rooted in a characteristic inner drive, their other-directed life orientation and regarding their work as a 'calling'. Next, each of the three main themes are conceptualised in sub-themes grounded in verbatim data.

*3.1. Volunteer Resilience Rooted in a Unique Inner Drive*

The volunteers' unique inner drive is reflected in their self-determination and autonomous agency as well as in an innate desire to be productive and useful.

3.1.1. Being Self-Determined and Demonstrating Autonomous Agency

The volunteers' narratives reveal a characteristic self-determination and autonomous agency. They take responsibility for and are in control of their own lives and choices, as opposed to being directed by external forces, and this drives them to make their own decisions. Self-determination is demonstrated by PR4 who is active in shaping his own life and takes responsibility by acting persistently on his motives: *" ... I became aware of the organisation, and then one day I came here, to XXX's office ... and then I came for a second time with the same person, and then I just decided okay, I want to continue with this, you know"*. Similarly, PR5 makes her internal locus of control apparent when she takes responsibility for her decision to volunteer: *" ... you work under pressure, I am not working under pressure. When I am tired, or God wants to speak to me, I just stand and listen"*. Both PR4 and PR5 made a personal decision to become involved in volunteering. This personal and informed decision was based on a willing engagement that was free from external coercion.

Volunteer self-determination was not only evident in relation to the volunteering environment. PR1's self-determination is revealed in the way she approached her life from an early age:

> *"You grow up knowing what you want and where you want to go. Because most of the children that I grew up with, their parents taking care of them and doing everything to them, today like they are still depending on their parents in such a way that everything the parents have to take decisions for them and even if a person is matured like me, they are waiting for their parents to take a decision for them ... somebody has to come to a point where you have to take decisions on yourself"*.

In addition to being self-determined, the volunteers are also autonomous agents, voicing a proclivity and capacity to make their own choices. This agency on the part of the volunteer is demonstrated by the free yet deliberate choices they made to engage in volunteer work. PR7 resolved: *"Volunteer is to work with your own ability, you do not, somebody does not push you. I want to volunteer, I want it"*. PR3 highlights that volunteering is *"a matter of choice"* and the deliberate choice to become involved in volunteering is confirmed by PR4 who indicates that volunteering *"is something you want to do"*.

This theme indicates the volunteer's tendency to act independently, take deliberate action and apply freedom of choice. This innate predisposition of being self-determined and demonstrating autonomous agency acts as a general resistance resource, facilitating the volunteers' resilience in vigorously continuing the work they do.

### 3.1.2. A Desire to be Productive and Useful

Volunteers voice an innate desire to be busy and useful, despite their personal difficult circumstances and challenges. After retiring for health reasons, PR7 explains how she was not happy to sit at home and feel as if she was doing nothing: *"When I am at home I think about the patient because there are other patients there at the hospital, the patient who did not have the relatives, and they struggled a lot at the hospital and I decided to go there, not doing nothing"*. She (PR7) further emphasised that she could not sit at home knowing that she had the opportunity to contribute to the patients in the hospital: *"I do not want to sit at home doing nothing whereas there is somebody who want me to comfort her or him"*. Similarly, PR8 wanted to participate in new tasks as opposed to being inactive as a result of her health challenges: *"When I am busy staying at home I start thinking now, I am just sitting here, I do not do nothing ... I start to think man, no man this sickness is going to kill me because I do nothing, I must start now, I'm going to rise up and ... I must go and tell the people about something, encourage people at hospital"*. Despite her discomfort, PR8 is adamant that she needs to be productive explaining that she *"cannot sit here every day thinking of this pain, there is some other people there at hospital, they have got this pain also, I must go and say to him, no man God will help you, I must go and encourage the person"*. PR1 describes how she constantly strives to do something: *"I just made sure that all of my spare time I spend it in something, doing something. Either I am studying or I am helping somebody or I am doing something"*.

The volunteers' need to be busy is complemented by the desire to be useful in their daily lives. This is evident from PR2, who lost her job after being declared incapacitated: *" ... ek wil nou eerder in XXX, [met] sieklike mense gaan [tyd] spandeer as om by die huis te sit"* (I would rather spend time with the ill patients than to sit at home). The extent of PR2's desire to be useful is visible in the variety of activities she is actively involved in:

> *"Mondays I work with the SAP (South African Police Service) and when there is accidents or robberies or everything . . . Monday is this time. And when I am not busy, I pray for the people in the NPO... And then I—Tuesday is NPO. Wednesday is NPO. Thursday is 'ouetehuis' (old age home)—all the old people. And Friday is me and my husband come to NPO".*

In addition to the independent deliberate action and freedom of choice that acts as a resource for the volunteers' coping and positive adaptation, their desire to be productive provides them with psychological resilience in that they are driven to take action, to be industrious and engaged in useful activity.

### 3.2. Volunteer Resilience Stemming from An Other-Directed Life Orientation

Volunteers' other-directedness is characterised by being people-centred and having a religious orientation to life.

### 3.2.1. Being People-Centred through Care, Compassion and Empathy

Volunteers unanimously report a strong people-centred orientation to life. This people-centredness is primarily rooted in an intrinsic desire to care for those in need. As described by PR1, this intrinsic need is inherent to her personality: *"I find that helping people is one of my, I don't know if I can call it a weakness or a . . . because I can give what I have and remain with nothing by helping someone"*. This seems to suggest that the needs of others are more important than her own, illustrating how deeply ingrained and important this orientation is: *"I cannot live like this while others they are suffering outside. I rather use the small that I am having and do something"*. Similarly, PR3 indicates that he has a *"heart for people. I like you know helping people in a way that I can. Ja (yes) if it means buy you food I will buy you food"* and also explains that he is *"the type of person who sort of you know wants to do something for the people you know"* because *"there is something inside of me that needs to do good"*. Although a people-centred orientation to life is also inherent to PR3's personality, he explains how it is further entrenched in him through his African values:

> *"Among the Africans, when somebody has lost a spouse or a child, then we go there and then, you know by going there it is the same as saying "Listen, I am here if there be any need, I am willing to get involved", and they ask you to go and fetch water, they ask you to go and fetch wood and so forth. And during the circumstances of their mourning, then you provide some kind of help".*

Apart from an intrinsic need to care for others, having compassion is also central to the volunteer's people-centred orientation to life. PR4 says that *"I am a person that have sympathy and empathy with other people, you know, in their time of suffering"*. PR5's compassion, which is founded in experiencing her own suffering, drives her to ease the anguish of others spiritually: *" . . . in my heart there was a, I don't know how to speak it, uh, a heart for that sick people because I come from there. I was so sick, I was feeling the love, to love them and show them that Jesus is the only way, there is no other way than Jesus"*. Similarly, PR8's compassion motivates her: *" . . . but my spirit inside, I have got some, I have compassion with people, I want to encourage people with the words of God, you know"*. Being compassionate enables the volunteers to provide a support system for the hospital patients. This entails for example P6 acting as a family member who can listen to their fears and just be a presence next to their hospital bed. Likewise, PR4's compassion drives him to encourage and listen to the patients and give them hope: *" . . . and sit down with them and just listen to them, listen to their fears or frustrations and things, and then you come there and you listen and you see in their eyes and you hear in their voice . . . and just give them hope"*.

Empathy also characterises the volunteer's people-centredness and it stems from their own experiences. PR2 who underwent a back operation, explains how she is able to understand the pain, suffering and difficulty the patients are feeling and experiencing because of her own medical history: *" . . . when I get out of the 8 weeks, I will tell God 'I know now how the people in the NPO, in the hospital feel'. Because I cannot tell you I know how you are feeling if you are not going through this"*. PR2 further mentions how she prays *"for everybody in this hospital, hospitals in the whole world because I know when I lay in the bed how they are feeling"*. Similarly, PR5 is also aware of what it entails to be a patient in the hospital: *"I come from there. I was so sick"*. Additionally, PR7 not only understands what it feels like to be a patient, but also how volunteer services were valuable to her: *"I was at hospital five years back, and when I was in hospital there comes a pastor and talk with me about God, and said God is love, and I take that message and restore my, my soul"*.

Being people-centred presents a unique other-directed orientation to life in general and presents as an intrapersonal strength or resource fundamental to the participants' resilience in this volunteering context. The establishment of a disposition for well-being in the volunteer, such a people-centredness, is based on being driven by one's care and compassion for others and the ability to feel empathy. The volunteer's other directedness is also exemplified by its focus on a higher spiritual source, as discussed next.

### 3.2.2. Religiously Rooted and Focused on God

A profound religious attitude, namely, a belief in and commitment to God, seems to be at the core of the volunteer's positive adaptation. The volunteers' focus on God is apparent as they have made it explicit that their effort is directed towards God himself and that they conduct volunteering as a service to Him: *"I work for God. I did not work for me, I work for God, who created me"* (PR7) and according to PR5, *"to volunteer, I don't think it is a volunteer, it is just the work from God"*. PR5 also indicates that *"I'm going to the hospital for the purpose of God, for the sick people"*. PR2 further emphasises that *"I work for God, not for XXX, for God"*. Similarly, PR4 sees volunteering as a *"service in the kingdom of God"*.

The volunteers in this study were all specifically vocal about being directed by Christian principles and teachings. These engendered their serving behaviour, as explained by PR3: *"You know actually it's one of the Christian principles where Jesus says, "If you want to be number one, start by serving ... if you want to be number one, be a slave to everyone ... I have been created to serve and to do good work"*. Similarly for PR6, to conduct service-oriented activities means to feel God's power flowing through him: *"I know that if I want to be anointed I need to give something, so I need to give my service, that is why I am giving my service voluntarily"*. Christian practices and teachings furthermore direct the volunteers in their work role, as PR1 explains: *"There is something that I am relying on—it is the Bible. Most of the things that I do, I do according to the Bible"*. Acts of worship such as prayer and ministry provide patients with hope and encouragement as described by PR8: *"I am going to encourage them with the words of God and pray"*. PR2 recounts: *" ... and I sit on the chair next to her and I pray for her and I give her one scripture, and I read in the Bible for her and ... I ask her if I can lay my hands on her"*. While volunteering, the volunteers express the virtue of love, which for them is central to Christianity: *"I am trying to practise the Bible, that is what I have to give out to the patients ... That love is to show them that there is a purpose for everything, there is an end out of everything"* (PR1). PR6 further explains that he tells the patients he loves them because *"when you start ministering you can see they need the love of God, because I believe Christianity is more about love than anything else"*.

Although the volunteer's actions are predominantly focused on God and carried out in His service, signalling their commitment to Him, the volunteers also depend on God during times of difficulty, such as relying on Him for guidance to solve challenging problems faced in the workplace. This is noticeable in PR1's explanation of dealing with a challenging patient while volunteering: *"You must ask God for a descending spirit that will help you to choose and to separate things and to do the right decision. So, sometimes when things like this are happening, I just ask God to help me how to come out of this or how to solve this"*. In this way, God acts as a support resource, fostering the belief that they will be able to deal with the demands posed by life. PR5 has faith that God will assist her with the challenges she is experiencing: *"I see my children, don't have anything to eat or clothes, they said, mama (mom), we want money to go to school. I don't have money, I am not working but I trust God. Because we are living by God's grace, we are living by God's grace"*. He (God) is further the source of the volunteer's strength and gratitude. PR2 explains that *"when I feel down, and I can say "I am not so bad", because the people in NPO in hospital is lying down, they don't got legs, they have stomach cancer and everything you know. And then I say, 'thank you God that I can make a difference, that you pick me up every morning'".*

In addition to being people-centred in a caring, compassionate and empathetic manner, other-directedness is evident in being rooted in and deriving strength from a strong religious attitude, in this study context, specifically, a Christian religion.

### 3.2.3. Work that is a "Calling"

The volunteer's resilience is strengthened by a specific orientation to work, namely the need to conduct work that is regarded as a calling. This entails work that one is passionate about and intrinsically motived to do. Over and above their religious calling, this calling orientation to work specifically stood out in the narrative of PR1: *"being a volunteer I find that it is a calling. It is a calling and it is a passion"* which *"you will do it with passion and love"* and *"I found it being a calling like when you are called for something, when you are called for being a Pastor, when you are called to be a doctor, you are with that thing inside of you"*. PR3 shares this view: *"I think it's a calling"*. A calling work orientation is described as being passionate about your work or having an intense love for it: *"a calling it is like something that you have a passion. Something that you have love when you do it. You have that, you do it whole heartedly with love. That is why you will be able to come here without somebody giving you an allowance, compensation to come here. You come here voluntarily. You come here using your own time for someone's life"* (PR1). PR5 exemplifies her work passion by emphasising how much she *"love[s] this job. I love this job"*. This type of calling work orientation stems from an inner desire and motivation on the part of the volunteer to conduct such work and is reflected in the words of PR3: *"[Y]ou see to be a volunteer, it springs from the heart"*. In PR6's explanation of how he started volunteering, he notes that volunteer work is intrinsically motivated: *"I wanted to volunteer, do the volunteer work but I never really have the volition, you know, if you know what I mean, I did not have the oomph to go and do it, because the passion was not ignited"* and how this changed for him to become a passion: *"but, from the time to 2012, when I came back from overseas, it was on my heart"*.

Two of the volunteers related to being called by God to volunteer, linking their work orientation to a religious calling. PR6 said: *"So ja (yes), it was not until that one day I had a dream, it was a vision actually, in the morning, and God said to me go to XXX . . . he said go and pray for the sick, and I went"*. PR2 had a similar experience during her recovery from a back operation: *"And I got a voice from God, go to NPO and do something for the sick people because they are laying in the bed and they can do nothing for themselves"*.

The volunteers' resilience is supported by the unique way in which they approach their work environment, that is, from a calling work orientation. A calling orientation presents a unique orientation to work that reflects an approach based on being intrinsically motivated to do work one has a passion for.

## 4. Discussion

This article set out to explore volunteers' psychological resilience from a salutogenic perspective. The findings highlight a characteristic predisposition or work-life orientation that supports and sustains positive adaptation as reflected in the will to continue volunteering.

In this study, volunteers from a low income context with limited material resources, portray a disposition or work-life orientation that is characterised by (1) a peculiar inner drive, (ii) an other-directedness and (iii) a "calling" work orientation. The three elements of their work-life orientation echo the sub-component dynamics of SOC and demonstrate how these dynamics are reflective of intrapersonal GRRs that promote and sustain their psychological resilience.

The first element of the volunteers' predisposition is their inner drive. The volunteers' unique inner drive is operationalised in their self-determined nature and autonomous agency as well as in a desire to be productive and useful. These intrapersonal characteristics are similar to volunteering studies that have linked autonomy orientation to engagement in pro-social activities, job satisfaction and intentions to sustain volunteer work [64,65]. Autonomy is also conceptually related to self-determination, independence, self-regulated behaviour and acting volitionally, according to one's own will [66–68]. Self-determination and autonomy are furthermore important for optimal functioning and well-being [69,70] and, therefore, proposed here as important GRRs in the psychological resilience of volunteers in this study.

Volunteers' inner drive reflects a predisposition to cognitively appraise difficult circumstances in a way that demonstrates a pervasive sense that life is comprehensible, manageable and meaningful. Their self-determination and autonomous agency support the SOC sub-component of *comprehensibility* in that volunteers take responsibility for their responses towards stressful stimuli in their external environment. Rather than blaming or questioning external forces, their appraisal of circumstances results in taking responsibility and acting decisively and of their own volition. Furthermore, the volunteers' innate desire to be productive and useful is related to the SOC sub-component of *manageability*. Responding to their circumstances by taking action shows that the volunteers believe in their capacity to meet life's demands. In being actively engaged and industrious through the volunteering work they do, volunteers confirm and build their self-efficacy and sense of feeling useful. This need to feel useful is fundamental to the volunteers' orientation to be productive and to contribute, and therefore also relates to the motivational aspect of SOC namely, *meaningfulness*.

The second element in the volunteers' predisposition, is their other-directedness. Having a life orientation directed towards servicing others or being in the service of religious beliefs and God, secondly also resemble the SOC components of *meaningfulness, manageability* and *comprehensibility*. In the data, their self-reflections and descriptions portray volunteers to be characteristically caring, compassionate and empathetic. This is congruent to studies showing that volunteers typically have pro-social personality characteristic such as other-oriented empathy and helpfulness which motivate them to volunteer [8]. The motivational effect of these characteristics is revealed in the dynamic that when enacting them, life is regarded as *meaningful*. Activities such as volunteering are therefore experienced as meaningful to engage in because the volunteer is then congruent to the authentic self. Their other-directedness also resemble the behavioural SOC sub-component of *manageability* as these characteristics provide intrapersonal resources that enable the volunteers to positively and actively respond to and act in their environment. Other-directedness thus seems to be a strength resource underlying psychological resilience in this study context, especially since expressing concern for the welfare of others, is directly related to psychological well-being [71] and empathy is considered a character strength [72]. Acting congruently to these pro-social traits, demonstrates volunteers' cognitive appraisal of their circumstances as *comprehensible*, because they do not shy away from difficulties. They rather engage in life, despite its difficulties, by creating opportunities in which they can authentically enact their other-directed character traits and needs. Their other-directedness is also pertinently rooted in being religious and focused on God. Believing in God and enacting the religious call to serve others provide the volunteers with the potential for innate well-being, as it fosters *meaningfulness* strength resources such as hope and gratitude. Other-directed characteristics such as caring, compassion, empathy and a religious belief system thus act as intrapersonal strengths or GRRs that enable the volunteer to remain resilient in the face of adverse circumstances.

Volunteers' psychological resilience is lastly rooted in the third element, namely a specific "calling" work orientation, which is conceptualised as having passion and an intrinsic motivation for the work. Volunteers with a religious identity (such as in this study context) have been shown to be motivated to volunteer as a way of following a calling [73]. Theory suggests that experiencing a calling to work results in positive effects such as work and life satisfaction, finding work meaningful, being more motivated and experiencing engagement with work [74]. Engaging in work that is aligned with a calling generally enhances well-being [75–77] and in this study context affirms the volunteers' SOC through *meaningfulness*. Loving the volunteer work they do and being passionate about it, shows that the volunteers find meaning in answering the call to volunteer. The call to volunteer is, however, not only experienced in terms of their religion. Volunteering is experienced in general as a deep internal motivation to find purpose and meaning in life.

Research generally confirms that volunteering contributes to well-being [14,17,27,78–80]. This study contributes to the body of knowledge by explaining the dynamic that builds the psychological resilience of volunteers. The unique disposition of volunteers described in this study context, predisposes the volunteer to appraise and respond to stressful life circumstances in a way that builds their psychological resilience and leads to active coping and positive adaptation. Having GRRs, being aware of them and having the ability to use them buffer the risk of poor mental health and distress [81]. In this sense, the act of volunteering may be regarded as an SRR which the volunteer accesses in order to congruently enact their intrapersonal character strengths. Research generally cites the characteristics and values of volunteers as aspects that motivate their decision to volunteer. From this study, understanding how volunteers sustain their well-being despite the difficult circumstances they work and live in, augments the unique person characteristics that underscore the motivation to volunteer—not as motivational factors per se, but as their innate well-being potential. Developing an understanding of the unique intrapersonal GRRs that enable volunteers to resile, may enable organisations and government to better manage and retain this valuable resource.

The small sample and qualitative nature of the study present with in-depth, rich and contextual understanding, yet the study is limited with regard to generalisation. Moreover, the study had an idiographic purpose and did not investigate the potential GRRs present in the organisational or societal context of the study. Future research should explore the resources required to facilitate volunteer well-being and specifically investigate possible developmental interventions to promote volunteer resilience.

## 5. Conclusions

Due to their non-profit service agreement, volunteers do not receive the same benefits as full-time employees. Although organisations spend some resources on their recruitment, training and management, this return on investment is too often overlooked [4]. Investing in the development and well-being of volunteers seems to be an area of need, especially in African contexts that are subject to limited resources and lower socio-economic conditions of volunteers. The findings of this study have several implications for investing in volunteer well-being, based on the premise that volunteers will extend their services in the longer term if they constructively cope with and positively adapt to their limited work circumstances. Whether or not one volunteers or gets paid for your services, endorsing your character strengths is directly related to higher job and life satisfaction and indirectly to greater well-being [82]. It is, therefore, proposed that organisations employing the services of volunteers create developmental opportunities to identify and endorse volunteer character strengths such as their pro-social nature, inner drive and need to work with passion and purpose.

**Author Contributions:** Conceptualisation: A.B. and A.F.; methodology: A.B. and A.F.; data curation: A.F.; investigation: A.F.; formal analysis: A.D. and A.F.; validation: A.B.; resources: A.F.; writing—original draft preparation: A.F.; writing—review and editing: A.B. and A.F.; supervision: A.B.; project administration: A.F. All authors have read and agreed to the published version of the manuscript.

**Funding:** This research received no external funding.

**Acknowledgments:** The authors would like to acknowledge the NPO that allowed and enabled the study to be conducted as well as the volunteers who willingly participated and offered their stories. Without them the study would not have been possible.

**Conflicts of Interest:** The authors declare no conflict of interest.

## References

1. Butcher, J.; Einolf, C.J. *Perspectives on Volunteering*; Springer: Cham, Switserland, 2017; pp. 1–297.
2. Detollenaere, J.; Willems, S.; Baert, S. Volunteering, income and health. *PLoS ONE* **2017**, *12*, 1–11. [CrossRef] [PubMed]
3. Carter Kahl, S. Making the Invisible Visible: Capturing the Multidimensional Value of Volunteerism to Nonprofit Organizations. Ph.D. Thesis, University of San Diego, San Diego, CA, USA, 2019.

4.  Manetti, G.; Bellucci, M.; Como, E.; Bagnoli, L. Investing in volunteering: Measuring social returns of volunteer recruitment, training and management. *Volunt. Int. J. Volunt. Nonprof. Organ.* **2015**, *26*, 2104–2129. [CrossRef]

5.  Dawson, C.; Baker, P.L.; Dowell, D. Getting into the 'Giving Habit': The dynamics of volunteering in the UK. *Volunt. Int. J. Volunt. Nonprof. Organ.* **2019**, *30*, 1006–1021. [CrossRef]

6.  Pietersen, W.J. The Value of Social Networks to Community Volunteers from High Risk Communities. Master's Thesis, University of Pretoria, Pretoria, South Africa, 2017.

7.  Sajardo, A.; Serra, I. The economic value of volunteer work: Methodological analysis and application to Spain. *Nonprofit. Volunt. Sect. Q.* **2010**, *40*, 873–895. [CrossRef]

8.  Rodell, J.B.; Breitsohl, H.; Schröder, M.; Keating, D.J. Employee volunteering: A review and framework for future research. *J. Manag.* **2016**, *42*, 55–84. [CrossRef]

9.  Omoto, A.M.; Snyder, M. Considerations of community: The context and process of volunteerism. *Am. Behav. Sci.* **2002**, *45*, 846–867. [CrossRef]

10.  Meier, S.; Stutzer, A. Is volunteering rewarding in itself? *Economica* **2008**, *75*, 39–59. [CrossRef]

11.  Snyder, M.; Omoto, A.M. Volunteerism: Social issues perspectives and social policy implications. *Soc. Iss. Pol. Rev.* **2008**, *2*, 1–36. [CrossRef]

12.  Crook, J.; Weir, R.; Willms, D.; Egdorf, T. Experiences and benefits of volunteering in a community AIDS organization. *J. Assoc. Nurs. AIDS Care* **2006**, *17*, 39–45. [CrossRef]

13.  Fuertes, F.C.; Jiménez, M.L. V Motivation and burnout in volunteerism. *Psychol. Spain* **2000**, *4*, 75–81.

14.  Mellor, D.; Hayashi, Y.; Firth, L.; Stokes, M.; Chambers, S.; Cummins, R. Volunteering and well-being: Do self-esteem, optimism, and perceived control mediate the relationship? *J. Soc. Ser. Res.* **2008**, *34*, 61–70. [CrossRef]

15.  Vecina, M.L.; Fernando, C. Volunteering and well-being: Is pleasure-based rather than pressure-based prosocial motivation that which is related to positive effects? *J. Appl. Soc. Psychol.* **2013**, *43*, 870–878. [CrossRef]

16.  MacKenzie, S.; Baadjies, L.; Seedat, M. A phenomenological study of volunteers' experiences in a South African waste management campaign. *Volunt. Int. J. Volunt. Nonprof. Organ.* **2015**, *26*, 756–776. [CrossRef]

17.  Kiyange, F. Volunteering in hospice and palliative care in Africa. In *The Changing Face of Volunteering in Hospice and Palliative Care*; Scott, R., Howlett, S., Eds.; Oxford University Press: Oxford, UK, 2018; pp. 142–156.

18.  Grandi, F.L.; Lough, B.J.; Bannister, T. The state of volunteering infrastructure globally. *Voluntaris* **2019**, *7*, 22–43.

19.  Trialogue Business in Society Handbook, 21st ed. Available online: https://trialogue.co.za/publications/csi-handbook-2018-free-download/ (accessed on 26 February 2020).

20.  Statistics South Africa (StatsSA) Volunteer Activities Survey. Available online: http://www.statssa.gov.za/publications/P02113/P021132010.pdf (accessed on 22 January 2020).

21.  VSO; RAISA. *Volunteer Management Needs Assessment among South African Civil Society Organisations*; Unpublished Research Report; VSO; RAISA: Pretoria, South Africa, 2011; Available online: http://www.vosesa.org.za/sadcconference/papers/12.pdf (accessed on 18 February 2020).

22.  Baillie Smith, N.; Laurie, M. South–South volunteering and development. *Geogr. J.* **2018**, *184*, 158–168. [CrossRef]

23.  Prince, R.; Brown, H. *Volunteer Economies: The Politics and Ethics of Voluntary Labour in Africa*; James Currey: Woodbridge, UK, 2016; pp. 1–254.

24.  Niyimbanira, F.; Krugell, W. The characteristics of volunteers in South Africa. *J. Econ. Financ. Sci.* **2017**, *10*, 424–436.

25.  Russell, B. Measuring the Contribution of Volunteering to the Sustainable Development Goals: The Measurement of Volunteering in the Global South. Available online: https://socialsurveys.co.za/wp-content/uploads/2018/08/Russell-2016-ISTR-Measurement-Volunteering-Global-South.pdf (accessed on 18 January 2020).

26.  Doyle, W.R.; Skinner, B.T. Does postsecondary education result in civic benefits? *J. Higher Educ.* **2017**, *88*, 863–893. [CrossRef]

27.  Son, J.; Wilson, J. Education, perceived control, and volunteering. *Sociol. Forum* **2017**, *32*, 831–849. [CrossRef]

28.  Wilson, J.; Mantovan, N.; Sauer, R.M. The economic benefits of volunteering and social class. *Soc. Sci. Res.* **2020**, *85*, 1–32. [CrossRef]

29. Handy, F.; Mook, L. Volunteering and volunteers: Benefit-cost analyses. *Res. Soc. Work Pract.* **2011**, *21*, 412–420. [CrossRef]

30. Lewig, K.A.; Xanthopoulou, D.; Bakker, A.B.; Dollard, M.F.; Metzer, J.C. Burnout and connectedness among Australian volunteers: A test of the Job Demands-Resources model. *J. Vocat. Behav.* **2007**, *71*, 429–445. [CrossRef]

31. Brudney, J.L.; Meijs, L.C.P.M. It ain't natural: Toward a new (natural) resource conceptualization for volunteer management. *Nonprof. Volunt. Sect. Q.* **2009**, *38*, 564–581. [CrossRef]

32. Folwell, A.; Kauer, T. 'You see a baby die and you're not fine': A case study of stress and coping strategies in volunteer emergency medical technicians. *J. Appl. Commun. Res.* **2018**, *46*, 723–743. [CrossRef]

33. IJntema, R.C.; Burger, Y.D.; Schaufeli, W.B. Reviewing the labyrinth of psychological resilience: Establishing criteria for resilience-building programs. *Consult. Psychol. J. Pract. Res.* **2019**, *71*, 288–304. [CrossRef]

34. Infurna, F.J.; Luthar, S.S. Re-evaluating the notion that resilience is commonplace: A review and distillation of directions for future research, practice, and policy. *Clin. Psychol. Rev.* **2018**, *65*, 43–56. [CrossRef] [PubMed]

35. Choi, K.W.; Stein, M.B.; Dunn, E.C.; Koenen, K.C.; Smoller, J.W. Genomics and psychological resilience: A research agenda. *Mol. Psychiatr.* **2019**, *24*, 1–9. [CrossRef] [PubMed]

36. Antonovsky, A. *Health, Stress, and Coping*; Jossey-Bass: San Francisco, CA, USA, 1979.

37. Mittelmark, M.B.; Bauer, G.F. The meanings of salutogenesis. In *The Handbook of Salutogenesis*; Mittelmark, M.B., Sagy, S., Eriksson, M., Bauer, G.F., Pelikan, J.M., Lindström, B., Espnes, G.A., Eds.; Springer: Cham, Switserland, 2017; pp. 7–14.

38. Eriksson, M. The sense of coherence in the salutogenic model of health. In *The Handbook of Salutogenesis*; Mittelmark, M.B., Sagy, S., Eriksson, M., Bauer, G.F., Pelikan, J.M., Lindström, B., Espnes, G.A., Eds.; Springer: Cham, Switserland, 2017; pp. 91–96.

39. Eriksson, M.; Lindström, B. Salutogenesis. *J. Epidemiol. Commun. Health* **2005**, *59*, 440–442.

40. Bernabé, E.; Tsakos, G.; Watt, R.G.; Suominen-Taipale, A.L.; Uutela, A.; Vahtera, J.; Kivimäki, M. Structure of the sense of coherence scale in a nationally representative sample: The Finnish Health 2000 survey. *Qual. Life Res.* **2009**, *18*, 629–636. [CrossRef]

41. Feldt, T.; Leskinen, E.; Koskenvuo, M.; Suominen, S.; Vahtera, J.; Kivimäki, M. Development of sense of coherence in adulthood: A person-centered approach.The population-based HeSSup cohort study. *Qual. Life Res.* **2011**, *20*, 69–79. [CrossRef]

42. Basinska, M.A.; Andruszkiewicz, A.; Grabowska, M. Nurses' sense of coherence and their work related patterns of behaviour. *Int. J. Occupat. Med. Environ. Health* **2011**, *24*, 256–266. [CrossRef]

43. Ablett, J.R.; Jones, R.S.P. Resilience and well-being in palliative care staff: A qualitative study of hospice nurses' experience of work. *Psycho-Oncol. J. Psychol. Soc. Behav. Dimens. Cancer* **2007**, *16*, 733–740. [CrossRef] [PubMed]

44. Idan, O.; Eriksson, M.; Al-Yagon, M. The salutogenic model: The role of generalized resistance resources. In *The Handbook of Salutogenesis*; Mittelmark, M.B., Sagy, S., Eriksson, M., Bauer, G.F., Pelikan, J.M., Lindström, B., Espnes, G.A., Eds.; Springer: Cham, Switzerland, 2017; pp. 57–69.

45. Laverty, S.M. Hermeneutic phenomenology and phenomenology: A comparison of historical and methodological considerations. *Int. J. Qual. Meth.* **2003**, *2*, 21–35. [CrossRef]

46. Geldenhuys, D.J. Social constructionism and relational practices as a paradigm for organisational psychology in the South African context. *SA J. Industr. Psychol.* **2015**, *41*, 1–10. [CrossRef]

47. Kafle, N.P. Hermeneutic phenomenological research method simplified. *Bodhi Interdiscip. J.* **2011**, *5*, 181–200. [CrossRef]

48. Leeds-Hurwitz, W. Social theories: Social constructionism and symbolic interactionism. In *Engaging Theories in Family Communication: Multiple Perspectives*; Braithwaite, D.O., Baxter, L.A., Eds.; Sage Publications: Thousand Oaks, CA, USA, 2006; pp. 229–242.

49. Crowther, S.; Ironside, P.; Spence, D.; Smythe, L. Crafting stories in hermeneutic phenomenology research: A methodological device. *Qual. Health Res.* **2017**, *27*, 826–835. [CrossRef]

50. Galbin, A. An introduction to social constructionism. *Soc. Res. Rep.* **2014**, *26*, 82–92.

51. Braun, V.; Clarke, V. *Successful Qualitative Research: A Practical Guide for Beginners*; Sage Publications: London, UK, 2013; pp. 1–373.

52. Yin, R.K. *Qualitative Research from Start to Finish*, 2nd ed.; The Guilford Press: New York, NY, USA, 2016; pp. 1–348.

53. Davidsen, A.S. Phenomenological approaches in psychology and health sciences. *Qual. Res. Psychol.* **2013**, *10*, 318–339. [CrossRef]

54. Maphumulo, W.T.; Bhengu, B.R. Challenges of quality improvement in the healthcare of South Africa post-apartheid: A critical review. *Curationis* **2019**, *42*, 1–9. [CrossRef]

55. Ngulube, P.; Ngulube, B. Application and contribution of hermeneutic and eidetic phenomenology to indigenous knowledge research. In *Handbook of Research on Theoretical Perspectives on Indigenous Knowledge Systems in Developing Countries*; Ngulube, P., Ed.; IGI Global: Hershey, PA, USA, 2016; pp. 127–155.

56. Patton, M.Q. *Qualitative Research and Evaluation Methods: Integrating Theory and Practice*, 4th ed.; Sage Publications: Thousand Oaks, CA, USA, 2015.

57. Monette, D.R.; Sullivan, T.J.; DeJong, C.R. *Applied Social Research: A Tool for the Human Services*, 8th ed.; Brooks/Cole, Cengage Learning: Belmont, CA, USA, 2011; pp. 1–543.

58. Fusch, P.I.; Ness, L.R. Are we there yet? Data saturation in qualitative research. *Qual. Rep.* **2015**, *20*, 1408–1416.

59. Churchill, S.D. Explorations in teaching the phenomenological method: Challenging psychology students to "grasp at meaning" in human science research. *Qual. Psychol.* **2018**, *5*, 207–227. [CrossRef]

60. Dahlberg, K.; Dahlberg, H.; Nyström, M. *Reflective Lifeworld Research*, 2nd ed.; Studentlitteratur: Lund, Sweden, 2008; pp. 1–370.

61. Beck, M.; Martinsen, B.; Birkelund, R.; Poulsen, I. Raising a beautiful swan: A phenomenological-hermeneutic interpretation of health professionals' experiences of participating in a mealtime intervention inspired by protected mealtimes. *Int. J. Qual. Stud. Health Well-being* **2017**, *12*, 1–12. [CrossRef] [PubMed]

62. Lindseth, A.; Norberg, A. A phenomenological hermeneutical method for researching lived experience. *Scand. J. Caring Sci.* **2004**, *18*, 145–153. [CrossRef] [PubMed]

63. Giorgi, A. The theory, practice, and evaluation of the phenomenological method as a qualitative research procedure. *J. Phenomenol. Psychol.* **1997**, *28*, 235–260. [CrossRef]

64. Boezeman, E.J.; Ellemers, N. Intrinsic need satisfaction and the job attitudes of volunteers versus employees working in a charitable volunteer organization. *J. Occupat. Organ. Psychol.* **2009**, *82*, 897–914. [CrossRef]

65. Haivas, S.; Hofmans, J.; Pepermans, R. Volunteer engagement and intention to quit from a self-determination theory perspective. *J. Appl. Soc. Psychol.* **2013**, *43*, 1869–1880. [CrossRef]

66. Roth, G.; Deci, E.L. Autonomy. In *The Encyclopedia of Positive Psychology*; Lopez, S.J., Ed.; Blackwell: West Sussex, UK, 2013; pp. 78–82.

67. Ryff, C.D. Happiness is everything, or is it? *Explor. Mean. Psychol. Well-Being* **1989**, *57*, 1069–1081.

68. Wehmeyer, M.L.; Little, T.D.; Sergeant, J. Self-determination. In *Oxford Handbook of Positive Psychology*; Snyder, C.R., Lopez, S.J., Eds.; Oxford University Press: New York, NY, USA, 2009; pp. 357–366.

69. Ryan, R.M.; Deci, E.L. Autonomy Is No Illusion: Self-Determination Theory and the Empirical Study of Authenticity, Awareness, and Will. In *Handbook of Experimental Existential Psychology*; Greenberg, S.L.K., Pyszczynksi, T., Eds.; The Guilford Press: New York, NY, USA, 2004; pp. 455–485.

70. Ryan, R.M.; Deci, E.L.; Vansteenkiste, M. Autonomy and autonomy disturbances in self-development and psychopathology: Research on motivation, attachment, and clinical process. In *Developmental Psychopathology: Theory and Method*; Cicchetti, D., Ed.; Wiley: New York, NY, USA, 2016; pp. 385–438.

71. Ryff, C.D. Psychological well-being revisited: Advances in the science and practice of eudaimonia. *Psychother. Psychosomat.* **2013**, *83*, 10–28. [CrossRef]

72. Peterson, C.; Seligman, M.E.P. *Character Strengths and Virtues: A handbook and Classification*; Oxford University Press: New York, NY, USA, 2004; pp. 1–789.

73. Grönlund, H. Identity and volunteering intertwined: Reflections on the values of young adults. *Volunt. Int. J. Volunt. Nonprof. Organ.* **2011**, *22*, 852–874. [CrossRef]

74. Wrzesniewski, A.; Dekas, K.; Rosso, B. Calling. In *The Encyclopedia of Positive Psychology*; Lopez, S.J., Ed.; Blackwell: West Sussex, UK, 2013; pp. 115–118.

75. Duffy, R.D.; Bott, E.M.; Allan, B.A.; Torrey, C.L.; Dik, B.J. Perceiving a calling, living a calling, and job satisfaction: Testing a moderated, multiple mediator model. *J. Counsel. Psychol.* **2012**, *59*, 50–59. [CrossRef]

76. Duffy, R.D.; Dik, B.J. Research on calling: What have we learned and where are we going? *J. Vocat. Behav.* **2013**, *83*, 428–436. [CrossRef]

77. Wrzesniewski, A.; McCauley, C.; Rozin, P.; Schwartz, B. Jobs, careers, and callings: People's relations to their work. *J. Res. Pers.* **1997**, *31*, 21–33. [CrossRef]

78. Krause, N.; Rainville, G. Volunteering and psychological well-being: Assessing variations by gender and social context. *Pastor. Psychol.* **2018**, *67*, 43–53. [CrossRef]
79. McMunn, A.; Nazroo, J.; Wahrendorf, M.; Breeze, E.; Zaninotto, P. Participation in socially-productive activities, reciprocity and wellbeing in later life: Baseline results in England. *Ageing Soc.* **2009**, *29*, 765–782. [CrossRef]
80. Tabassum, F.; Mohan, J.; Smith, P. Association of volunteering with mental well-being: A lifecourse analysis of a national population-based longitudinal study in the UK. *BMJ Open* **2016**, *6*, 1–9. [CrossRef] [PubMed]
81. Vinje, H.F.; Langeland, E.; Bull, T. Aaron Antonovsky's development of salutogenesis, 1979 to 1994. In *The Handbook of Salutogenesis*; Mittelmark, M.B., Sagy, S., Eriksson, M., Bauer, G.F., Pelikan, J.M., Lindström, B., Espnes, G.A., Eds.; Springer: Cham, Switzerland, 2017; pp. 25–40.
82. Littman-Ovadia, H.; Steger, M. Character strengths and well-being among volunteers and employees: Toward an integrative model. *J. Posit. Psychol.* **2010**, *5*, 419–430. [CrossRef]

International Journal of
*Environmental Research
and Public Health*

*Article*

# Health Assets, Vocation and Zest for Healthcare Work. A Salutogenic Approach to Active Coping among Certified Nursing Assistant Students

**Natura Colomer-Pérez** [1,2,*], **Elena Chover-Sierra** [1,3], **Vicente Gea-Caballero** [4,5] **and Joan J. Paredes-Carbonell** [6,7]

1    Department of Nursing, University of Valencia, 46010 Valencia, Spain; elena.chover@uv.es
2    Development and Advising in Traffic Safety (DATS) Research Group, INTRAS (Instituto de Investigación en Tráfico y Seguridad Vial), 46022 Valencia, Spain
3    Consorcio Hospital General Universitario de Valencia, Medicina Interna, 46014 Valencia, Spain
4    Escuela de Enfermería La Fe, Centro adscrito Universitat de València, 46026 València, Spain; vicentegeacaballero@gmail.com
5    Grupo de Investigación GREIACC, Instituto de Investigación Sanitaria IIS La Fe, 46026 València, Spain
6    Centre de Salut Pública d'Alzira, Conselleria de Sanitat Universal i Salut Pública, Alzira, 46600 València, Spain; paredes_joa@gva.es
7    Fundació per al Foment de la Investigació Sanitària i Biomèdica de la Comunitat Valenciana (FISABIO), 46035 València, Spain
*    Correspondence: natura.colomer@uv.es

Received: 16 March 2020; Accepted: 18 May 2020; Published: 20 May 2020

**Abstract:** People's health assets (HA) mapping process and design dynamization strategies for it are paramount issues for health promotion. These strategies improve the health heritage of individuals and communities as both the salutogenic model of health (SMH) and health assets model (HAM) defend. Connecting and mobilizing HA and strengthens the 'sense of coherence' (SOC) are both related to enhancing stress active and effective coping strategies. This study aims to describe the HA present in a population of certified nursing assistant students ($n = 921$) in Spain and then to explore their relationships with the SOC, the motivation to choose healthcare studies and their academic performance. A great variety of HA were identified and mapped. Findings showed that individuals with greater motivation towards self-care and 'caring for others' as internal HA, possessed higher SOC levels and a strong vocation for healthcare work. Differences in HA were identified according to gender, age and employment situation. Consistent connections between the care–relation factor and vocational factor with interpersonal and extrapersonal HA were reported. Evidence and results substantiated the salutogenic and asset-based approach as a proper strategy to strengthen SOC, dynamize their HA map, reinforce the *sense of calling and* enable Certified Nurse Assistant (CNA) students to buffer against caregiving-related stress and thrive in their profession.

**Keywords:** salutogenic model of health; health assets model; asset-based approach; nurse; certified nurse assistant; vocation; active coping

## 1. Introduction

### 1.1. Salutogenesis and the Health Asset Framework as Models to Cope with Stress

Even though the formal Salutogenic Model of Health (SMH) has not received enough attention in research and practice since its origins, there is a renewed global interest in seeking for future directions for the concept. It emerges from the need for a better understanding of the theory and its implications addressing the full spectrum of the human health experience [1]. Likewise, the concept of health assets is becoming increasingly popular and it was explored in different settings and populations throughout the world [2]. Thus, health promotion from a positive health paradigm is grounded in two frameworks: Aaron Antonovsky's salutogenic theory and the Asset–based community development (ABCD) approach [3,4]. The identification of people's health resources and assets and the design of dynamization strategies for them is a paramount issue by improving the health heritage of individuals and communities as it is defended by the SMH and the Health Assets Model (HAM) [4–8].

In this sense, the salutogenic approach promotes the concept that when people can make sense of the world that surrounds them, they will also notice a correspondence between their actions and the effects these actions will have on their environment [9,10]. In this regard, there are closer connections between salutogenesis and a health asset-based model. The salutogenic assumption seeks to explore the origin and stability of health by understanding how it can be created and determines the optimum conditions for its development [11,12]. On the other side, the HAM establishes that: the more possibilities someone has to experience and accumulate positive effects of a series of assets throughout their life, the higher the chances of achieving health goals are [4]. To this effect, HAM implements community intervention's methodologies (such as social participation and action research) to develop a salutogenic health promotion strategy. This strategy is usually addressed in four phases: identification of protective factors, participatory asset mapping, connection and dynamization of health assets (HA) and finally, evaluation [13,14].

In parallel, those approaches also maintain interesting and close linkages to 'Stress and Coping' Lazarus and Folkman's theory. These authors address the existence of internal, interpersonal (conceived as social support) and external factors to individuals that buffer adverse effects of stress, allowing people to develop mechanisms to regulate emotional responses to stressful circumstances and having a high impact on well-being. Then, stress is a two-way process that involves stressors produced by the environment and the individual subjective responses to them by using primary and secondary cognitive appraisals [15,16]. Hence, these factors may be identified, equated and assimilated on many occasions to the salutogenic 'general resistant resources' (GRR) defined by Antonovsky [9,10] because they also include psychological traits, coping strategies, social and cultural factors and social support. Moreover, these factors contribute to increasing people's resilience, enabling them to solve problems adaptively, assessing stressful events as meaningful, predictable and manageable [17]. All in all, the theory posits that life experiences shape the sense of coherence (SOC)—the core element of the SMH—that helps to mobilize resources to cope with stressors and manage tension successfully [18].

In exploiting the perspective's whole meaning, the present research assumes the notion of these resistant resources consolidated as health assets. Consequently, to operationalize all these factors for health-promoting purposes, both the SMH and HAM advocate for categorizing the intrapersonal, interpersonal and extrapersonal health elements which operate as protective and promoting factors to buffer against life's stressors [19]. A health asset itself can be defined as any factor (or resource) which enhances the ability of individuals, groups, communities, populations, social systems and institutions to maintain and sustain health and well-being and to help to reduce health inequities. These assets can operate at the level of the individual (for example, abilities, competences and talents), group and community (including the role of supportive networks and population as protective or promoting factors to buffer against life's stresses) and eventually an organizational or institutional level (for example making use of external financial, physical or even environmental resources) [19]. According to the management of stress, literature advocates that all those resources not only immediately help people to cope better with stress and surviving [1]; but also, over time, personal and environmental resources can help with recovery and healing [20–23], even from early life adversities in adult populations [24].

### 1.2. Salutogenic Active Coping and Zest for Work in Healthcare Professionals

Advancing and empowering the SMH and HAM to understand better the ways of coping productively with stress seems to be a paramount purpose. Furthermore, this challenge must be primarily tackled in health care professionals and their prior academic and formative context.

Regarding the case of CNA nursing students and their future job demands envisaged, stress is a psychosocial factor that influences the academic performance and well-being of this group [25]. Nursing students not only face academic, but also face pressure at work during their training period [26]. Like previous findings, care behaviors correlated negatively with depression, distress and emotional exhaustion and positively correlated with coping strategies and a positive attitude to one's role at work [27]. Not surprisingly, the negative consequences of not having adequate coping strategies to undertake the inherent demands of nursing degree, as well as the future professional life, have an impact on their health and mental well-being. Furthermore, this situation is also directly related to professional performance [28].

Overall, it was shown that those using a greater variety of health assets can develop a greater sense of coherence (SOC) that will also allow them to promote active and effective stress coping strategies [18]. Concretely, healthcare students and workers with strong SOC may perceive and appraise the demands of their work environment as challenging rather than threatening, according to Antonovsky's research on health-promoting factors at work [9]. In addition, active coping is a valuable asset, especially in very demanding situations that nurses have to face up every day; therefore, resilient professionals are vital to the proper functioning of a health system [29]. Given those facts, researchers and professors suggest that daring to strengthen and reinforce the salutogenic capacity of the students must be expanded as part of the professional training in healthcare professional's degrees in order to promote and maintain the engagement and the zest for healthcare work [30]. More recently, it was observed the impact of the motivational factor in job engagement mediated by a sense of calling. This calling–vocation match brings forth from introspection, sensibility and reflection, which produces a working situation that for the most part, feels deeply gratifying and meaningful to the individual, resulting in zest for work and vitality [31].

In the spirit of the whole latest reflections on the salutogenic paradigm, for a better conceptualization of salutogenic orientation, it is necessary to encourage alternative approaches, including qualitative research [1].

Conversely, the health asset literature is underdeveloped, and its sustained credibility depends on future research dealing with definitional, theoretical and evaluative issues, being, therefore, imperative accomplishing more research to deeply apprehend the health assets model in a global context [2]. Thus, the pursuit of a better understanding between the salutogenic perspective (measuring SOC) and a health–asset approach (observing reported health assets) is the primary purpose of this study, to tempt a potential and early incorporation of a salutogenic orientation in healthcare-providers' studies. To this effect, the first phase of the present study has explored the salutogenic paradigm among nursing assistant (CNA) students in a region of Spain. Based on those findings, it seems that possessing a strong SOC appears to contribute towards improved resistance to stress, which in part, may also justify the motivation for studying a career that is pleasing and obtaining high academic performance despite being a profession with high demands and marked stressors [32]. Additional analyses of this research also have confirmed that CNA students referring a good practice on self-care and the willingness to caring for others (described as an internal health asset) also display an optimal zest for work in the nursing discipline [33].

This current study faces the last phase of the research seeking the opinions of CNA students about the HA that provides opportunities for well-being and health and determines a Health–asset Map articulated by participants using mixed-methods. Subsequently, it intends to explore, thorough a quantitative approach, the relationships between those HA, the SOC, the sense of calling (vocation-motivation variable) to choose healthcare studies as a career in concert with the academic performance for this certification in public education and vocational centers (Comunitat Valenciana, Spain).

## 2. Materials and Methods

### 2.1. Study Design and Sample

Mix-method study (qualitative and quantitative study: cross-sectional, analytical and exploratory) was carried out in 2016. Participants were enrolled—at data collection time-, in the last semester of certification of nursing assistant (CNA) from the total of public upper secondary schools providing vocational education and training (VET) certifications at Comunitat Valenciana (Spain). The study was aimed at the entire student population ($n = 1150$) enrolled in the region. With an IC = 95% and an error = 5%, a minimum sample of $n = 289$ was required.

### 2.2. Data Collection

Sociodemographic data collected were: (a) gender (male, female), (b) age (categorized: <30, 30–45, >45), (c) employment status (employed, unemployed), (d) income level (net income of the student's household, understood as the level of income received, from among the following options: high, medium/high, medium, medium/low, low, does not know/does not answer), (e) public secondary education centers in which CNA studies are taught, (f) geographical emplacement of the center (rural, urban, large city), (g) self-reported academic performance: students were asked about their academic record at the end of the last semester—when they already knew their global marks- and the responses were: fail (<5), pass (5–5.9), good (6–6.9), remarkable (7–7.9), outstanding (8–8.9), with distinction (9–10); in Spain, the academic record is scored in a scale of 0–10, with 10 being the highest score to reach and below 5 is considered as failed), (h) motivation of choice of studies (vocational, could not be enrolled in other studies, seek for better employment option, unmotivated). Some opened questions served to identify HA (intrapersonal, interpersonal, extrapersonal), defined as things/people/places that increased their well-being. SOC levels (a global orientation of the personality that facilitates the solution of problems in an adaptive way in stressful situations to which people are subjected throughout their lives) were assessed by the orientation-to-life questionnaire—13 items (OLQ-13 or SOC-13) [34]. This 13-item questionnaire also measures the dimensions of *comprehensibility* (with 5 items), *manageability* (with 4 items) and *meaningfulness* (with 4 items). The SOC-13 scale has shown good internal consistency, with a Cronbach's alpha between 0.70 and 0.92 [34–36].

### 2.3. Procedure

Professors from all educative centers attending Nursing Assistant public certifications were first contacted to mail them the questionnaire. Students completed a self-administered online questionnaire (with internet protocol—IP-response restriction) during their schedule's classes collecting qualitative data: HA; and quantitative data: the sense of coherence scale (SOC), factors related (motivation to study this career and self-reported academic performance) and sociodemographic variables. The questionnaire included information about the study and the contact details of the principal investigator. There were no exclusion criteria, and permission to participate in the study and consent to use the data were required. The qualitative analysis was carried out to identify the different types of HA, categorizing them into the categories already proposed. Subsequently, additional quantitative analysis was also carried out.

### 2.4. Data Analysis

#### 2.4.1. Qualitative Phase

The CNA student's HA mapping was underpinned according to the fundamentals of HAM methodology [3,13,37–39]. Most of these authors propose six categories of health assets: people, agencies or organizations (with or without profit), institutions, infrastructure or physical resources, economy and culture (including traditions, identity and sense of belonging). In this study, HA were collected and categorized in four HA groups, according to recommendations and results of previous research in this field [40]. First, the intrapersonal HA, which corresponds to an individual level; second, the interpersonal HA; third, the extrapersonal HA I—as institutions, organizations, etc.-; and finally, the extrapersonal HA II—as infrastructures, indoor/outdoor spaces, etc.-, which correspond to a community level.

In order to dump qualitative data collected from the questionnaire regarding to HA, the procedure was developed in 3 operationalization's phases and conducted by the main researcher and a different extra researcher. Phase (1) first, consisted of an information's transcription of the given responses from the open-ended asset questions, through a thematic analysis. This analysis was employed to codify information, also using the word economy, making significant groupings of the answers whenever possible, and trying to preserve the literalness of the discourse. Phase (2) was a reflection stage to prepare emerging subcategories for the four HA types that included sensitizing concepts. This means that these subcategories were aroused as a result of raising significant reference frames from the thematic analysis. Thus, a total of 30 HA's subcategories were built and identified with a subsequent numeric code: 3 subcategories for the Intrapersonal HA, 7 subcategories for the Interpersonal HA, 8 subcategories for the Extrapersonal HA I and 12 subcategories for the Extrapersonal HA II. Phase (3) was a proceeding of classifying and reconversion of each thematic content in its related HA subcategory—concretely, into its code number- with the purpose to prepare the database for the posterior statistical analysis.

In parallel and once again following the HAM methodology, the graphical students' HA map was built as a reflection of the literal qualitative data collected at the open HA questionnaire.

## 2.4.2. Quantitative Phase

Both the population characteristics and the subcategories of the HA classification that emerged after the qualitative analysis were analyzed in this phase. Descriptive statistics were applied to obtain frequencies and percentages in case of qualitative variables or means (M) and standard deviations (SD) to describe the quantitative ones.

chi-squared test was used to analyze the relationship among the HA identified and some population characteristics. Differences in SOC scores (global and for each dimension), according to the HA subcategory, were analyzed using the nonparametric Kruskal–Wallis test. In case of finding differences in SOC scores among groups, post hoc analysis using Bonferroni correction was performed to identify between which groups these differences occur.

In all cases, statistical significance was set at $p$-value $< 0.05$.

## 2.5. Ethical Considerations

In the case of underage students, prior authorization was obtained from parents or legal guardians to participate in the study. At the time the questionnaire was administered, the following measures were taken to ensure the anonymity and protection of the study participants: the professors—who were instructed to give the relevant indications to answer questionnaire right and accurately informed students that their participation was voluntary. They were also informed that not participating in the study did not imply grievances for them. The first screen of the online questionnaire informed about the legal details of the research. Data were anonymized and processed according to the recommendations of the State Data Protection Agency based on Organic Law 15/1999 and the European Directive on Data Protection 95/46/EC. Furthermore, permissions were also requested and obtained from each educational center and the competent organism in the area of education in the region (05ED01Z/2016/106/S) Resolution of February 25th, 2016 of the Autonomous Secretariat of Education and Research of the Conselleria d'Educació, Investigació, Cultura i Esport.

## 3. Results

### 3.1. Characteristics of the Participating Students

921 students answered the questionnaire voluntarily, 751 were women (81.54%) and 170 were men (18.46%). This number of responses meant 80.09% of the population of CNA in the Valencian Community. The average age of the participants was 28.52 (SD = 11.43, Range = 16–57).

The characteristics of the studied population studied are shown in Table 1.

**Table 1.** Characteristics of the studied population.

| Characteristics | | N | % |
|---|---|---|---|
| | Under 30 | 577 | 62.65 |
| Age group | 30–45 | 219 | 23.78 |
| | Over 45 | 125 | 13.57 |
| Gender | Male | 150 | 18.46 |
| | Female | 771 | 81.54 |
| | Rural areas | 66 | 7.17 |
| Geographic context | Urban areas | 520 | 56.46 |
| | Large cities | 335 | 36.37 |
| | Low | 283 | 30.68 |
| | Medium/Low | 261 | 28.36 |
| Familiar income | Medium | 297 | 32.19 |
| | Medium/High | 64 | 6.99 |
| | High | 16 | 1.78 |
| Employment situation | Employed | 222 | 24.11 |
| | Unemployed | 699 | 75.89 |
| | Vocational motivation | 444 | 48.21 |
| | Impossibility of access to other studies | 21 | 2.28 |
| Career choice motivation | Seeking better work | 316 | 34.31 |
| | No motivation | 25 | 2.71 |
| | Other | 115 | 12.49 |
| | Fail | 25 | 2.71 |
| | Pass | 119 | 12.92 |
| | Good | 242 | 26.27 |
| Academic performance | Remarkable | 303 | 32.89 |
| | Outstanding | 179 | 19.43 |
| | With distinction | 53 | 5.75 |

### 3.2. Qualitative Analysis: Mapping the HA Identified by CNA Students

The qualitative analysis of the answers offered by the participants allowed the researchers to design a qualitative map of HA identified by the CNA students, which is shown in Figure 1.

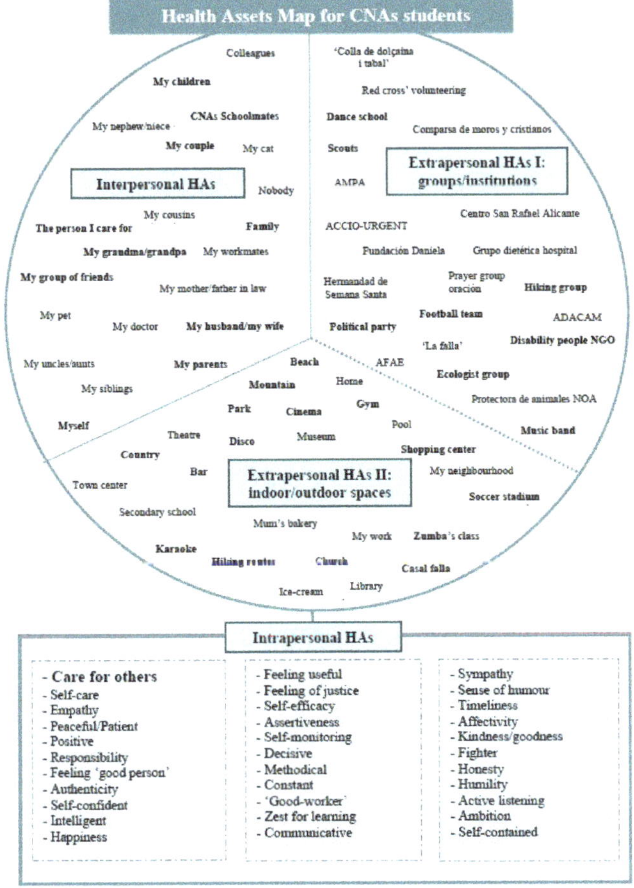



**Figure 1.** Map of health assets (HA) articulated by CNA students.

### 3.3. Quantitative Analysis of HA Identified by CNA Students

The quantitative analysis of the students' responses is presented in Table 2, which shows the frequencies of each one of the different HA identified by the participants.

**Table 2.** Quantitative description of health assets (HA) identified by CNA students.

| HA Category | HA Subcategory | N | % |
|---|---|---|---|
| Intrapersonal HA | Selfcare | 14 | 1.52 |
| | 'Caring for others' | 82 | 8.91 |
| | Others | 825 | 89.57 |
| Interpersonal HA | Couple | 126 | 13.68 |
| | Friends | 59 | 6.41 |
| | Ascendant relatives | 165 | 17.91 |
| | Descendant relatives | 113 | 12.27 |
| | Extended family | 274 | 29.78 |
| | Family + Friends | 143 | 15.52 |
| | Others | 41 | 4.45 |

**Table 2.** *Cont.*

| HA Category | HA Subcategory | N | % |
|---|---|---|---|
| Extrapersonal HA I (Groups/Institutions) | Volunteering | 47 | 5.10 |
| | Educative institutions | 33 | 3.58 |
| | Sporting institutions | 142 | 15.42 |
| | Recreational/Leisure institutions | 95 | 10.31 |
| | Religious institutions | 14 | 1.52 |
| | Musical/artistic institutions | 63 | 6.84 |
| | Social group | 396 | 42.99 |
| | No assets identified | 131 | 14.22 |
| Extrapersonal HA II (Indoor/Outdoor spaces) | Mountain (natural space) | 125 | 13.57 |
| | Beach (natural space) | 131 | 14.22 |
| | Both mountain and beach | 77 | 8.36 |
| | Urban spaces | 59 | 6.41 |
| | Sporting spaces | 63 | 6.84 |
| | Recreational/leisure spaces | 239 | 25.95 |
| | Spiritual/religious spaces | 6 | 0.65 |
| | Educative spaces | 15 | 1.63 |
| | Cultural spaces | 14 | 1.52 |
| | Home | 106 | 11.51 |
| | Working place | 10 | 1.08 |
| | No assets identified | 76 | 8.25 |

According to HA identified in this study, it is observed that in the case of intrapersonal ones, a great variety of them was identified by the CNA students. However, 9% of students refer to the fact of "caring for others" as one of the prior inner assets. Regarding interpersonal HA, the couple and the family nucleus (more or less extensive) were identified by up to 89% of students and the group of close friends by 6.4% of them. Finally, and concerning extrapersonal HA, a high number of groups and institutions were identified. On one hand, it is essential to highlight the social group (friends, classmates)—which was identified as an asset by 43% of students—and sporting institutions, identified by 15.4% of students. In addition, a wide variety of physic spaces was also identified by these students; however, they highlighted the recreational spaces (26%) and natural outdoor spaces (beach by 14.2% and mountain by 13.6%) and or specific spaces to do sports activities (6.8%). Interestingly, another 11.5% also identified their own home as an asset for well-being and health.

When analyzing the HA identified by the participants based on specific descriptive characteristics of the population, several differences are found, some of them statistically significant (chi-squared test). Table 3 shows these differences according to the gender, age and employment situation of CNA students.

Table 3. HA identified by CNA students according to their gender, age group and employment situation.

| HA Category | HA Subcategory | Male | Female | | Under 30 | 30–45 | Over 45 | | Employed | Unemployed | |
|---|---|---|---|---|---|---|---|---|---|---|---|
| | | % | % | $p$ | % | % | % | $p$ | % | % | $p$ |
| Intrapersonal HA | Self-care | 2.35 | 1.33 | 0.302 | 1.04 | 2.28 | 2.4 | 0.629 | 1.8 | 1.43 | 0.559 |
| | 'Caring for others' | 6.47 | 9.45 | | 9.01 | 8.22 | 9.6 | | 7.21 | 9.44 | |
| Interpersonal HA | Couple | 14.11 | 13.58 | 0.001 | 12.82 | 16.44 | 12.8 | 0.000 | 11.71 | 14.31 | 0.029 |
| | Friends | 12.94 | 4.93 | | 6.76 | 6.39 | 4.8 | | 6.31 | 6.44 | |
| | Ascendant relatives | 23.53 | 16.64 | | 21.66 | 15.07 | 5.6 | | 12.16 | 19.74 | |
| | Descendant relatives | 7.65 | 13.31 | | 3.29 | 22.37 | 36 | | 17.12 | 10.73 | |
| | Extended family | 24.71 | 30.89 | | 32.93 | 22.83 | 27.2 | | 32.43 | 28.89 | |
| Extrapersonal HA I Group/institutions | Sporting institutions | 22.94 | 13.71 | 0.021 | 14.9 | 18.72 | 12 | 0.003 | 15.31 | 15.45 | 0.030 |
| | Recreational/Leisure inst. | 11.76 | 9.99 | | 10.05 | 9.59 | 12.8 | | 8.55 | 10.87 | |
| | Social group | 32.94 | 45.27 | | 46.45 | 34.7 | 41.6 | | 39.18 | 44.21 | |
| | Volunteering | 5.29 | 5.06 | | 3.29 | 7.82 | 8.8 | | 7.66 | 4.29 | |
| Extrapersonal HA II Indoor/Outdoor spaces | Mountain spaces | 15.29 | 13.18 | 0.075 | 10.05 | 18.72 | 20.8 | 0.000 | 12.61 | 13.88 | 0.012 |
| | Beach spaces | 11.76 | 14.78 | | 15.42 | 14.15 | 8.8 | | 15.31 | 13.87 | |
| | Recreational/Leisure space | 25.29 | 26.09 | | 30.5 | 19.18 | 16.8 | | 22.07 | 27.18 | |
| | Home | 11.76 | 11.45 | | 12.48 | 8.67 | 12 | | 8.56 | 12.44 | |

### 3.4. Relationship between SOC and HA

The results on the SOC-13 obtained by this population and its relationship with academic performance can be read in a previously published study [32]. However, a summary is exposed below: the mean score (M) for total SOC measurement was 56.38 (SD = 12.24). Regarding the SOC dimensions, the average score for each subscale was: (i) *manageability* Mean = 16.45 (SD = 4.53); (ii) *comprehensibility* Mean = 19.27 (SD = 5.642; 30) and (iii) *meaningfulness*, Mean = 20.65 (SD = 4.48; 23).

When studying the relationship between the intrapersonal HA identified and the scores in SOC, it is found that those students who outlined aspects related to 'taking care of others' got higher scores in SOC than those who identified other introspective features of behavior, although these differences were not statistically significant (Kruskal–Wallis test). All these differences are shown in Table 4.

**Table 4.** SOC scores, according to intrapersonal HA identified by CNA students.

| HA Subcategory | SOC (Global) | | | Manageability | | | Comprehensibility | | | Meaningfulness | | |
|---|---|---|---|---|---|---|---|---|---|---|---|---|
| | Mean | SD | *p* | Mean | SD | *p* | Mean | SD | *p* | Mean | SD | *p* |
| Selfcare | 56.07 | 7.79 | | 16.36 | 4.38 | | 17.71 | 4.41 | | 22 | 2.93 | |
| 'Caring for others' | 56.89 | 12.68 | 0.95 | 15.82 | 4.75 | 0.48 | 19.56 | 5.61 | 0.54 | 21.51 | 4.61 | 0.08 |
| Others | 56.34 | 12.26 | | 16.52 | 4.51 | | 19.27 | 5.66 | | 20.55 | 4.49 | |

In the case of the relationship with interpersonal HA, shown in Table 5, it is found higher values of SOC in those who identified their children as health and well-being generating factors than those who identified other members of their family or their group of friends. These differences were statistically significant (Kruskal–Wallis test).

**Table 5.** SOC scores, according to interpersonal HA identified by CNA students.

| HA Subcategory | SOC (Global) | | | Manageability | | | Comprehensibility | | | Meaningfulness | | |
|---|---|---|---|---|---|---|---|---|---|---|---|---|
| | Mean | SD | *p* | Mean | SD | *p* | Mean | SD | *p* | Mean | SD | *p* |
| Couple | 55.47 | 13.75 | | 15.91 | 5 | | 18.80 | 6.10 | | 20.75 | 4.55 | |
| Friends | 54.39 | 12.35 | | 16.02 | 4.83 | | 19.05 | 5.27 | | 19.32 | 5 | |
| Ascendant relatives | 54.96 | 10.99 | | 16.02 | 4.43 | | 19.15 | 5.10 | | 19.79 | 4.17 | |
| Descendant relatives | 60.76 | 11.83 | 0.003 | 17.79 | 4.36 | 0.004 | 21.04 | 5.76 | 0.035 | 21.94 | 4.24 | 0.001 |
| Extended family | 56.24 | 12.08 | | 16.43 | 4.40 | | 19.14 | 5.45 | | 20.68 | 4.51 | |
| Family + Friends | 56.65 | 12.32 | | 16.71 | 4.43 | | 18.77 | 6.12 | | 21.17 | 4.22 | |
| Others | 55.68 | 11.73 | | 16.02 | 4.17 | | 19.39 | 5.43 | | 20.27 | 5.21 | |

The post hoc study to analyze intergroup differences found in some cases statistically significant differences ($p < 0.007$, with Bonferroni correction) in SOC scores obtained by students who identified their descendant relatives (the highest values) as HA and those who identified the other interpersonal HA. Table 6 shows mean differences in SOC scores, when comparing them among groups, considering as a reference those who identified their descendent relatives as HA.

**Table 6.** Post hoc analysis. Differences in SOC scores according to interpersonal HA identified by CNA, students, considering as reference the scores obtained by those who consider their descendent relatives as HA.

| HA Subcategory | SOC (Global) * Mean Differences | *p* | Manageability * Mean Differences | *p* | Comprehensibility * Mean Differences | p | Meaningfulness * Mean Differences | *p* |
|---|---|---|---|---|---|---|---|---|
| Couple | 5.29 | 0.002 | 1.88 | 0.001 | 2.24 | 0.003 | 1.19 | 0.055 |
| Friends | 6.37 | 0.002 | 1.76 | 0.013 | 1.99 | 0.034 | 2.62 | 0.001 |
| Ascendant relatives | 5.8 | 0.000 | 1.77 | 0.000 | 1.89 | 0.004 | 2.15 | 0.000 |
| Extended family | 2.15 | 0.001 | 1.36 | 0.002 | 1.9 | 0.002 | 1.26 | 0.015 |
| Family + Friends | 4.11 | 0.008 | 1.08 | 0.029 | 2.27 | 0.002 | 0.77 | 0.141 |
| Others | 5.08 | 0.014 | 1.77 | 0.004 | 1.35 | 0.097 | 1.67 | 0.084 |

* Mean differences are the differences between the scores obtained by CNA students who identified descendent relatives (reference group) as HA and the scores obtained by those who identified the HA shown in each row.

In this post hoc analysis, other statistically significant differences were found when comparing scores obtained in other groups such as in the case of people who identified as HA their parents and people who identified their family and friends as HA, but only in the scores of SOC *meaningfulness* dimension ($p = 0.003$).

Eventually, concerning extrapersonal HA I (groups or spaces), as shown in Table 7, individuals that referred to using open sporting spaces as a HA stood out for higher scores in SOC. In contrast, those who referred to cultural spaces (theatres, cinemas, museums, etc.) showed an average of around 6 points less in SOC. Higher values of SOC with students who used to accomplish volunteer actions and those who identified educational institutions as a health and well-being asset were also perceived. These differences were not statistically significant in all cases (Kruskal–Wallis test).

**Table 7.** SOC scores, according to extrapersonal HA identified by CNA students.

| HA Category/Subcategory | SOC (Global) | | | Manageability | | | Comprehensibility | | | Meaningfulness | | |
|---|---|---|---|---|---|---|---|---|---|---|---|---|
| | Mean | SD | *p* | Mean | SD | *p* | Mean | SD | *p* | Mean | SD | *p* |
| Groups/institutions (HA I) | | | | | | | | | | | | |
| Volunteering | 57.96 | 13.77 | | 16.02 | 4.52 | | 20.77 | 6.05 | | 21.17 | 4.49 | |
| Educative institutions | 57.73 | 13.08 | | 15.70 | 5.47 | | 19.88 | 6.22 | | 22.15 | 4.44 | |
| Sporting institutions | 57.58 | 12.20 | | 16.80 | 4.62 | | 19.93 | 5.40 | | 20.85 | 4.54 | |
| Recreational/leisure institutions | 57.35 | 11.85 | 0.038 | 16.94 | 4.47 | 0.204 | 19.64 | 5.27 | 0.002 | 20.77 | 4.08 | 0.138 |
| Religious institutions | 49.93 | 12.44 | | 14.71 | 4.39 | | 14.36 | 6.12 | | 20.86 | 4.27 | |
| Musical/artistic institution | 53.87 | 13.05 | | 15.79 | 4.94 | | 18.13 | 5.95 | | 19.95 | 4.52 | |
| Social group | 56.77 | 11.87 | | 16.70 | 4.32 | | 19.32 | 5.60 | | 20.75 | 4.48 | |
| No assets identified | 54.21 | 12.10 | | 15.84 | 4.59 | | 18.56 | 5.47 | | 19.81 | 4.68 | |
| Indoor/Outdoor Spaces (HA II) | | | | | | | | | | | | |
| Mountain (natural space) | 58.19 | 12.36 | | 16.94 | 4.40 | | 20.12 | 5.82 | | 21.13 | 4.29 | |
| Beach (natural space) | 55.75 | 11.05 | | 16.04 | 4.57 | | 18.81 | 4.82 | | 20.90 | 4.23 | |
| Both mountain and beach | 58.88 | 11.51 | | 17.09 | 4.48 | | 20.13 | 5.93 | | 21.66 | 3.94 | |
| Urban/rural spaces | 56.36 | 14.77 | | 17.02 | 5.01 | | 18.76 | 6.34 | | 20.58 | 5.56 | |
| Sporting spaces | 60.37 | 12.98 | | 18.02 | 4.73 | | 21.06 | 5.87 | | 21.29 | 4.73 | |
| Recreational/leisure spaces | 54.54 | 11.63 | 0.037 | 15.84 | 4.25 | 0.093 | 18.69 | 5.52 | 0.075 | 20.02 | 4.64 | 0.099 |
| Spiritual/religious spaces | 60.83 | 15.20 | | 17.33 | 6.28 | | 20.50 | 7.18 | | 23 | 2.68 | |
| Educative spaces | 55.27 | 12.96 | | 16 | 5.37 | | 17.33 | 6.14 | | 21.93 | 3.95 | |
| Cultural spaces | 54.50 | 13.51 | | 15.79 | 4.89 | | 18.07 | 7.24 | | 20.64 | 3.05 | |
| Home | 55.92 | 11.90 | | 16.27 | 4.35 | | 19.41 | 5.09 | | 20.25 | 4.39 | |
| Working place | 59.70 | 14.21 | | 17.70 | 3.33 | | 20.60 | 6.45 | | 21.40 | 5.38 | |
| No assets identified | 54.87 | 12.50 | | 16.14 | 4.78 | | 18.72 | 5.78 | | 20 | 4.38 | |

Post hoc analysis to study differences among CNA students according to extrapersonal HA I did not show statistically significant differences in SOC global scores. Nevertheless, in the case of the comprehensibility dimension, students who considered religious institutions as a HA got the lowest scores; these significant differences were found when comparing their scores with those achieved by the students who identified other extrapersonal HA I, such as sporting or educational institutions or their social group ($p < 0.006$, with Bonferroni correction). Furthermore, when comparing SOC global scores, according to HA II identified, no statistically significant differences were found.

### 3.5. HA According to Motivation for Choosing Nursing Studies

In light of exploring the calling for CNA's work in our population, 48.2% of students showed a vocational orientation when choosing this career, which was also significantly related to their better self-reported academic records. Furthermore, they scored higher in SOC (both, globally and three dimensions separately), as published previously [32] and those were also the ones who most frequently identified the concept *care for others* within intrapersonal HA (85.7%).

When analyzing the interpersonal HA referred according to this motivational variable, it is observed that the students with a vocational orientation identified their own families as a prior factor providing well-being to them. As for both categories of extrapersonal HA prioritized by these vocation-motivated students, the most valued were to be optimally included in a social group (44.4%), natural spaces (27.7%) and recreational/leisure spaces (24.3%). A lower priority appears in cultural (1.8%) and religious institutions (0.7%).

## 4. Discussion

The purpose of this research was to describe the HA identified by a sample of CNA students and to establish a relationship with essential resources to deal with their learning (and life) environment. It was also interesting to explore if having a coherent life meaning—and the presence of attaining their personal goals through an academic achievement- showed relations with vocational factors like taking care of others or owning a *sense of calling* for nursing studies, as the results confirmed.

Concerning intrapersonal HA related to relevant values as 'caring for others,' this study has connected the willingness for care asset-value with consistent scores of SOC, especially were women who referred the chance to 'taking care of others' as an internal resource generating well-being. This aspect and other ones related to patience and fondness provided in care were also referred to as protective assets by informal caregivers of Alzheimer's patients in a study developed by Agulló-Cantos [41]. Another study, which mapped internal HA thorough an intervention with resilient practices (mindfulness) in also informal caregivers determined that this technique—even it cannot be used by itself- could help them to manage a multitude of stressful situations and guarantee a good level of care for another person and even maintain their health and well-being [42]. Reviewing literature in educational contexts, a study mapping HA explained better academic achievement as a factor of well-being for students in a regular schooling experience [5], which justifies the importance of attending self-efficacy from a salutogenic perspective as the present study has observed. Linking to the essence of the salutogenic framework, several works have examined the relationships between SOC and school-related stress [43–45]. In the same line of this research, SOC correlated significantly and positively with school marks, school performance, achievement and success [32,46]. The present study determines that students relating this 'caring for others' asset-value also scored higher in SOC and those variables were significant related to students with a powerful sense of calling as a reason to perform this career. These findings reflect the proposal of Vinje [47] matching vocational element, giving meaningfulness to the chosen profession along with assuming and integrating healthcare strategies as combined, synergic and protective factors in health for nurses. This author even suggests that this chain of phenomena acts like a real snowball providing enthusiasm for the profession called zest for healthcare work, which explains the commitment to the practice of nursing and genuine job engagement [29,30].

In terms of highlighting the interpersonal HA pointed by CNA students, the ubiquity of positive social values provided by family and social network is consistent with findings from other studies which have found positive identity and positive social values [17] and the protective effects of supportive family relationships [48,49], as resources to cope with stressful life events. Concretely, women with children in the moment of the data collection were those who identified them as the main interpersonal assets in their lives. In contrast, men students referred to friendship as their mostly identified HA.

Observing the availability of social support resources in adolescents, they stressed the importance of family, friends and neighbors and the feeling of being supported and taken care of by their parents, community and friends constituting an essential social resource that contributed to resilience [50–52]. Regarding the school context, support from teachers and support from classmates seem to be critical elements during adolescence [6,53]. Besides that, stronger SOC scores were reported by students identifying their children as an interpersonal HA, as the post hoc results also confirmed, these students also obtained higher scores than individuals who identified another interpersonal HA. Based on the findings from the literature review, also those who identified other relatives and friends as HA scored slightly lower in SOC. These findings concur with a study conducted in Spain by Malagón [54], that reported higher SOC levels for nurses accounting for the satisfactory and supportive nuclear familiar network.

Finally, and discussing the extrapersonal HA, it was observed more frequency of response concerning using sporting institutions among CNA male students. In contrast, women identified preferences for educational institutions as healthy resources. These data lead to a reflection about the expression of a society that tends to perpetuate the assignment of roles and stereotypes to women and men in a differentiated way and the need to work in breaking these stereotypes from a salutogenic educational environment. Another interesting relation was observed between higher SOC scores and the fact to undertake some volunteer activities and referred to the frequent use of educational and cultural institutions. On the other hand, those that frequented religious institutions as a coping strategy scored quite lower in SOC at the time of data collection. Even though, in this case, the post hoc analysis did not find statistically significant differences. All in all, for this typology of assets, the results are consistent with those identified as HA in previous researches: neighborhood and community network, sociocultural environment and heritage, sporting and leisure time spaces and natural environments as the most relevant among the population [3,4,40,55,56].

At this stage and analyzing the specific HA map articulated by Valencian CNA students, some aspects must be tackled, such as the consistent connections of the care–relation factor and vocational factor with real health assets reported. Therefore, is the salutogenic and asset-based approach a facilitating strategy to allow CNA nursing students to strengthen SOC, reinforce their *sense of calling*, delve into the *zest for healthcare work* and consequently, enable them to buffer against work-related caregiving stress and thrive in their professions? Even more, could this holistic and positive health-promoting paradigm be early conducted at nursing VET schools and university nursing schools? The findings sustain partial aspects of this postulate. One research focused on undergraduate nursing education emphasized the importance of the role of some personal HA as self-efficacy, emotional intelligence and develop nursing professionalism as inherent aspects to be included in educational strategies for these healthcare students [57]. Mayer and Boness [58] suggested that educational contexts can play a crucial role in creating consistency and a safe and respectful learning environment that promotes social support and enhances SOC. In this regard, a systematic review from the UK released that an Asset–based community development proved a useful 'lens' to view research in schools on the interaction of education and health improvement, having confirmed that there are promising areas for health gain from using schools as 'health assets' [59]. Lindström & Eriksson [12] reflected on the importance of introducing the salutogenic framework in educational science by starting a discussion about the content of health education and health literacy expanding towards healthy learning, with emphasis on positive health promotion.

Then, the commitment seems to point towards the convenience for a decisive introduction of salutogenic orientation in nursing curricula. However, this academic engagement should probably be extended to the rest of health sciences' disciplines as a core component of the study plans [57]. In addition, asset-based interventions must also be implemented in educative programs (as well as in continuing education) and the challenge goes through advocating for teachers being specially trained in salutogenic approaches. That is worthwhile because the salutogenic framework seems to provide a better understanding of the ways to tackle workload and cope with professional stressors, promotes positive health and well-being among future caregivers professionals and could improve, all in all, their efficacy as healing agents.

The implications of our study go in that direction. On one hand, it contributes to the theoretical development of the Salutogenic Model of Health and the Health Assets Model; both have shown to be adequate constructs to improve the health and well-being of any individual from a positive perspective, also in students [2,10]. Furthermore, this is relevant to the extent that this study contributes to reinforcing the growing existing evidence [1,4,6,13], based on non-theoretical population studies. According to the transference to a practical level, it is proposed a transversal implementation of the salutogenic approach in nursing curricula. It is also believed that the implementation of this approach at earlier ages (school) may help to encourage vocational choice in health-related studies, which we have observed to be associated with better academic achievement and higher SOC levels [32]. Salutogenic educative orientation expedites us to train students in order to reinforce their SOC and mainly, dynamize optimal HA against adversities and experiences that caring profession itself will make them live, in addition to improving their academic performance [59]. In this way, we will better prepare future healthcare professionals to be able to adapt to frequently hostile work environments and their internal threats, with a resilient capacity that will enable them to emerge strengthened from these experiences. This approach would reinforce their self-efficacy and self-esteem [17] and all of these would entail an indirect benefit for the health system itself and society [29].

*Study Limitations*

One of the weaknesses of the present study is the individual recollection of HA data, but the strict bases of building up HA maps point at the recommendation of releasing it throughout participatory processes. Nor should be forgotten the importance of devolution of the results from mapping actions to the population in order to provoke reflection and raise awareness about their health potential. This aspect was not developed in this study due to the exploratory design. Another limitation is related to the cross-sectional study design, which hampers the ability to make causal inferences. Although these types of measures help enrich the perception of reality, they may, however, induce bias; while considering that a bias exists regarding the homogeneity of the sample in favor of female participants. On this, it is also convenient to highlight that one of the strengths of the present study is the use of mixed-methods combining the qualitative HA identification with another quantitative phase focused on exploring the relations between all HA categories reported with Antonovsky's sense of coherence. This groundbreaking approach could allow obtaining new information of great interest in the fulfillment of the objectives of global salutogenic strategies.

## 5. Conclusions

The HA map articulated by CNA students was built and described. An implement that shall provide future opportunities for health and well-being among participants through empowering students in order to access and mobilize their resources and increasing their control over their health and its wider determinants. The relationships between primary HA, high SOC levels, a strong vocation for healthcare in concert with a heightened academic performance for this certification in public VET centers (Comunitat Valenciana, Spain) were confirmed. These consistent connections between the care–relation factor and vocational factor with specific HA reported, substantiate the salutogenic and asset-based approach. Ultimately, a salutogenic educative strategy root in: strengthen CNA student's SOC, dynamize their HA map, reinforce their *sense of calling* which delves into the *zest for healthcare work*, consequently enable them to buffer against work-related caregiving stress and seems to be the proper to thrive in their profession.

**Author Contributions:** Conceptualization, N.C.-P., J.J.P.-C. and V.G.-C.; methodology, N.C.-P., J.J.P.-C., E.C.-S. and V.G.-C.; formal analysis, E.C.-S. and N.C.-P.; data curation, E.C.-S. and N.C.-P.; writing—original draft preparation, E.C.-S. and N.C.-P.; writing—review and editing, N.C.-P., J.J.P.-C., E.C.-S and V.G.-C.; supervision, N.C.-P., J.J.P.-C. and V.G.-C. All authors have read and agreed to the published version of the manuscript.

**Funding:** This research received no external funding.

**Conflicts of Interest:** The authors declare no conflict of interest.

## References

1. Bauer, G.F.; Roy, M.; Bakibinga, P.; Contu, P.; Downe, S.; Eriksson, M.; Espnes, G.A.; Jensen, B.B.; Juvinya Canal, D.; Lindström, B.; et al. Future Directions for the Concept of Salutogenesis: A Position Article. *Health Promot. Int.* **2019**, 1–9. [CrossRef] [PubMed]
2. Van Bortel, T.; Wickramasinghe, N.D.; Morgan, A.; Martin, S. Health Assets in a Global Context: A Systematic Review of the Literature. *BMJ Open* **2019**, *9*, e023810. [CrossRef] [PubMed]
3. Kretzmann, J.P.; McKnight, J.L. *Building Communities from the Inside Out: A Path Toward Finding and Mobilizing a Community's Assets*; The Asset-Based Community Development Institute, Institute for Policy Research: Evanston, IL, USA, 1993; pp. 140–145.
4. Morgan, A.; Ziglio, E. Revitalising the Evidence Base for Public Health: An Assets Model. *Promot. Educ.* **2007**, *14*, 17–22. [CrossRef] [PubMed]

5.  Pérez-Wilson, P.; Hernán, M.; Morgan, A.R.; Mena, A. Health Assets for Adolescents: Opinions from a Neighbourhood in Spain. *Health Promot. Int.* **2015**, *30*, 552–562. [CrossRef]
6.  Rivera de los Santos, F.; Ramos Valverde, P.; Moreno Rodríguez, C.; Hernán García, M. Análisis Del Modelo Salutogénico En España: Aplicación En Salud Pública E Implicaciones Para El Modelo De Activos En Salud. *Rev. Española Salud Pública* **2011**, *85*, 129–139. [CrossRef]
7.  Mittelmark, M.B.; Bauer, G.F. The meanings of salutogenesis. In *The Handbook of Salutogenesis*; Mittelmark, M.B., Ed.; Springer: Berlin/Heidelberg, Germany, 2017.
8.  Idan, O.; Eriksson, M.; Al-Yagon, M. The salutogenic model. In *The Handbook of Salutogenesis.*; Mittelmark, M.B., Ed.; Springer: Berlin/Heidelberg, Germany, 2017; p. 57.
9.  Antonovsky, A. *Unraveling the Mystery of Health: How People Manage Stress and Stay Well*; Jossey-Bass: San Francisco, CA, USA, 1987.
10. Antonovsky, A. The Salutogenic Model as a Theory to Guide Health Promotion. *Health Promot. Int.* **1996**, *11*, 11–18. [CrossRef]
11. Hernan, M.; Morgan, A.; Mena, A.L. *Formación En Salutogénesis Y Activos Para La Salud*; Escuela Andaluza de Salud Pública: Sevilla, Spain, 2010.
12. Lindström, B.; Eriksson, M. From health education to healthy learning: Implementing salutogenesis in educational science. *Scand. J. Public Health* **2011**, *39* (Suppl. 6), 85–92.
13. Cofiño, R.; Aviñó, D.; Benedé, C.B.; Botello, B.; Cubillo, J.; Morgan, A.; Paredes-Carbonell, J.J.; Hernán, M. Promoción De La Salud Basada En Activos: ¿cómo Trabajar Con Esta Perspectiva En Intervenciones Locales? *Gac. Sanit.* **2016**, *30*, 93–98. [CrossRef]
14. Cassetti, V.; Powell, K.; Barnes, A.; Sanders, T. A Systematic Scoping Review of Asset-Based Approaches to Promote Health in Communities: Development of a Framework. *Glob. Health Promot.* **2019**, *7*, 1–25. [CrossRef]
15. Lazarus, R.S.; Folkman, S. *Stress, Appraisal, and Coping*; Springer Publishing Company: New York, NY, USA, 1984.
16. Lazarus, R.S. Emotions and Interpersonal Relationships: Toward a Person-Centered Conceptualization of Emotions and Coping. *J. Personal.* **2006**, *74*, 9–46. [CrossRef]
17. Lindström, B.; Eriksson, M. Contextualizing Salutogenesis and Antonovsky in Public Health Development. *Health Promot. Int.* **2006**, *21*, 238–244. [CrossRef] [PubMed]
18. Mittelmark, M.; Bull, T.; Bouwman, L.I. Emerging Ideas Relevant to the Salutogenic Model of Health. In *The Handbook of Salutogenesis*; Mittelmark, M.B., Ed.; Springer: Berlin/Heidelberg, Germany, 2017.
19. Morgan, A. Revisiting the Asset Model: A Clarification of Ideas and Terms. *Glob. Health Promot.* **2014**, *21*, 3–6. [CrossRef]
20. Todahl, J.L.; Walters, E.; Bharwdi, D.; Dube, S.R. Trauma Healing: A Mixed Methods Study of Personal and Community-Based Healing. *J. Aggress. Maltreatment Trauma* **2014**, *23*, 611–632. [CrossRef]
21. Sakallaris, B.R.; Macallister, L.; Voss, M.; Smith, K.; Jonas, W.B. Optimal Healing Environments. *Glob. Adv. Health Med.* **2015**, *4*, 40–45. [CrossRef] [PubMed]
22. Parkin, S. Salutogenesis: Contextualising Place and Space in the Policies and Politics of Recovery from Drug Dependence (UK). *Int. J. Drug Policy* **2015**, *33*, 21–26. [CrossRef] [PubMed]
23. Boscherini, G. A Sense of Coherence: Supporting the Healing Process. *Archit. Des.* **2017**, *87*, 108–113. [CrossRef]
24. Dube, S.R.; Rishi, S. Utilizing the Salutogenic Paradigm to Investigate Well-being among Adult Survivors of Childhood Sexual Abuse and Other Adversities. *Child Abus. Negl.* **2017**, *66*, 130–141. [CrossRef]
25. Rudman, A.; Gustavsson, J.P. Burnout during Nursing Education Predicts Lower Occupational Preparedness and Future Clinical Performance: A Longitudinal Study. *Int. J. Nurs. Stud.* **2012**, *49*, 988–1001. [CrossRef]
26. Pulido-Martos, M.; Augusto-Landa, J.M.; Lopez-Zafra, E. Sources of Stress in Nursing Students: A Systematic Review of Quantitative Studies. *Int. Nurs. Rev.* **2012**, *59*, 15–25. [CrossRef]
27. Chana, N.; Kennedy, P.; Chessell, Z.J. Nursing Staffs' Emotional Well-being and Caring Behaviours. *J. Clin. Nurs.* **2015**, *24*, 2835–2848. [CrossRef]
28. Lee, Y.E.; Kim, E.; Park, S.Y. Effect of Self-Esteem, Emotional Intelligence and Psychological Well-being on Resilience in Nursing Students. *Child Health Nurs. Res.* **2017**, *23*, 385–393. [CrossRef]
29. Bakibinga, P.; Vinje, H.F.; Mittelmark, M.B. Self-Tuning for Job Engagement: Ugandan Nurses' Self-Care Strategies in Coping with Work Stress. *Int. J. Ment. Health Promot.* **2012**, *14*, 3–12. [CrossRef]

30. Vinje, H.F.; Ausland, L.H.; Langeland, E. The Application of Salutogenesis in the Training of Health Professionals. In *The Handbook of Salutogenesis*; Mittelmark, M.B., Ed.; Springer International Publishing: Berlin/Heidelberg, Germany, 2016.

31. Jenny, G.J.; Bauer, G.F.; Vinje, H.F.; Vogt, K.; Torp, S. The application of salutogenesis to work. In *The Handbook of Salutogenesis*; Mittelmark, M.B., Ed.; Springer: Berlin/Heidelberg, Germany, 2016.

32. Colomer-Pérez, N.; Paredes-Carbonell, J.J.; Sarabia-Cobo, C.; Gea-Caballero, V. Sense of Coherence, Academic Performance and Professional Vocation in Certified Nursing Assistant Students. *Nurse Educ. Today* **2019**, *79*, 8–13. [CrossRef] [PubMed]

33. Colomer-Perez, N.; Paredes-Carbonell, J.J.; Sarabia-Cobo, C.M.; Gea-Caballero, V. Self-Care Agency and Sense of Coherence Implications for a Salutogenic Model of Care in Certified Nursing Assistant Students. *Health Educ. Behav.*. under review.

34. Antonovsky, A. The Structure and Properties of the Sense of Coherence Scale. *Soc. Sci. Med.* **1993**, *36*, 725–733. [CrossRef]

35. Lizarbe-Chocarro, M.; Guillén-Grima, F.; Aguinaga-Ontoso, I.; Canga Armayor, N. Validación Del Cuestionario De Orientación a La Vida (OLQ-13) De Antonovsky En Una Muestra De Estudiantes Universitarios En Navarra. *An. Sist. Sanit. Navar.* **2016**, *39*, 237–248.

36. Eriksson, M.; Lindström, B. Validity of Antonovsky's Sense of Coherence Scale: A Systematic Review. *J. Epidemiol. Community Health* **2005**, *59*, 460–466. [CrossRef]

37. Morgan, A.; Ziglio, E.; Davies, M. *Health Assets in a Global Context*; Springer: New York, NY, USA, 2010.

38. Blickem, C.; Dawson, S.; Kirk, S.; Vassilev, I.; Mathieson, A.; Harrison, R.; Bower, P.; Lamb, J. What is Asset-Based Community Development and how might it Improve the Health of People with Long-Term Conditions? A Realist Synthesis. *SAGE Open* **2018**, *8*, 1–13. [CrossRef]

39. McKnight, J.L.; Russell, C. *The Four Essentian Elements of an Asset-Based Community Developement Process*; De Paul University: Chicago, IL, USA, 2018.

40. Paredes-Carbonell, J.J.; Agulló-Cantos, J.M.; Vera-Remartínez, E.J.; Hernán-García, M. Sentido De Coherencia Y Activos Para La Salud En Jóvenes Internos En Centros De Menores. *Rev. Española Sanid. Penit.* **2013**, *15*, 87–97. [CrossRef]

41. Agulló-Cantos, J.M.; García-Alandete, J.; Paredes-Carbonell, J.J. Activos Para La Salud En Cuidadores Familiares De Enfermos De Alzheimer: Desarrollo De Un Mapa De Activos Para La Salud. *Glob. Health Promot.* **2019**, 175797591984307–1757975919843076. [CrossRef]

42. Alpuente, A.C.; Cintas, F.A.; Foà, C.; Cosentino, C. Mapping Caregivers' Health Assets. A Self-Care Project using Salutogenesis and Mindfulness. *Acta Bio Med. Atenei Parm.* **2018**, *89*, 70–77.

43. Natvig, G.K.; Albrektsen, G.; Qvarnstrøm, U. Associations between Psychosocial Factors and Happiness among School Adolescents. *Int. J. Nurs. Pract.* **2003**, *9*, 166–175. [CrossRef] [PubMed]

44. Moksnes, U.K.; Rannestad, T.; Byrne, D.G.; Espnes, G.A. The Association between Stress, Sense of Coherence and Subjective Health Complaints in Adolescents: Sense of Coherence as a Potential Moderator. *Stress Health* **2011**, *27*, e157–e165. [CrossRef]

45. Torsheim, T.; Aaroe, L.E.; Wold, B. Sense of Coherence and School-Related Stress as Predictors of Subjective Health Complaints in Early Adolescence: Interactive, Indirect or Direct Relationships? *Soc. Sci. Med.* **2001**, *53*, 603–614. [CrossRef]

46. Kristensson, P.; Öhlund, L.S. Swedish Upper Secondary School Pupils' Sense of Coherence, Coping Resources and Aggressiveness in Relation to Educational Track and Performance. *Scand. J. Caring Sci.* **2005**, *19*, 77–84. [CrossRef] [PubMed]

47. Vinje, H.F. Thriving Despite Adversity: Job Engagement and Self-Care among Community Nurses. Ph.D. Thesis, Bergen University, Germany, 2007.

48. Oliva Delgado, A.; Jiménez Morago, J.M.; Parra Jiménez, A.; Sánchez-Queija, I. Acontecimientos Vitales Estresantes, Resiliencia Y Ajuste Adolescente. *Rev. Psicopatol. Psicol. Clínica* **2008**, 53–62. [CrossRef]

49. Reina, M.D.C.; Oliva-Delgado, A.; Parra-Jiménez, Á. Percepciones De Autoevaluación: Autoestima, Autoeficacia Y Satisfacción Vital En La Adolescencia. *Psychol. Soc. Educ.* **2017**, *2*, 55. [CrossRef]

50. Gonzalez, M.P.; Rey Yedra, L. La Escuela Y Los Amigos: Factores Que Pueden Proteger a Los Adolescentes Del Uso De Sustancias Adictivas. *Ensen. Investig. Psicol.* **2006**, *11*, 23.

51. Moyano Díaz, E.; Ramos Alvarado, N. Bienestar Subjetivo: Midiendo Satisfacción Vital, Felicidad Y Salud En Población Chilena De La Región Maule. *Universum Talca* **2007**, *22*, 177–193. [CrossRef]

52. Vera-Remartínez, E.J.; Paredes-Carbonell, J.J.; Aviñó Juan-Ulpiano, D.; Jiménez-Pérez, M.; Araujo Pérez, R.; Agulló-Cantos, J.M.; Mora Notario, A. Sentido De Coherencia Y Mapa De Activos Para La Salud En Jóvenes Presos De La Comunidad Valenciana En España. *Glob. Health Promot.* **2017**, *24*, 112–121. [CrossRef]

53. García-Moya, I.; Rivera, F.; Moreno, C.; López, A. Calidad De La Relación Entre Los Progenitores Y Sentido De Coherencia En Sus Hijos Adolescentes. El Efecto De Mediación De La Satisfacción Familiar. *An. Psicol.* **2013**, *29*, 482–490. [CrossRef]

54. Malagón Aguilera, M.C. El Sentido De La Coherencia Y El Compromiso Laboral De Las Enfermeras En El Ámbito Sociosanitario De Girona. Ph.D. Thesis, Universitat de Barcelona, Barcelona, Spain, 2016.

55. Restrepo, H.E.; Málaga, H. *Promoción De La Salud: Cómo Construir Vida Saludable*; Organización Panamericana de la Salud: Washington, DC, USA, 2001.

56. Paakkari, L.; Torppa, M.; Välimaa, R.; Villberg, J.; Ojala, K.; Tynjälä, J. Health Asset Profiles and Health Indicators among 13- and 15-Year-Old Adolescents. *Int. J. Public Health* **2019**, *64*, 1301–1311. [CrossRef] [PubMed]

57. Kim, H.; Ryu, S.; Jang, K. Effect of Structured Pre-Simulation Preparation and Briefing on Student's Self-Confidence, Clinical Judgment, and Clinical Decision-Making in Simulation. *Contemp. Nurse* **2019**, *55*, 317–329. [CrossRef] [PubMed]

58. Mayer, C.; Boness, C. Interventions to Promoting Sense of Coherence and Transcultural Competences in Educational Contexts. *Int. Rev. Psychiatry* **2011**, *23*, 516–524. [CrossRef]

59. Caan, W.; Cassidy, J.; Coverdale, G.; Ha, M.; Nicholson, W.; Rao, M. The Value of using Schools as Community Assets for Health. *Public Health* **2014**, *129*, 3–16. [CrossRef]

International Journal of
*Environmental Research
and Public Health*

*Review*

# Sense of Coherence in Nurses: A Systematic Review

**Giuseppe Michele Masanotti** [1,*], **Silvia Paolucci** [2], **Elia Abbafati** [2], **Claudio Serratore** [2] and **Michela Caricato** [2]

1   Director of Experimental Centre of Research for Health Promotion and Health Education (CeSPES),
    06129 Perugia, Italy
2   School of Specialization in Hygiene and Preventive Medicine, University of Perugia, 06129 Perugia, Italy;
    silvia.paolucci01@gmail.com (S.P.); elia.abbafati@gmail.com (E.A.); claudio.serratore@gmail.com (C.S.);
    michelacaricato@gmail.com (M.C.)
*   Correspondence: giuseppe.masanotti@unipg.it; Tel.: +39-07-55-858-034

Received: 11 February 2020; Accepted: 8 March 2020; Published: 13 March 2020

**Abstract:** *Background*: Nurses experience high levels of distress due to the nature of their work and workplaces; Antonovsky's salutogenic theory shows that individual and work-related factors can influence human health. The aim of this paper is to analyze the possible correlations with different work-related and individual variables, which influence or are influenced by Sense of Coherence (SOC) and verify the possible use of SOC scales to prevent negative health determinants in workplaces. *Methods*: Electronic databases were searched with selected studies compared for sample, sample size, study design and basic results. Cross-sectional studies were reviewed for correlations between individual physical and mental health, distress, burnout, job satisfaction and SOC, with intervention studies used to assess the possible impact of training on nurses' SOC. *Results*: The review found several correlations between SOC and different work-related variables; but also with several individual characteristics. *Conclusion*: The review found that SOC was predictor of depressive state, burnout, job dissatisfaction among female nurses; therefore, SOC could be a health promoting resource.

**Keywords:** nurses; salutogenesis; Antonovsky

---

## 1. Introduction

Nurses are in the front line in the psychologically demanding everyday-care of patients, which can often undermine their emotional balance, influencing both their physical and mental wellbeing [1]. Moreover formal caregivers are frequently burdened with an excessive workload, high working pressure and demands, spending more time at work than on other dimensions of their lives [2,3]. All together these factors may contribute to the creation of a stressful working environment, which requires great coping abilities.

According to the salutogenic theory [4], one of the most critical determinants of the capacity to cope successfully with distress is the Sense of Coherence (SOC), which shapes the individual experience of a stressful event and allows it to be perceived as meaningful, manageable and comprehensible. This can be achieved by mobilizing efficiently the so-called GRRs (generalized resistance resources), which include biological, material and psychosocial factors, triggering a virtuous cycle and in turn strengthening the personal SOC [5]. Similarly, the Conservation of Resources theory (COR) states that the stress can result from circumstances involving loss of valued resources, and that the desire to preserve the individual resources is the basis of the coping ability [6].

Transposing Antonovsky's theory on the working context, the SOC can be modified, in a positive or negative way, by the nature of the current working environment. This re-adaptation explains how "job resources" are an integral part of the GRRs and participate in the modeling of the worker's SOC, which consequently influences how the so-called job demands (hours and pattern of work, workload, relations among the colleagues and every organizational aspects of a job that require continuous physical and/or psychological effort) are perceived, appraised, faced and overcome [5].

The SOC scale, named initially by Antonovsky's "Orientation to Life Questionnaire", investigating the three dimensions of SOC (Meaningfulness, Manageability and Comprehensibility) is available, to date, in two versions: the original form of 29 items (SOC-29) and the shorter version of 13 items (SOC-13). Each item is scored on a 7-point Likert scale, ranging 29–203 and 13–91, respectively, with higher score corresponding to a more developed SOC.

The aim of this paper is to analyze the possible correlations with different work-related and individual variables, which influence or are influenced by SOC and verify the possible use of SOC-29 or SOC-13 to prevent negative health determinants in workplaces.

## 2. Materials and Methods

### 2.1. Research Methods

A systematic search was conducted up to January 2019 on major healthcare databases: PubMed, Web of Science and Scopus. The following terms were included: sense of coherence, nurse, nurses, nursing, nursing staff, formal caregiver, formal caregivers; no additional filters used. Additional articles were retrieved from the consultation of relevant authors and paper's bibliography.

### 2.2. Articles Selection

Two independent reviewers selected the studies according to the following inclusion criteria: (1) original articles, and (2) administration of the SOC questionnaire to a sample of formal nurses.

Exclusion criteria applied were: (1) SOC questionnaire not administered, (2) language other than English, (3) sample different from working nurses (nurse teachers, unemployed nurses), (4) impossibility to retrieve a specific SOC value for the nurse sample, (5) absence of both mean SOC value and type of SOC questionnaire, and (6) use of other SOC questionnaire other than SOC-29 or SOC-13. Disagreements on article selection were resolved by consensus.

### 2.3. Data Extraction and Synthesis

Extraction of paper's data was independently performed by the reviewers through a pre-set table and consensus was reached, upon common revision, for each item inserted therein.

Selected papers were subsequently divided into three categories, based on whether the field of investigation of the Sense of Coherence was work-related or within the individual's sphere; articles assessing SOC variation upon interventions were categorized separately. The categories were named "Work-Related Variables", "Individual Variables" and "Interventions".

## 3. Results

A total of 876 papers were obtained. After duplicates removal, 535 records were screened initially by title and abstract and then by full text assessment. This process led to the exclusion of $n = 454$ and $n = 42$ articles respectively, yielding a total of 39 records included in the present review (Figure 1).

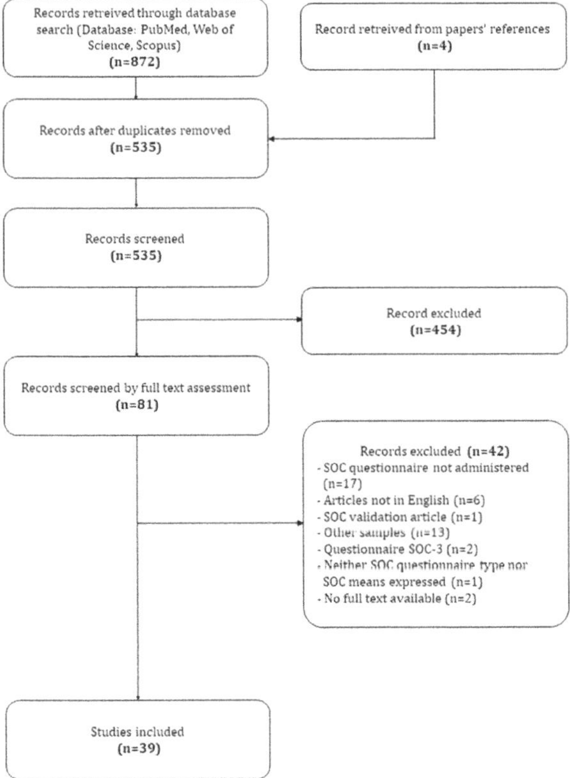

**Figure 1.** Flowchart of the searching and screening of literatures.

Data extraction from the included studies were performed and, according to the variables assessed, they were allocated to the three categories mentioned above: "Work-Related Variables", "Individual Variables", "Interventions" (Figure 2). Features of the same article, falling into more than one category were assessed separately. Table 1 summarizes the articles.

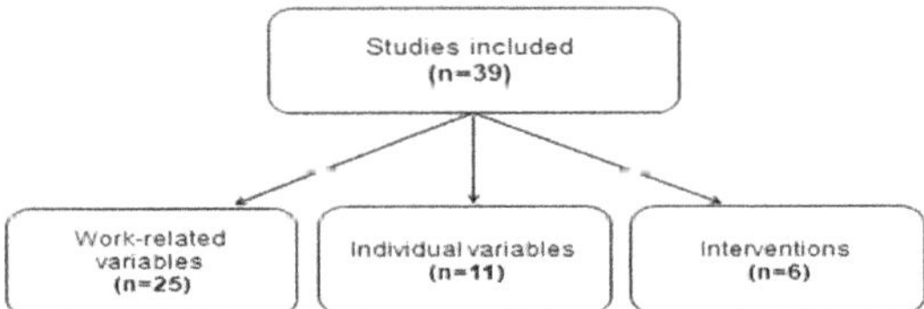

**Figure 2.** Flowchart of the allocation into the three categories: "Work-related Variables", "Individual Variables", "Interventions"

**Table 1.** Included study.

| Author | Sample-Size/sex | Sample | Country | Mean Age (SD) | Mean Experience (SD) | SOC Scale | Mean SOC (SD) | Study Design | Other Used Instruments | Sense of Coherence (SOC)-Related Results |
|---|---|---|---|---|---|---|---|---|---|---|
| Kowitlawkul et al., 2018 | 1040 nurses 955 females (91.8%) 69 males (6.6%) 16 missing (1.5%) | Registered nurses and enrolled nurses working in inpatient and outpatient departments at a tertiary hospital (in the hospital for ≥6 months) | Singapore | 30.6 (8.5) | – | SOC-13 | 50/50 group: 55.7 (9.5) 60/40 group: 56.0 (9.6) 70/30 group: 54.5 (9.2) 80/20 group: 52.4 (10.8) | Descriptive quantitative study | Work life balance (WLB) Job satisfaction questionnaire Social support questionnaire WHO-Quality of Life (WHO-QOL)-BREF-26 questionnaire | - Division in four groups depending on the proportion of time spent on work and private life (50/50 and below, 60/40, 70/30, 80/20 and above) -"50/50 and below" and "60/40" groups' SOC scores are higher the other two groups-lowest SOC scores are in the "80/20 and above" group -SOC and social support are significant predictors for all QoL domains -A unit increase in SOC results in a 6–12% increase in likelihood of having high QoL for all domains |
| Kuraoka et al., 2018 | 63 nurses 60 females (95.2%) 3 males (4.8%) | Nurse managers in the first 3 years of a supervisory role, working in acute-care hospitals Participation in an experiential learning-based program | Japan | 45.2 (3.4) | 1.4 (0.6) | SOC-13 | Pre-test: 57.17 (10.28) Post-test: 54.97 (10.4) | Quasi-experimental study | Experiential Learning Inventory on the Job (ELI) Ad hoc questionnaire for Knowledge of Experiential Learning Social Support Questionnaire (SSQ) Nurse Managers Competence Inventory (NMCI) | -Mean SOC scores reduction after participation to the experiential learning-based program |
| Schäfer et al., 2018 | 27 nurses 44 females (65.4%) 18 males (34.6%) | Nurses working in ICU and anesthesiology unit | Germany | 39 (10) | – | SOC-13 | 43.19 (9.63) | Cross sectional study | Resilience Scale (RS-11) Scale for assessment of internal and external control beliefs (IE-4) ICD 10 symptom rating (ISR) PTSD Checklist (PCL-5) | -No differences in resilience, nor SOC or LOC (locus of control) between physicians and nursing staff -SOC, resilience, and internal and external LOC correlate with ISR score and PTSD symptoms -SOC is a significant predictor of mental health problems and of symptom severity -Resilience, internal and external LOC account for 59% of SOC variance and have a significant indirect effects on symptom severity measures mediated by SOC on symptoms severity measures |

**Table 1.** *Cont.*

| Author | Sample-Size/sex | Sample | Country | Mean Age (SD) | Mean Experience (SD) | SOC Scale | Mean SOC (SD) | Study Design | Other Used Instruments | Sense of Coherence (SOC)-Related Results |
|---|---|---|---|---|---|---|---|---|---|---|
| Debska et al., 2017 | 164 nurses (sex unknown) | Nurses working at inpatient chemotherapy wards | Poland | 43.07 (7.99) | In chemotherapy ward: 11.77 (7.65) | SOC-29 | 125.05 (18.30) | Cross-sectional study | Meister Questionnaire | -Highest SOC scores for the Manageability subscale (45.15), lower for Comprehensibility and Meaningfulness subscales (41.18 and 38.73, respectively) - Inverse correlation between SOC and Monotony (r = −0.398, p < 0.001), Unspecific Load (r = −0.370, p < 0.001), - Mental Load (r = −0.378, p < 0.001) and Work experience (r = −0.19, p = 0.016) -No correlation between SOC and work-experience -Inverse correlation between age and Manageability level (r = −0.193, p = 0.014) -No correlation between overall SOC and age, nor educational level |
| Fusz et al., 2017 | 518 nurses 482 females (93.1%) 36 males (6.9%) | Nurses working in eight different hospitals | Hungary | 42.44 (9.59) | – | SOC -13 | 61.75 | Cross-sectional study | Ad hoc questionnaire for sense of quality, quality of sleep, frequency of psychosomatic symptoms, and work schedule regularity | -Higher SOC score for day-shift workers (65.84), compared to shift workers (61.02) (t = 2.933; p = 0.004) -Lower SOC score for irregular work schedule workers (58.19), compared to flexible work schedule workers (63.17, p = 0.04) -Lower scores in self-heath-assessment for irregular work schedule workers, compared to flexible work schedule workers (p = 0.019) |
| Stock et al., 2017 | 12 nurses 10 females (83.33%) 2 males (16.67%) | Nurses self-described as thriving, experienced (10+ y) or recently retired (<1 y) | Hawaii | 55–64 y: 33.33% | 10–21 y: 33% >34 y: 67% | SOC-13 | 73.53 | Mixed method, exploratory-descriptive study | Ad hoc questionnaire for thriving view and experience | Possible link between SOC and self-described thriving nurses Nurses have a high SOC and the use of GRRs |

**Table 1.** *Cont.*

| Author | Sample-Size/sex | Sample | Country | Mean Age (SD) | Mean Experience (SD) | SOC Scale | Mean SOC (SD) | Study Design | Other Used Instruments | Sense of Coherence (SOC)-Related Results |
|---|---|---|---|---|---|---|---|---|---|---|
| Ando et al., 2016 | 130 nurses 102 females (78.5%) 28 males (21.5%) | Nurses from psychiatric and internal medicine wards at a national hospital | Japan | 40–49 y: 35.4% | 18.7 | SOC-13 | 53.5 (9.7) | Cross-sectional study | Moral Distress Scale for Psychiatric Nurses (MDS-P) General Heath Questionnaire - 12 (GHQ-12) Job Satisfaction Scale (JS) | Inverse correlation between MDS-P and SOC Inverse correlation between MDS-P and JS "Unethical conduct by caregivers" negatively correlates with Manageability ($r = -0.28$, $p < 0.01$), Comprehensibility ($r = -0.22$, $p < 0.01$) and Meaning ($r = -0.017$, $p < 0.05$) "Low staffing" negatively correlated with Comprehensibility ($r = -0.30$, $p < 0.001$) and Manageability ($r = -0.22$, $p < 0.01$) "Acquiescence to patients' rights violations" negatively correlates with Manageability ($r = -0.31$, $p < 0.01$), Comprehensibility ($r = -0.28$, $p < 0.01$) and Meaning ($r = -0.22$, $p < 0,01$). "Acquiescence to patients' rights violations" (standard $\beta = -0,26$, $p < 0.01$) and "Meaning" of SOC (standard $\beta = 0.35$, $p < 0.001$) influence Job Satisfaction more than other variables. |
| Lindmark et al., 2016 | 165 nurses (sex unknown) | Dental nurses working in the Public Dental Service | Sweden | – | – | SOC-13 | 68.4 (11.0) | Cross-sectional study | Salutogenic Health Indicator Scale (SHIS) Work Experience Measurement Scale (WEMS) | Clinical coordinators have higher SOC score (75.1; SD 10.0), compared with all other professions, including dental nurses Dental hygienists have higher scores for meaningfulness (23.8; SD 3.3), and dentists have higher scores for manageability (22.1; SD 3.6), compared with dental nurses (22.5; SD 3.4 and 20.3; SD 4.0, respectively) |

**Table 1.** *Cont.*

| Author | Sample-Size/sex | Sample | Country | Mean Age (SD) | Mean Experience (SD) | SOC Scale | Mean SOC (SD) | Study Design | Other Used Instruments | Sense of Coherence (SOC)-Related Results |
|---|---|---|---|---|---|---|---|---|---|---|
| Vifladt et al., 2016 | 143 nurses 126 females (83.1%) 13 males (9.1%) 4 unknown (2.8%) | Nurses working in ICUs in six hospitals for ≥ 3 months | Norway | 41–50 y: 37.4% | ≥21 y: 26.1% | SOC-13 | 72.2 (8.96) | Cross-sectional study | Hospital Survey on Patient Safety Culture (HSOPSC) Bergen Burnout Indicator | Positive safety culture negatively correlates with burnout and positively correlates with SOC (r = −0.451, r = 0.393, respectively, $p <$ 0.001). Inverse correlation between SOC and burnout (r = −0.577, $p <$ 0.001) Positive correlation between SOC and safety culture at the unit level (β = 0.014, 95%CI = 0.005-0.024, $p =$ 0.003) and hospital level (β = 0.017, 95%CI = 0.008–0.026, $p <$ 0.001) |
| Yoshida et al., 2016 | 430 nurses 416 females (96.7%) 14 males (3.3%) | Public health nurses (after the earthquake, tsunami and Fukushima Daiichi Nuclear Power Station accident following the Great East Japan Earthquake in 2011) | Japan | ≥ 50 y: 35.6% | <10 y: 22.8% >10 y: 71.2% | SOC-13 | 43.0 (7.7) | Cross-sectional study | Ad hoc questionnaire for anxiety | Division in two groups based on anxiety score on a 10-point Likert scale: 1–5 'anxiety (−)' and 6–10 'anxiety (+)' No difference in mean SOC between the anxiety (+) and the anxiety (−) groups Higher ratio of nurses <40 years of age in the anxiety (+) group ($p <$ 0.001) Higher ratio of nurses with <10 years of working experience, staff positions and nursing licenses in the anxiety (+) group ($p <$ 0.001) |
| Hochwälder et al., 2015 | 1012 nurses (all females) | Nurses working at three hospitals and two primary health care settings, ≥ 30 y of age, with no negative life events during the previous year | Sweden | 46.6 (8.9) | – | SOC-29 | 150.21 (20.28) Low SOC (<144) group $n$ = 322): 126.35 (13.82) Moderate SOC (144–160) group $n$ = 348: 152.20 (4.84) High SOC (>160) group $n$ = 342: 170.63 (7.49) | Longitudinal study | Ad hoc questionnaire for Controllable and Uncontrollable Negative Life events | No significant age differences between Low, Moderate and High SOC groups Higher dropout rate in the Low SOC group No significant difference between SOC groups and mean number of uncontrolled negative life events High and Moderate SOC groups have fewer controllable negative life events compared to the Low SOC group ($p <$ 0.001) |

**Table 1.** *Cont.*

| Author | Sample-Size/sex | Sample | Country | Mean Age (SD) | Mean Experience (SD) | SOC Scale | Mean SOC (SD) | Study Design | Other Used Instruments | Sense of Coherence (SOC)-Related Results |
|---|---|---|---|---|---|---|---|---|---|---|
| Kretowicz et al., 2015 | 310 nurses 301 females (97.1%) 9 males (2.9%) | Nurse managers in selected medical units | Poland | 45.7 (6.7) | Mean nursing experience 24.7 (7.2) y Mean managerial experience 8.8 (6.7) y | SOC -29 | 147.00 (20.47) | Cross-sectional study | – | No correlation between age and SOC Correlation between educational background and SOC (Kruskal-Wallis test 9.04; $p = 0.029$) and Meaningfulness (Kruskal-Wallis test 12.82; $p = 0.005$) No correlation between SOC and general working experience, managerial working experience, position at work and characteristics of employment Higher SOC in nurses employed as strategic managers (U Mann-Wallis test -2.74; $p = 0.006$) |
| Makabe et al., 2015 | 1202 nurses 1116 females (93%) 86 males (7%) | Nurses working in three hospitals | Japan | 37 (11) | 15 (12) | SOC-13 | 50/50 group (A): 55.7 (9.5) 60/40 group (B): 56.0 (9.6) 70/30 group (C): 54.5 (9.2) 80/20 group (D): 52.4 (10.8) | Cross sectional study | Work-Life Balance (WLB) Work-Life Balance Satisfaction Job Satisfaction Scale WHO Quality of Life (26-item) | Division in four groups depending on the proportions of percentages of time spent on work and private life - WLB status [50/50 and below (A), 60/40 (B), 70/30 (C), 80/20 and above (D)] Group A has a higher SOC score than all other groups (ANCOVA $p < 0.001$) Group D has a lower SOC score than all other groups (ANCOVA $p < 0.001$) |
| Miyata et al., 2015 | 1425 nurses 1333 females (94%) 92 males (6%) | Nurse staff ($n =$ 1248%–88%) and nurse managers ($n = 177$–12%) working in 10 hospitals | Japan | 35.5 (9.9) | 12.8 (9.5) | SOC-13 | median: 50 (IQR 45–55) | Cross-sectional study | Recognition behavior scale | No significant correlation between SOC and marital status Positive correlation between SOC and good mental health status (OR = 4.07, 95%CI = 2.53–6.53), good physical health status (OR = 1.08, 95% CI = 1.09–2.89), overall work experience (OR = 1.05, 95%CI= 1.04–1.07), with $p < 0.001$, and recognition behaviors by the nurse manager (OR = 1.02 (95% CI = 1.01–1.04), $p = 0.006$ |

**Table 1.** *Cont.*

| Author | Sample-Size/sex | Sample | Country | Mean Age (SD) | Mean Experience (SD) | SOC Scale | Mean SOC (SD) | Study Design | Other Used Instruments | Sense of Coherence (SOC)-Related Results |
|---|---|---|---|---|---|---|---|---|---|---|
| Kikuchi et al., 2014 | 347 nurses (all females) | Nurses working at a general hospital (intensive care, pediatrics, surgery, oncology, and emergency medicine) | Japan | 33.7 (9.2) | – | SOC-13 | 54.2 (11.9) | Cross-sectional study | Brief Job Stress Questionnaire (BJSQ) K6 short screening questionnaire Ten-item Personality Inventory (TIPI-J) | Age negatively correlates with depressive state ($r = -0.18$, $p = 0.00$) and positively to SOC ($r = 0.27$, $p = 0.00$) Inverse correlation between SOC and shift work ($r = -0.16$, $p = 0.00$), job rank ($r = -0.14$, $p = 0.01$), and overtime hours ($r = -0.33$, $p = 0.00$). SOC and depressive state correlate negatively with almost all job stressors and personality traits Inverse correlation between SOC and Neuroticism ($r = -0.49$, $p = 0.00$). Inverse correlation between SOC and depressive state ($r = -0.67$, $p = 0.00$ Positive correlation between SOC and job and life satisfaction ($r = 0.47$, $p = 0.00$) |
| Kikuchi et al., 2014 | 348 nurses (all females) | Nurses working at a general hospital (intensive care, pediatrics, surgery, oncology, and emergency medicine) | Japan | 34.4 (9.0) | – | SOC-29 | 124.4 (21.2) | Cross-sectional study | K6 short screening questionnaire Effort-reward imbalance (ERI) scale | SOC ($\beta = -0.46$, $p < 0.001$), over-commitment, effort-esteem ratio, and age significantly correlate with the depressive state Age correlates positively with SOC ($r = 0.12$, $p < 0.05$) |
| Sarid et al., 2012 | 36 nurses (sex unknown) | Nurses working both clinical and administrative roles in one major regional hospital, ≥ 5 y of experience Intervention group ($n = 20$): Participation in a Cognitive-Behavioral course Control group ($n = 16$) | Israel | 50.6 (10.7) | – | SOC-13 | Intervention group T1: 70.83 (7.67) T2: 75.05 (6.7) Control group T1: 72.07 (8.9) T2: 69.51 (7.64 | Pre-post test design, with control | Perceived stress scale (PSS) Profile of Mood states (POMS) | In the intervention group at T2 higher SOC, more vigor, less perceived stress, and less fatigue, compared to T1 At T1 no significant difference in mean SOC scores between the groups At T2 significant difference in mean SOC scores between the two groups $F(p) = 10.44$ ($p < 0.05$) |

**Table 1.** *Cont.*

| Author | Sample-Size/sex | Sample | Country | Mean Age (SD) | Mean Experience (SD) | SOC Scale | Mean SOC (SD) | Study Design | Other Used Instruments | Sense of Coherence (SOC)-Related Results |
|---|---|---|---|---|---|---|---|---|---|---|
| Ando et al., 2011 | 28 nurses (sex unknown) | Nurses working in geriatric wards, ≥ 20 y of experience, without severe mental problems Intervention group ($n$ = 15): participation in mindfulness based therapy sessions Control Group ($n$ = 13) | Japan | – | – | SOC-13 | Post-intervention Intervention group: 52 Control group: 53 | Pre-post test design, with control | General Health Questionnaire (GHQ) Functional Assessment of Chronic Illness Therapy (FACIT-sp) | - Significant decrease in GHQ after the intervention (improvement in general health) - Significant increase in SOC after the intervention, no change in the control group - Meaningfulness: higher scores and significant increase after intervention: Higher scores and significant increase after intervention, compared to comprehensibility and manageability - No effect on spirituality |
| Basinska et al., 2011 | 331 nurses (all females) | Nurses working shifts in three general care hospitals | Poland | 34.15 (6.61) | – | SOC-29 | 136.46 (21.43) | Descriptive quantitative study | Work Related Patterns of Behavior and Experience Questionnaire (AVEM) | Division in work related behavior type groups: type G-healthy, type S-frugal, type A-risk (overburdened), type B-burnout Positive correlation between SOC and healthy type G and the frugal type S ($r$ = 0.50, and $r$ = 0.20 respectively, $p < 0.001$) Inverse correlation between SOC and burnout type B and the overburdened type A ($r$ = −0.57 $p < 0.001$ and $r$ = −0.13 $p < 0.05$ respectively) SOC explains 28% of the variability of type B, 21% of type G and 7% of type S |
| Hochwälder et al., 2011 | 1012 nurses (all females) | Nurses working in hospitals and primary health care setting, > 30 y of age Group 0: no negative life events Group 1: ≥1 negative life event in the previous year | Sweden | 46.90 (8.85) | 46.9 (8.85) | SOC-29 | At T1: 150.21 (20.28) At T2: 150.89 (20.38) | Cross sectional study | Ad hoc questionnaire for Negative Life Events | SOC stable from T1 to T2 No strong evidence that negative life events lower SOC No evidence that negative life events lower SOC more in persons with an initially low or moderate SOC that in persons with an initially high SOC. SOC for those who experience a negative life event, is initially weaker, than those who do not experience any negative life events |

**Table 1.** *Cont.*

| Author | Sample-Size/sex | Sample | Country | Mean Age (SD) | Mean Experience (SD) | SOC Scale | Mean SOC (SD) | Study Design | Other Used Instruments | Sense of Coherence (SOC)-Related Results |
|---|---|---|---|---|---|---|---|---|---|---|
| Sarid et al., 2010 | 36 nurses (sex unknow) | Nurses working in all hospital wards, > 5 y of experience, both clinical and administrative roles Study group (n = 20) participation in a CBI course Control group (n = 16) | Israel | 50.6 (10.7) | – | SOC-13 | – | Pre-post test design | Perceived Stress Scale (PSS) Profile of Mood States (POMS) | No differences in SOC, perceived stress and mood states between the groups at T1 Intervention group at T2: increased levels of SOC and vigor and decreased levels of perceived stress and fatigue Mean changes in SOC were -4,20 (SD = 1.28) for the study group and 2,11 (SD = 1.46) for the control group, F = 10.44, $p < 0.05$ |
| Takeuchi et al., 2010 | 138 nurses (all females) | Nurses working at three hospitals, who are also mothers and/or wives | Japan | 36.2 (8.0) | 12.5 (8.3) | SOC-13 | 56.7 (9.9) | Descriptive quantitative study | Job Content Questionnaire (JCQ, 6 items) Cumulative fatigue symptoms index (CFSI-18) Center for Epidemiologic Studies Depression (CES-D) scale Work-to-Family conflicts scale | Inverse correlation between work-family conflict and SOC (β = −0.233, partial r = −0.340, $p < 0.01$) Inverse correlation between SOC and cumulative fatigue (β = −0.397, $p < 0.001$) and depression (β = −0.517, $p < 0.001$). The interaction of SOC and WFC influences depression (β = 0.214, $p < 0.05$), SOC has a buffering effect on WFC with respect to depression. |
| Ida et al., 2009 | 502 nurses (all females) | Nurses working at a major university hospital | Japan | 32.4 (9.9) | – | SOC-29 | – | Cross-sectional study | Job Content Questionnaire (JCQ-12) Ad hoc questionnaire for medical errors and nurses career levels | Positive correlation between Comprehensibility and professional experience (r = 0.125, $p < 0.001$) Positive correlation between SOC and workplace adaptability (r = 0.335, $p < 0.01$), and job satisfaction (r = 0.280, $p < 0.01$) Inverse correlation between SOC and organization environment (r = −0.611, $p < 0.01$) and health risk (r = −0.364, $p < 0.01$) SOC affects sickness-absence (OR = 0.982, 95%CI = 0.970–0.995) |

**Table 1.** *Cont.*

| Author | Sample-Size/sex | Sample | Country | Mean Age (SD) | Mean Experience (SD) | SOC Scale | Mean SOC (SD) | Study Design | Other Used Instruments | Sense of Coherence (SOC)-Related Results |
|---|---|---|---|---|---|---|---|---|---|---|
| Van der Colff et al., 2009 | 818 nurses 791 females (97.4%) 21 males (2.6%) 6 sex unknown | Nurses working in hospital wards, psychiatric wards, community/occupational services and nursing management. | South Africa | 40 | 19 | SOC-29 | 137.95 (20.46) | Cross sectional study | Nursing Stress Inventory (NSI) Coping Orientation for Problem Experienced Questionnaire (COPE) Maslach Burnout Inventory- Human Services Survey (MBIHSS) Utrecht Work Engagement Scale (UWES) | Correlation between SOC and burnout subscores. Emotional exhaustion (−0,49), Depersonalization (−0,47) and Personal accomplishment (0.34) ($p < 0.05$) Positive correlation between SOC and Engagement (0.42) Negative correlation between SOC and Nursing stress inventory (NSI) variables: Lack of organizational support (−0.23), job demands (−0.28) and nursing-specific demands (−0.15) ($p < 0.05$) SOC correlates with coping strategies: Approach coping (0.0), Seeking emotional/social support (0.18) and Turning to religion (0,11), Avoidance (−0.35) and Focus on and ventilation of emotions (−0.29) ($p < 0.05$) |
| Hall-Lord et al., 2006 | 71 nurses (sex not specified) | Nurses participating in an educational program | Sweden | 39.2 (8.05) | 16.7 (7.35) | SOC-13 | – | Cross-sectional study | Ad hoc questionnaire for assessment of pain and distress of patients Five-factor personality inventory (FFPI) | Patients' age and type of illness seems to influence nurses' assessments of pain and distress Nurses with high emotional stability and high SOC scores assess pain and distress for acute patients as less intense and assess it more intense for chronic patients |

**Table 1.** *Cont.*

| Author | Sample-Size/sex | Sample | Country | Mean Age (SD) | Mean Experience (SD) | SOC Scale | Mean SOC (SD) | Study Design | Other Used Instruments | Sense of Coherence (SOC)-Related Results |
|---|---|---|---|---|---|---|---|---|---|---|
| Engström et al., 2005 | 33 nurses 31 females (94%) 2 males (6%) | Nurses working in a residential home for persons with dementia Experimental group (n = 17): participation in IT support project Control group (n = 16) | Sweden | 41 | Experimental group -in nursing care 12 (8) -in dementia care 5 (3) Control group -in nursing care 15 (9) -as district nurse 9 (7) | SOC-13 | Intervention group -baseline: 72 (12) -at 6 months: 69 (11) -at 12 months: 75 (9) Control group -baseline: 68 (10) -at 6 months: 69 (9) -at 12 months: 65 (13) | Quasi-experimental non-equivalent groups design | Satisfaction with Work Questionaires (SWQ) Life Satisfaction Questionnaire (LSQ) | Perception of psychosocial job satisfaction and quality of care improve in the experimental group No significant within-subject effect for the total SOC scale and meaningfulness subscale, split on the experimental and control groups Interaction effect for the factors family relation, close friend relation (LSQ), total SOC scale and meaningfulness subscale |
| Höge et al., 2004 | 160 nurses (sex not specified) | Nurses working at two hospitals | Germany | – | – | SOC-13 | 68.1 (10.83) | Cross-sectional study | Work load screening TAA-KH-S Negative Affectivity Scale (NAS) Maslach Burnout Inventory (Emotional exhaustion subcategory) Irritation Strain questionnaire Short Form Health Survey (SF-12) | Inverse correlation between SOC and Negative Affectivity (r = −0.61) Inverse correlation between SOC and overall strain (rs = −0.33) and overall stressors (rs = −0.34) |
| Shimizu et al., 2004 | 285 nurses (all females) | Nurses working at a hospital Intervention group (n = 62): participation in an assertiveness training Reference group (n = 196) | Japan | Intervention group: 44 (7.1) Reference group: 27.8 (5.9) | – | SOC-13 | Intervention group T1: 57.4 (10.5) T2: ΔSOC 2.2 (8.3) Comparison group T1: 52.7 (9.4) T2: ΔSOC 1.5 (8.6) | Pre-post test design, with control | Rosemberg's self-esteem scale (SES) | -No significant difference between the ΔSOC of the intervention group and that of the reference group. -Improvement in SES at six months after the intervention |

**Table 1.** *Cont.*

| Author | Sample-Size/sex | Sample | Country | Mean Age (SD) | Mean Experience (SD) | SOC Scale | Mean SOC (SD) | Study Design | Other Used Instruments | Sense of Coherence (SOC)-Related Results |
|---|---|---|---|---|---|---|---|---|---|---|
| Cilliers et al., 2003 | 105 nurses (all females) | Nurses working in large hospitals, ≥ 5 y of experience | South Africa | range: 28-57 | | SOC-29 | 141.28 (16.44) | Cross-sectional study | Maslach Burnout Inventory (MBI) Personal Views Survey (HAR) Self-Control Schedule (LR) | Inverse correlation between burnout and salutogenic functioning (SOC, HAR and LR) Inverse correlation between SOC and emotional exhaustion (r = −0.21, $p < 0.01$) and depersonalization (r = −0.25, $p < 0.01$) Positive correlation between SOC and personal accomplishment (r = 0.35, $p < 0.001$) |
| Yam et al., 2003 | 29 nurses 28 females (96.5%) 1 male (3.4%) | Nurses working in critical care in public and private hospitals | Hong Kong | 32 | 7.6 | SOC-13 | – | Cross-sectional study | Critical Care Nursing stress scale (CCNSS) Perceived Stress Scale (PSS) | Inverse correlation between SOC and CCNSS (r = −0.20, $p = 0.30$) Inverse correlation between SOC and PSS (r = −0.64, $p < 0.001$) |
| Michael et al., 2001 | 233 nurses 225 females (96.56%) 8 males (3.43%) | Nurses working in operating suites in private and public hospitals | Australia | 41 | 11 years in operating suite | SOC-13 | 66.75 (9.77) | Mixed method trianglulated study | Ad hoc questionnaire for work-related traumatic events Ad hoc questionnaire for social support | nurses who did not experience a traumatic event had a higher SOC (t(231) = −3.12, $p < 0.005$) |
| Tselebis et al., 2001 | 79 nurses 62 females (78.5%) 17 males (21.5%) | Nurses working in general internal medicine, general surgery and respiratory medical wards in a major hospital | Greece | 37.7 (5.5) | 11.0 (6.3) | SOC-13 | 63.60 (11.70) | Cross-sectional study | Maslach Burnout Inventory (MBI) Beck's Depression Inventory (BDI) | No differences in SOC between sexes nor marital status Inverse correlation between SOC and BDI (r = −0.58, $p < 0.05$) Correlation between SOC and MBI categories, negative correlation with sentimental exhaustion (r = −0.55, $p < 0.05$) and depersonalization (r = −0.45, $p < 0.05$), positive correlation with personal achievement (r = 0.44, $p < 0.05$) |

**Table 1.** *Cont.*

| Author | Sample-Size/sex | Sample | Country | Mean Age (SD) | Mean Experience (SD) | SOC Scale | Mean SOC (SD) | Study Design | Other Used Instruments | Sense of Coherence (SOC)-Related Results |
|---|---|---|---|---|---|---|---|---|---|---|
| Levert et al., 2000 | 94 nurses 67 females (71.3%) 27 male (28.7%) | Nurses working in psychiatric units | South Africa | 39 | – | SOC-13 | 60.61 (12.42) | Cross-sectional study | Maslach Burnout Inventory (MBI) Work Load and Lack of Collegial Support. Role Conflict and Role Ambiguity | Correlation between SOC and MBI's components: emotional exhaustion r = 0.41 ($p < 0.0001$) and depersonalization r = 0.36 ($p > 0.001$) SOC (t = 4.48; p.OOO) and work load (t = 4.50; p.OOO) together explain the majority of the variance in emotional exhaustion (36.6%) SOC (t = 3.51; p.OOl) and work load (t = 2.61; p.Oll) together explain the majority of the variance in depersonalization (21.3%) |
| Berg et al., 1999 | 22 nurses 16 females (72.7%) 6 male (27.3%) | Nurses working in psychiatric ward, during a year of systematic clinical supervision | Sweden | 39.7 (7.1) | 13.5 (8.3) | SOC-29 | T1: 146.6 (21.2) T2: 153.6 (18.3) | Pre-post-test design | Creative Climate questionnaire (CCQ) Work-related strain inventory (WRSI) Satisfaction with nursing care and work (SNCW) | Not significant Improvement in SOC scores at T1 and T2 Inverse correlation between SOC and WRSI (r = −0.48, $p < 0.05$) Inverse correlation between SOC and SNCW, factor involvement (r = −0.46, $p < 0.05$) |
| Shiu et al., 1998 | 20 nurses (all females) | Public health nurses (females, ≥ 1 children, promoted to nursing officer or in charge of a center, unit or team Work perturbed by interruption signals | Hong Kong | 39 | – | SOC-29 | 135.75 (12.27) | Cross-sectional study | Experience sampling diary (ESD) | Positive correlation between SOC and perceived goal progress and perceived control Low SOC group (scores <136) has a lower positive affect [t(323)= - 6.79, p < 0.001] and higher negative affect [t(321.98)= 2.88, p < 0.005], than the high SOC group (≥136) in response to interruption signals. |

**Table 1.** *Cont.*

| Author | Sample-Size/sex | Sample | Country | Mean Age (SD) | Mean Experience (SD) | SOC Scale | Mean SOC (SD) | Study Design | Other Used Instruments | Sense of Coherence (SOC)-Related Results |
|---|---|---|---|---|---|---|---|---|---|---|
| Pålsson et al., 1996 | 33 nurses (all females) | District nurses working in 10 primary health care districts Supervisory group (n=21): participation in a training program Comparison group (*n* = 12) | Sweden | Supervisory group: 49.0 (7.1) Comparison group: 46.3 (8.3) | Supervisory group -in nursing care 24.0 (8.2) -as district nurse 16.1 (6.5) Comparison group -in nursing care 21.8 (6.9) -as district nurse 14.0 (7.3) | SOC-29 | Supervisory group T1: 148 (17.5) T2: 151 (16.6) Comparison group T1: 154 (13.6) T2: 153 (17.3) | Pre-post test design, with control | Karolinska Scales of Personality (KSP) Burnout Measure Empathy Construct Rating Scale (ECRS) | Burnout, empathy and SOC scores are concordant in respect to the KSP variables Inverse correlation between SOC and KSP variables: somatic anxiety ($r = -0.44$, $p < 0.05$), impulsiveness ($r = -0.40$, $p < 0.05$), monotony avoidance ($r = -0.42$, $p < 0.05$), detachment ($r = -0.44$, $p < 0.05$), hostility ($r = -0.37$, $p < 0.05$) and psychasthenia ($r = -0.40$, $p < 0.05$) Positive correlation between SOC and KSP variable, socialization ($r = 0.44$, $p < 0.01$) Inverse correlation between SOC and burnout ($r = -0.69$, $p < 0.001$) Positive correlation between SOC and empathy ($r = 0.76$, $p < 0.001$) No significant change in burnout, empathy and SOC over time within the groups nor between the groups at T1 or at T2 |
| Lewis 1994 | 49 nurses (all female) | Nurses working in dialyses units | USA | 39.6 | – | SOC-29 | 148.7 | Cross-sectional study | Perceived Stress Scale Nursing Stress Scale Coping Resources Inventory Myers-Briggs Type Indicator (MBTI) Maslach burnout inventory (MBI) | Inverse correlation between SOC and personal stress ($r = -0.715$) Inverse correlation between SOC and work stress ($r = -0.436$) Inverse correlation between SOC and burnout ($r = -0.395$) Positive correlation between SOC and total coping resources ($r = 0.667$) |

**Table 1.** *Cont.*

| Author | Sample-Size/sex | Sample | Country | Mean Age (SD) | Mean Experience (SD) | SOC Scale | Mean SOC (SD) | Study Design | Other Used Instruments | Sense of Coherence (SOC)-Related Results |
|---|---|---|---|---|---|---|---|---|---|---|
| Langius et al., 1992 | 57 nurses (all females) | Nurses participating in an in-house training program (total 5 groups) Sample 1 ($n = 35$: SOC through VAS format) Sample 2 ($n = 22$: SOC and SMI through VAS format | Sweden | Sample 1: 41 Sample 2: 43 | – | SOC-29 | Sample 1: 152 (17) Sample 2: 143 (17) | Cross-sectional study | Self-Motivation Inventory (SMI) | No differences in SOC scores means between the participating groups, who had questionnaire administered through different formats. Negative correlation between SOC and SMI ($r = -0,685$, $p < 0.001$) |
| Lewis 1991 | 238 nurses 224 females (94%) 14 males (6%) | Nurses working in dialysis units | USA | 36.1 | – | SOC-29 | 143.1 | Cross-sectional study | Nursing stress scale Maslach Burnout inventory (MBI) | Men show lower SOC scores than women Inverse correlation between SOC and overall stress ($r = -0.39$) Correlation between SOC and burnout subscales (emotional exhaustion $r = -0.57$, depersonalization $r = -0.54$, personal accomplishment $r = 0.53$) No relationship between SOC and age, marital status, educational level, position, years in nursing, number of patients or shift unit, shift length or hours worked per week. |

### 3.1. Work-Related Variables

#### 3.1.1. Job Characteristics

Debska et al. observed among the nurses highest SOC scores for the Manageability subscale (45.15), followed by the Comprehensibility and Meaningfulness subscales. They showed an inverse correlation between SOC and the dimensions of mental load investigated by the Meister questionnaire, such as Monotony, Unspecific Load and Mental Load [6]. The relationship between SOC and general working experience, position at work and employment characteristics was unclear, while some authors found no correlation [7,8], an inverse correlation between SOC and work experience was found by Debska et al. [6], in contrast Miyata et al. [9] observed a positive correlation. Among nurses there is a wide variety of work schedule such as regular, irregular, flexible, etc. Fusz et al. showed that day-shift workers had higher SOC score than shift workers, and that lower SOC value was found among irregular workers (58.19), compared to flexible work schedule workers [10], while Kikuchi et al. observed an inverse correlation between SOC and shift work, job rank, and overtime hours [11]. Several studies found differences of SOC between different professionals, because there was higher SOC in nurses employed as strategic managers [7], Lindmark et al. likewise showed that clinical coordinators have higher SOC score, compared with all other professions, for example dental hygienists have higher scores for meaningfulness, and dentists have higher scores for manageability, compared with dental nurses [12].

Ando et al. described the relationship between the moral distress for nurses and several job characteristics, such as job satisfaction, SOC and mental health, finding an inverse correlation between Moral Distress Scale for Psychiatric Nurses (MDS-P) and SOC. Inverse correlations were found between subscales of the MDS-P and those of SOC [13]. Positive correlation was observed between SOC and workplace adaptability [14], and job satisfaction [11,14,15]. Moreover, Ida et al. identified SOC as an important factor affecting sickness-absence [14].

Lastly planning effective pain and distress management is a crucial part of the nurses' profession. Hall-lard et al. found that patient's age and type of illness seems to influence nurses' assessments of pain and distress, nurses with high emotional stability and high SOC scores assess pain and distress for acute patients as less intense and assess it more intense for chronic patients [16].

#### 3.1.2. Work-Life Balance

Some authors, investigating Work-Life Balance as the proportions of percentages of time spent at work and private life (50/50 and below, 60/40, 70/30, 80/20 and above), reported significantly higher SOC scores in "50/50 and below" and "60/40" groups, whereas the lowest SOC scores were associated with the "80/20 and above" group [2,17].

As far as the Quality of Life (QoL) is concerned, the "50/50 and below" reported the higher scores for overall QoL and physical health, while the "80/20 and above" group the lowest in the overall QoL, in the physical health domain and in the environment domain. No significant differences among the four groups were observed in terms of social support, job satisfaction, and the psychological and social relationship domains of the QoL [2].

#### 3.1.3. Work Related Trauma

Michael et al. investigated the effect of social and personal resources at work, related to trauma. They observed that nurses who did not report a traumatic event had the strongest SOC. This could be due to some causes, nurses with strong SOC did not perceive an event as traumatic, or in contrast, traumatic events influence the SOC [18].

#### 3.1.4. Social Support

Social support and SOC were found to be significant predictors ($p < 0.05$) for all QoL domains. Indeed, a unit increase in SOC results in a 6–12% increase in the likelihood of having high QoL for all domains, however social support had more influence on nurses' QoL than their ability to cope with stress [2].

### 3.1.5. Stress and Burnout

Yam et al. analyzed SOC and perceived stress with a sample of critical care nurses, finding that SOC was a protective factor in relation to stress perceptions arising from the work environment [19].

Höge et al. investigated the possible impact of SOC and negative affectivity on the relationship between work stressors and strain. They found a strong correlation between SOC and negative affectivity [20].

Berg et al. [21] observed that Work-Related Strain Inventory (WRSI), measuring the feeling of psychological strain in occupational setting, and factor involvement of the Satisfaction with Nursing Care and Work (SNCW) scale, negatively related to SOC.

Several studies negatively correlated SOC with overall stress [8,20] and work-related stress [8], especially workload [22]; in these studies, nurses' overload in the workplace was identified as a critical factor for stress development. Burnout and SOC were found to inversely correlate in several studies [22–25].

Moreover, burnout subscales were observed to logically relate to SOC. A stronger coping ability is associated with higher scores in personal accomplishment, lower levels of emotional exhaustion, and depersonalization [8,15,26,27]. Workload was considered a major contributing factor for burnout [8].

### *3.2. Individual Variables*

### 3.2.1. Individual Characteristics

Five studies investigated the correlation between SOC and age, with discordant results: two studies [11,28] revealed a positive association between SOC and age of the participants, whereas another three studies [7,8,29] did not find this relationship significant. Nevertheless, in the study conducted by Debska et al., although no significant correlation was found between total SOC and age, an inverse association between age and Manageability subscale was observed [6].

Although one study did not find any correlation with sex of the participants [26], an earlier study by Lewis [8] observed a stronger SOC in women, compared to men.

SOC was associated to marital status in the study conducted by Tselebis et al. [26], whereas the same correlation was not found in other studies [8,9].

Educational background, considered by Kretowicz et al. was found to be positively associated to overall SOC and Meaningfulness [7]. Two studies by Debska et al. and Lewis et al. have not proven the same correlation [6,8].

### 3.2.2. Individual Physical and Mental Health

The relationship between SOC and nurses' health was the focus of several studies. Miyata et at. associated positively SOC with good mental health status and good physical health status [9].

Schäfer et al. observed a significant increase from the cut-off value of nurses' scores in ICD-10-Symptoms Rating (ISR), evaluating general health problems, as well as symptoms burden, depression and eating disorder symptoms. Moreover, when compared to physicians, nurses reported higher ISR and symptoms burden scores, the same was not found for variables such as Resilience, SOC or LOC (Locus of Control). Furthermore, SOC, Resilience, and Internal and External LOC correlated with ISR scores and Post-Traumatic Stress Disorder (PTSD) symptoms, correlating SOC as a significant predictor of mental health problems and of symptom severity [30].

An inverse correlation was found between health risk and SOC, the latter significantly affecting sickness-absences, especially for experienced and expert nurses, for whom it is the only casual factor, among the other investigated variables [14].

Depression and SOC have been found to negatively, and strongly, correlate in several studies [11, 26,28,31]. Takeuchi et al. also considered the interaction of SOC and work-family conflicts (WFC) on the degree of nurses' depression and pointed out the buffering effect of SOC against depression, resulting from WFC [31].

Moreover, an inverse correlation was found between SOC and personal stress [22] and cumulative fatigue [31].

### 3.2.3. Personality Traits and Characteristics

Van der Colff et al. showed that SOC was correlated positively with different coping strategies, evaluated through the Coping Orientation for Problem Experienced (COPE) questionnaire, namely Approach Coping (seeking emotional/social support) and Turning to Religion; the correlation was inverse for Avoidance and Focus on and ventilation of emotions [15].

Overall a higher SOC score was associated with stronger total coping resources [22], thriving and the use of GRRs [1] and greater self-motivation, measured by the Self-Motivation Inventory (SMI) [32].

SOC was positively related to perceived progress goal as well as perceived control, both related to the perception of characteristic tasks of the job and life activities in which nurses were involved, evaluated upon interruption signals. Such signals were found to have a lower positive affect and higher negative affect in nurses with lower SOC [33].

Few studies concentrated on the relationship between SOC and personality traits. Kikuchi et al. revealed that SOC had a strong correlation with almost all personality traits, the strongest being the one with Neuroticism [11]. Höge et al. underlined the same concordant relationship between SOC and Negative Affectivity [20]. Similarly, SOC was found to correlate to the Karolinska Scale of Personality (KSP): negatively to Impulsiveness, Monotony Avoidance, Detachment, Hostility and Psychasthenia, and positively to Socialization and Empathy [23].

The KSP variable "Somatic Anxiety" was inversely related to SOC [23], but in contrast, no differences in mean SOC between the Anxiety (+) and the Anxiety (−) groups were found by Yoshida et al. [34].

### 3.2.4. Negative Life Events

In two different studies Hochwälder et al. investigated the association of negative life events on nurses' SOC [29,35]. There was no strong evidence that negative life events lower SOC in the sample population, but those who experienced a negative life event had initially a weaker SOC, compared to those who did not experience any negative life events [35]. Although there was not a significant correlation between SOC and the number of uncontrolled negative life events, those with high and moderate SOC reported fewer controllable negative life events compared to individuals with low SOC [29].

### 3.3. Intervention Studies

A total of six studies analyzed the effect of an intervention on nurses' Sense of Coherence. Only two studies observed a significant improvement in the SOC scores [36,37]. In the first study, nurses participated in a modified version of the Mindfulness-Based Stress Reduction (MBSR) program, lasting two weeks. After the intervention, it was observed a significant decrease in GHQ and its subscales (Physical Symptoms, Anxiety/Sleep, Social Activities and Depression), indicating an overall improvement in general health. Furthermore, SOC increase was significant, compared to the control group, as it was the increase in the meaningfulness subscale score, compared to comprehensibility and manageability scores [36].

Sarid et al. investigated the effect of Cognitive-Behavioral Intervention (CBI), comprised of 16 meetings, once a week, on nurses' SOC. At baseline the two groups did not differ in respect to SOC, perceived stress and mood states [37,38]. At T2 (four months after the beginning, upon completion of the program), nurses of the intervention group scored higher in SOC and vigor scales, whereas reported decreased level of perceived stress and fatigue. Such changes were not reported in the control group.

Nurses in the study conducted by Shimizu participated in an Assertive Training program. Although no significant changes in SOC were reported, the effects of the intervention were appreciable as an improvement in Self-esteem scores in the sample analyzed [39].

Berg 1999 and Pålsson 1996 both investigated the outcomes of systematic clinical supervision strategies on nurses. The two studies did not report significant changes in SOC after the intervention [21,23].

Only one study observed a reduction of mean SOC scores of nurse managers in early years of their supervisory roles, after the participation in a four-month experiential learning-based program [40].

The effect of an IT support project on SOC was considered in one study: no significant within-subject effects for the total SOC scale and meaningfulness subscale was observed both in the group receiving the intervention and the control group. However, IT support improved the perception of psychosocial job satisfaction and the quality of care; in this sense the study showed a significant interaction effect for the family relation factors, close friend relation (LSQ), total SOC scale and meaningfulness subscale [41].

## 4. Discussion

The nursing profession is characterized by taking care of patients and their families, it is a factor increasing the mental and emotional burden, and for this reason nurses' Sense of Coherence needs to be strong enough to deal with several stressful working experiences. Among others, most of the strain experienced by nurses derives from heavy workload, unsatisfactory work environment and work conditions, deep emotional involvement in others, organizational structure, lack of resources, inter-professional conflicts and professional uncertainty [21].

The majority of nurses spend more time at work than on their private life and report significantly higher SOC scores for those whose percentages were proportionally lower, and the lowest scores were for nurses with higher percentages of time spent on working activities [2,17].

Nurses face moral distress and feel so powerless because of the management policy of institutions [13]. The crucial role of institutions in cooperating in the hospital management is also correlated to a positive perception of safety, which in turn is correlated with absence of burnout and a strong Sense of Coherence [24].

The raising of SOC and organization environment reduces sickness-absence. Improving comprehensibility by enriching professionalism, recovering meaningfulness and manageability through optimizing work-life balance and social support may also raise SOC.

SOC and social support were found to be significant predictors for all QoL domains. Social support had the most relevant influence on nurses' QoL and is considered as a buffer in the stressful situations of healthcare working environment to help the individuals to cope. Cultivating social support could indeed help the individuals to improve their coping abilities and their general health status [2].

Occupational stress is a major contributing factor to burnout [15]. This correlation is also supported by studies, showing that individuals with high burnout levels are expected to possess poor stress coping abilities, specifically in the manageability dimension of the Sense of Coherence, which was found to be related to emotional exhaustion of burnout [15]. Burnout is defined as a syndrome of emotional exhaustion, depersonalization and decreased sense of self-achievement, unfortunately, occupational burnout affects a considerable proportion of nurses who face daily stress experienced at work [25].

Despite the relative stability of SOC after the third decade, it may be shaped progressively throughout the whole course of someone's life and the GRRs [42], which are mobilized by the Sense of Coherence, arise from the cultural, social and environmental conditions of living, in addition to idiosyncratic factors [43].

This concept could explain why the analysis of the correlation between SOC and individual characteristics (age, sex, marital status and educational background), taken into account only in few studies, yield discordant results.

An interesting point was explored by Kretowicz et al., who correlated positively SOC and educational background: as SOC is considered to have an educational value and the progress in the academic education could elevate it, it is reasonable to think that this relationship could influence task completion in managerial positions [7].

Furthermore, Antonovsky did not exclude a possible influence of negative life events on SOC, especially for those with low or moderate SOC [42]. Starting from this assumption, Hochwälder et al. in two different studies [29,35] have investigated this relationship: no association was found between negative life events and nurses' SOC, however those who experienced a negative life event had initially a weaker SOC, compared to those who did not experience any [35]. This result led the authors to consider a low SOC as a vulnerability factor, rather than considering a high SOC as a protective factor. This finding is in accordance with Antonovsky's assumption that a high SOC could prevent the experience of negative life events, both helping individuals to avoid potential stressors and not allowing them to perceive them invariably as negative [4].

A strong SOC is believed to be related to general well-being [15]. This relationship was confirmed by Miyata et al., who demonstrated in nurses a positive association between SOC and good mental health status and good physical health status [9].

Moreover, nurses working in hospitals reported, compared to the general population, higher burden of general health problems, as well as symptoms of burden, depression and eating disorders symptoms. SOC was found to be the most important predictor for general mental health problems and post-traumatic stress symptoms. SOC could play a crucial role in the development and course of these health issues, by shaping the perception and attitude toward aversive work experiences and stress [30].

The inverse relationship between SOC and health risk and the identification of SOC as a key determinant of sickness absences demonstrates how a poor coping ability, in the presence of powerful stressors, such as advanced career levels, could represent a health risk, due to a decreased ability to cope successfully with the stress. Nevertheless, Ida et al. advanced the possibility that raising SOC and the organization of the environment could produce a positive effect on sickness absences [14].

The inverse correlation between depression and SOC was strong in several studies [26,30,31].

Possession of a strong SOC allowed nurses to better manage occupational stress due to lack of organizational support and job demands, through the choice of appropriate coping strategies [15], to define themselves as thriving, with a positive use of GRRs [1] and to possess a greater self-motivation [32]. Self- motivation was found related to certain specific behavior attitudes (propensity toward physical activity and giving it value in respect to health) and the hypothesis of Langius et al. of a positive relation to SOC was confirmed by their investigation [32].

Only two studies investigated the correlation between anxiety and SOC. Yoshida et al. confronted two groups, divided based on anxiety presence, assessed by an ad hoc questionnaire: no differences were found among the two groups, possibly explained by the initially high SOC possessed by the group at hand [34]. Palsson et al.'s finding indicated that there is an inverse relationship between self-rated pathogenic anxiety and self-rated salutogenic Sense of Coherence [23].

Among the five studies investigating the effects of an intervention on the SOC, only two studies reported significant results. Stress coping strategies improvement, achieved by the MBSR therapy was demonstrated by a significant increase in SOC scores. Moreover, a significant increase in the Meaningfulness subscale of SOC indicated that, through the program, nurses were able to focus their attention on mind and body, allowing them to find meaning in their life and work activities [36].

The effects of CBI were significant in increasing SOC and vigor levels and in decreasing perceived stress and fatigue. CBI aims to raise the personal awareness on possible stress reaction, to learn how to self-talk in anxiety-producing situation, to gain a balance and awareness on perspective stressful events and to facilitate cognitive restructuring of stressful work situations. These intrinsic characteristics of the therapy explain the improvement in nurses' coping ability and the reduction of negative moods [37,38].

The only study reporting a decrease in mean SOC score investigated the changes produced by an experiential learning-based program. This result was explained by the overload experienced by nurse managers in early years of their supervisory roles when discussing their behavior and stressful situation encountered at work. Furthermore, it has been argued that SOC could possibly increase after an initial decrease, which was not evaluated, due to the short-term follow-up performed [40].

The other intervention studies did no show significant changes on SOC, these results are coherent with the initial description of SOC by Antonovsky, for whom SOC tends to remain stable in adulthood under normal circumstances and can be considered as a moderating factor on negative work environment variables [21]. Moreover, also the well-recognized difficulty in obtaining a significant SOC change in respect to high or low baseline is considered a determining factor of SOC stability in these studies [23].

## 5. Conclusions

SOC provides a solid theoretical basis for examining the organization of work [32].

Therefore it has been proposed that nursing management could focus on building a healthy work environment, which fosters SOC, rather than concentrating on resolving the effects of stress and its management at individual level [33].

Comprehensibility is improved by a clear view on roles and responsibilities and by open communication channels; sense of manageability is strengthened by appropriate workload and availability of resources [8,22]. Participation in decision making and the perspective of a clear career path are factors improving the sense of meaningfulness [8,33].

We found that SOC was a protective factor for depressive state, burnout, job dissatisfaction among female nurses, but there is no clear correlation with factors such as working experience or position at work. In addition, a higher SOC enhances a good mental and physical health status, acting as a health promoting resource, according to Antonovsky's theory [44,45].

**Author Contributions:** Conceptualization, G.M.M. and M.C.; Methodology, M.C. and S.P.; Formal Analysis, G.M.M. and M.C.; Investigation, E.A. and C.S.; Resources, M.C. and E.A.; Data Curation, S.P.; Writing—Original Draft Preparation, M.C.; Writing—Review & Editing, G.M.M. and M.C.; Visualization, G.M.M.; Supervision, G.M.M.; Project Administration, G.M.M. All authors have read and agreed to the published version of the manuscript.

**Funding:** This research received no external funding.

**Conflicts of Interest:** The authors declare no conflict of interest.

## References

1. Stock, E. Exploring salutogenesis as a concept of health and wellbeing in nurses who thrive professionally. *Br. J. Nurs.* **2017**, *26*, 238–241. [CrossRef]
2. Kowitlawkul, Y.; Yap, S.F.; Makabe, S.; Chan, S.; Takagai, J., Tam WW, S., Nurumal, M.S. Investigating nurses' quality of life and work-life balance statuses in Singapore. *Int. Nurs. Rev.* **2018**, *66*, 61–69. [CrossRef]
3. Ilić, I.M.; Arandjelović, M.Ž.; Jovanović, J.M.; Nešić, M.M. Relationships of Work-Related Psychosocial Risks, Stress, Individual Factors and Burnout—Questionnaire survey among emergency physicians and nurses. *Med. Pr. J.* **2017**, *68*, 167–178. [CrossRef]
4. Antonovky, A. *Health, Stress, and Coping*; Jossey-Bass: San Francisco, CA, USA, 1979.
5. Mittelmark, M.B.; Sagy, S.; Eriksson, M.; Bauer, G.F.; Pelikan, J.M.; Lindström, B.; Espnes, G.A. *The Handbook of Salutogenesis*; Springer: Berlin, Germany, 2017.
6. Hobfoll, S.E. Conservation of Resources: A New Attempt at Conceptualizing Stress. *Am. Psychol.* **1989**, *44*, 513–524. [CrossRef]
7. Dębska, G.; Pasek, M.; Wilczek-Rużyczka, E. Sense of coherence vs. mental load in nurses working at a chemotherapy ward. *Cent. Eur. J. Public Health* **2017**, *25*, 35–40. [CrossRef]

8.	Kretowicz, K.; Bieniaszewski, L. Determinants of sense of coherence among managerial nursing staff. *Ann. Agric. Environ. Med.* **2015**, *22*, 713–717. [CrossRef]

9.	Lewis, S.L.; Campbell, M.A.; Becktell, P.J.; Cooper, C.L.; Bonner, P.N.; Hunt, W.C. Work stress, burnout, and sense of coherence among dialysis nurses. *ANNA J.* **1992**, *19*, 545–554.

10.	Miyata, C.; Arai, H.; Suga, S. Characteristics of the nurse manager's recognition behavior and its relation to sense of coherence of staff nurses in Japan. *Collegian* **2015**, *22*, 9–17. [CrossRef] [PubMed]

11.	Fusz, K.; Tóth, Á.; Varga, B.; Rozmann, N.; Oláh, A.; Tudományegyetem, P. Different work schedules of nurses in Hungary and their effects on health. *Ideggyógyászati Szle. Clin. Neurosci.* **2017**, *70*, 136–139. [CrossRef] [PubMed]

12.	Kikuchi, Y.; Nakaya, M.; Ikeda, M.; Okuzumi, S.; Takeda, M.; Nishi, M. Sense of Coherence and Personality Traits Related to Depressive State. *Psychiatry J.* **2014**, *2014*, 738923. [CrossRef] [PubMed]

13.	Lindmark, U.; Wagman, P.; Rolander, B. Workplace health in dental care—A salutogenic approach. *Int. J. Dent. Hyg.* **2018**, *16*, 103–113. [CrossRef] [PubMed]

14.	Ando, M.; Kawano, M. Relationships among moral distress, sense of coherence, and job satisfaction. *Nurs. Ethics* **2018**, *25*, 571–579. [CrossRef] [PubMed]

15.	Ida, H.; Miura, M.; Komoda, M.; Yakura, N.; Mano, T.; Hamaguchi, T.; Yamazaki, Y.; Kato, K.; Yamauchi, K. Relationship between stress and performance in a Japanese nursing organization. *Int. J. Health Care Qual. Assur.* **2009**, *22*, 642–657. [CrossRef] [PubMed]

16.	Van der Colff, J.J.; Rothmann, S. Occupational stress, sense of coherence, coping, burnout and work engagement of registered nurses in South Africa. *SA J. Ind. Psychol.* **2009**, *35*, 1–10. [CrossRef]

17.	Hall-lord, M.L.; Larsson, B.W. Registered nurses' and student nurses' assessment of pain and distress related to specific patient and nurse characteristics. *Nurse Educ. Today* **2006**, *26*, 377–387. [CrossRef]

18.	Makabe, S.; Takagai, J.; Asanuma, Y.; Ohtomo, K.; Kimura, Y. Impact of work-life imbalance on job satisfaction and quality of life among hospital nurses in Japan. *Ind. Health* **2015**, *53*, 152–159. [CrossRef]

19.	Michael, R.; Jenkins, H.J. Recovery from work-related trauma by perioperative nurses. *Collegian* **2001**, *8*, 8–13. [CrossRef]

20.	Yam, B.M.C.; Shiu, A.T. Perceived stress and sense of coherence among critical care nurses in Hong Kong: A pilot study. *J. Clin. Nurs.* **2003**, *12*, 144–146. [CrossRef]

21.	Höge, T.; Büssing, A. The Impact of Sense of Coherence and Negative Affectivity on the Work Stressor-Strain Relationship. *J. Occup. Health Psychol.* **2004**, *9*, 195–205. Available online: http://doi.apa.org/getdoi.cfm?doi=10.1037/1076-8998.9.3.195 (accessed on 10 February 2020). [CrossRef]

22.	Berg, A.; Hallberg, I.R. Effects of systematic clinical supervision on psychiatric nurses' sense of coherence, creativity, work-related strain, job satisfaction and view of the effects from clinical supervision: A pre-post test design. *J. Psychiatr. Ment. Health Nurs.* **1999**, *6*, 371–381. [CrossRef]

23.	Lewis, S.L.; Bonner, P.N.; Campbell, M.A.; Cooper, C.L.; Willard, A. Personality, stress, coping, and sense of coherence among nephrology nurses in dialysis settings. *ANNA J.* **1994**, *21*, 325–335; discussion 336. [PubMed]

24.	Palsson, M.; Norberg, A.; Bjorvell, H. Burnout, Empathy and Sense of Coherence among Swedish District Nurses before and after Systematic Clinical Supervision. *Scand. J. Caring Sci.* **1996**, *10*, 19–26. [CrossRef] [PubMed]

25.	Vifladt, A.; Simonsen, B.O.; Lydersen, S.; Farup, P.G. The association between patient safety culture and burnout and sense of coherence: A cross-sectional study in restructured and not restructured intensive care units. *Intensive Crit. Care Nurs.* **2016**, *36*, 26–34. [CrossRef] [PubMed]

26.	Basinska, M.; Andruszkiewicz, A.; Grabowska, M. Nurses' sense of coherence and their work related patterns of behaviour. *Int. J. Occup. Med. Environ. Health* **2011**, *24*, 256–266. [CrossRef]

27.	Tselebis, A.; Moulou, A.; Ilias, I. Burnout versus depression and sense of coherence: Study of Greek nursing staff. *Nurs. Health Sci.* **2001**, *3*, 69–71. [CrossRef]

28.	Cilliers, F. Burnout and salutogenic functioning of nurses. *Curationis* **2003**, *26*, 62–74. [CrossRef]

29.	Kikuchi, Y.; Nakaya, M.; Ikeda, M.; Okuzumi, S.; Takeda, M.; Nishi, M. Relationship between depressive state, job stress, and sense of coherence among female nurses. *Indian J. Occup. Environ. Med.* **2014**, *18*, 32–35. [CrossRef]

30.	Hochwalder, J. Test of antonovsky's postulate: High sense of coherence helps people avoid negative life events 1. *Psychol. Rep.* **2015**, *116*, 363–376. [CrossRef]

31. Schäfer, S.K.; Lass-hennemann, J.; Groesdonk, H.; Volk, T.; Michael, T. Mental Health in Anesthesiology and ICU Staff: Sense of Coherence Matters. *Front. Psychiatry* **2018**, *9*, 440. [CrossRef]
32. Takeuchi, T.; Yamazaki, Y. Relationship between work—Family conflict and a sense of coherence among Japanese registered nurses. *Jpn. J. Nurs. Sci.* **2010**, *7*, 158–168. [CrossRef]
33. Langius, A.; Bjorvell, H.; Antonovsky, A. The Sense of Coherence Concept and its Relation to Personality Traits in Swedish Samples. *Scand. J. Caring Sci.* **1992**, *6*, 165–171. [CrossRef] [PubMed]
34. Shiu, A.T.Y. The Significance of Sense of Coherence for the Perceptions of Task Characteristics and Stress During Interruptions Amongst a Sample of Public Health Nurses in Hong Kong: Implications for Nursing Management. *Public Health Nurs.* **1998**, *15*, 273–280. [CrossRef] [PubMed]
35. Yoshida, K.; Orita, M.; Goto, A.; Kumagai, A.; Yasui, K.; Ohtsuru, A.; Hayashida, N.; Kudo, T.; Yamashita, S.; Takamura, N. Radiation-related anxiety among public health nurses in the Fukushima Prefecture after the accident at the Fukushima Daiichi Nuclear Power Station: A cross-sectional study. *BMJ Open* **2016**, *6*, e013564. [CrossRef] [PubMed]
36. Hochwa¨lder, J.; Forsell, Y. Is Sense of Coherence Lowered by Negative Life Events ? *J. Happiness Stud.* **2011**, *12*, 475–492. [CrossRef]
37. Ando, M.; Natsume, T.; Kukihara, H.; Shibata, H.; Ito, S. Efficacy of mindfulness-based meditation therapy on the sense of coherence and mental health of nurses. *Health* **2011**, *3*, 118–122. [CrossRef]
38. Sarid, O.; Berger, R.; Segal-engelchin, D. The impact of cognitive behavioral interventions on SOC, perceived stress and mood states of nurses. *Procedia. Soc. Behav. Sci.* **2010**, *2*, 928–932. Available online: http://dx.doi.org/10.1016/j.sbspro.2010.03.128 (accessed on 10 February 2020). [CrossRef]
39. Orly, S.; Rivka, B.; Rivka, E.; Dorit, S. Are cognitive—Behavioral interventions effective in reducing occupational stress among nurses? *Appl. Nurs. Res.* **2012**, *25*, 152–157. [CrossRef]
40. Shimizu, T.; Kubota, S.; Mishima, N.; Nagata, S. Relationship between Self-Esteem and Assertiveness Training among Japanese. *J. Occup. Health* **2004**, *46*, 296–298. [CrossRef]
41. Kuraoka, Y. Effect of an experiential learning—Based programme to foster competence among nurse managers. *J. Nurs. Manag.* **2018**, *26*, 1015–1023. [CrossRef]
42. Engström, M.; Ljunggren, B.; Lindqvist, R.; Carlsson, M. Staff perceptions of job satisfaction and life situation before and 6 and 12 months after increased information technology support in dementia care. *J. Telemed. Telecare* **2005**, *11*, 304–309. [CrossRef]
43. Antonovsky, A. *Unraveling the Mystery of Health. How People Manage Stress and Stay Well*; Jossey-Bass: San Francisco, CA, USA, 1987.
44. Eriksson, M.; Lindstrom, B. Validity of Antonovsky's sense of coherence scale: A systematic review. *J. Epidemiol. Community Health* **2005**, *59*, 460–466. [CrossRef] [PubMed]
45. Eriksson, M.; Lindström, B. Antonovsky's sense of coherence scale and the relation with health: A systematic review. *J. Epidemiol. Community Health* **2006**, *60*, 376–381. [CrossRef] [PubMed]

International Journal of
*Environmental Research
and Public Health*

*Article*

# Listening to Hospital Personnel's Narratives during the COVID-19 Outbreak

**Shir Daphna-Tekoah** [1,2,*], **Talia Megadasi Brikman** [1], **Eric Scheier** [1] and **Uri Balla** [1,3]

[1] Kaplan Medical Center, Rehovot 7610000, Israel; drtaliamb1@clalit.org.il (T.M.B.);
Eric.scheier@gmail.com (E.S.); uriballa@gmail.com (U.B.)

[2] Faculty of Social Work, Ashkelon Academic College, Ashkelon 78211, Israel

[3] Faculty of Medicine, The Hebrew University of Jerusalem, Jerusalem 91905, Israel

* Correspondence: shir.dt@gmail.com

Received: 20 July 2020; Accepted: 27 August 2020; Published: 3 September 2020

**Abstract:** Healthcare workers (HCWs) facing the COVID-19 pandemic are required to deal with unexpectedly traumatic situations, concern about contamination, and mounting patient deaths. As a means to address the changing needs of our hospital's HCWs, we conducted a narrative analysis study in the early stages of the covid-19 outbreak. A focus group of medical experts, conducted as the initial step, recommended that a bottom-up research tool be used for exploring HCWs' traumatic experiences and needs. We therefore conducted 450 semi-structured in-depth interviews with hospital personnel. The interviews were based on Maslow's Pyramid of Needs model, and the narratives were analyzed by applying the Listening Guide methodology. The interviewees expressed a need for physical and psychological security in the battle against Covid-19, in addition to the need for attachment and meaning. Importantly, we also found that the interview itself may serve as a therapeutic tool. In light of our findings, we recommended changes in hospital practices, which were subsequently implemented. Further research on HCWs' traumatic experiences and needs will provide evidence-based knowledge and may enable novel approaches in the battle against Covid-19. To conclude, the knowledge generated by listening to HCWs' narratives may provide suitable support programs for professionals.

**Keywords:** COVID-19 pandemic; healthcare professionals; first responders; listening guide

---

## 1. Introduction

### 1.1. The Setting—COVID-19 in Hospitals

> The moment that I was informed that we had become a COVID-19 department, I was devastated. This coronavirus is so frightening, and I knew that I could die from it. I am a person who needs to be in control, and I had lost control, I was so frightened. This entire new situation was scary—a situation of life or death. Moreover, I was in it. At the level of the team, we did not know what to expect, personally and collectively, as a department. I did not know what was expected from me as a social worker and what were the guidelines; everything was new. We created everything from the beginning, and I was scared.
>
> *Emma, a social worker in the hospital's Corona Department*

Studies on outbreaks of infectious diseases reveal the profound and broad-spectrum psychological impacts that disease outbreaks can inflict on healthcare professionals [1,2]. Experience with the SARS and Ebola virus outbreaks suggests that healthcare professionals are subject to extremely high levels of stress and emotional turbulence [3–5]. In dealing with the SARS virus, which is similar in some respects to the 2019 novel coronavirus (SARS-CoV-2), healthcare professionals were troubled by intrusive thoughts and images associated with SARS and exhibited symptoms of PTSD and substantial psychological distress [6], if not mental illness [7]. In addition, healthcare professionals feared that they would fall ill from SARS but were equally or more worried about infecting family members and other people [8]. Nonetheless, prior experience with pandemics, disasters, and major traumatic events has indicated that enhanced support for healthcare professionals enabled them to remain efficient and focused during these stressful events [9].

The on-the-job stress and burnout of healthcare workers (HCWs) employed in hospitals [10–12] is one of the key themes in studies of the trauma of these professionals in the corpus of literature on the cost of caring [13]. This recurring motif is not surprising, since the hospital environment is challenging in that it frequently demands holistic treatment of patients that integrates psychological elements with physical treatment: In the course of their routine hospital work, HCWs are required to deal with patients and their families who have experienced traumatic events and life-threatening episodes [14].

It was against this background that hospital HCWs the world over first encountered the global COVID-19 pandemic in December 2019 [15]. Facing this global public health event situated HCWs in an unbearable situation, with them being required to function under extreme physical and psychological pressure on both the professional and personal levels [16]. In parallel, they were often required to make impossible decisions, including how to deal with limited equipment, how to balance their own physical and mental welfare needs with those of their patients, and how to bring into line their responsibility to their patients with their responsibility to their family and friends. In addition, their wish to provide optimal care for severely ill patients was often constrained by inadequate resources. Thus, in the COVID-19 pandemic, frontline HCWs had to work under particularly intense stress levels in unprecedentedly difficult situations [17]—situations that may indeed cause stress, moral injury, and physical and mental health problems [4,17].

The unfolding pandemic has been compared to war, as described in "'The Art of War' in the Era of Coronavirus Disease 2019 (COVID-19)" by Maxwell et al. [18], who claim that the image of war is often used in the field of infectious diseases. Indeed, from the beginning of the pandemic, HCWs in the hospital setting have been faced with caring for patients with an incredibly contagious and life-threatening disease about which nothing was known and for which there was no known lifesaving treatment—a situation similar to a war. They were—and still are—handling life-and-death situations while simultaneously putting their own lives at risk. This factor has contributed to a real sense of danger among hospital staff, who have found themselves in the forefront of defense against the pandemic [19]. In parallel, HCWs are being required to deal with emerging challenges [20]. They are often required to develop, in an extremely short time, novel concepts and new interventions for unpredictable situations. The need for HCWs to be proactive results from the fast-growing numbers of critically ill patients, the lack of treatment modalities, and shortages of critical medical resources and staff. Like the general population, the hospital staff is struggling with the emotional stressors imposed by the pandemic. However, they are also faced with additional stressors and a rapidly evolving work environment that differs significantly from their pre-COVID routines [21,22].

Marchand-Senécal et al. [23] point out that specialized, dedicated COVID-19 teams could quickly be overwhelmed as numbers of cases increase markedly. They also hold that longer shifts and increased work intensity may lead to HCW fatigue and lapses in the use of the correct techniques for handling personal protective equipment (PPE). They illustrate this conclusion by citing initial reports indicating that about 4% of Chinese HCWs caring for COVID-19 patients were infected, with 15% of those HCWs being classified as severe or critical cases. It is thus not surprising that a survey of 1257 frontline nurses, physicians, and other HCWs who treated COVID-19 patients in hospitals in China found that the participants carried a psychological burden, with symptoms related to depression, anxiety, and distress [1]. Earlier studies on epidemic outbreaks have indeed revealed that medical personnel, particularly first responders, including physicians, nurses, ambulance personnel, and other HCWs, become emotionally affected and traumatized and display heightened stress and higher levels of depression and anxiety [24,25]. These findings are to be expected, since anxiety and the fear of being infected are aggravated as the risk of exposure is elevated. This heightened anxiety is exacerbated even further by the fear of transmission of the infection to loved ones. The need to maintain a sense of balance between professional duty, altruism, and personal fear for oneself and others thus often causes a mental conflict for HCWs [1]. In the guidelines published on 19 March 2020 by the World Health Organization, it was declared that, by virtue of their caring for and close contact with COVID-19 patients, the people most at risk of acquiring the disease are HCWs and that protecting HCWs is of paramount importance.

In light of the above, at the beginning of the COVID-19 outbreak in Israel, we arranged a meeting in our hospital of an ad-hoc group of experts from different disciplines in healthcare and with different types of expertise. The recommendations of this focus group and the results of the interviews that were constructed and conducted as an outcome of these recommendations are described below.

*1.2. Theoretical Framework*

The theory of how humans usually function in daily life, as embodied in the ideas of Abraham Maslow, appear to be remarkably relevant to massive crises, especially to the present global crisis resulting from the COVID-19 pandemic [26]. Maslow's theory establishes a hierarchy of human needs [27] and, as such, provides a framework to describe the needs of hospital personnel [28,29]. Maslow's theory divides human needs into five categories. The first category, forming the base of a pyramid of needs, comprises physiological needs, such as air, water, food, shelter, sleep, and clothing. For medical personnel, Hale and his colleges [28] extend this level to include the basic determinants of good physical and mental health and safety. The first category is followed, in order, by four more layers: safety needs, such as personal security, employment, resources, health, and property; love and belonging, which includes friendship, intimacy, family, and a sense of connection; esteem and respect, which includes self-esteem, status, recognition, strength, and freedom; and, finally, at the top of the pyramid, the desire to become the best that one can be, relates to personal growth [30,31]. Maslow described each level as a separate need that relies on the previous need. However, modern-day theorists have modified this conceptualization into an overall concept in which each need coexists with the others [28]. We used semi-structured interviews based on Maslow's pyramid to survey HCWs in our hospital. We then analyzed the interviews by choosing, from the umbrella of narrative analysis, the Listening Guide methodology to analyze the recorded interviews and their transcripts [32,33]. This qualitative research methodology was developed as an alternative analysis to conventional coding schemes used to analyze qualitative data [33]. It differs from other means of analysis in that it places emphasis on the psychological complexities of people through attention to voice as a manifestation of the psyche. The Listening Guide in its attention to voice—and silence—thus provides a way of exploring the interplay of inner and outer worlds and of bringing the inner world out into the open [33–35]. By focusing on different voices, on the dynamics and interplay of these voices within the interview transcript, and on the socio-cultural setting of the research, the Guide establishes a contextual framework for understanding and/or interpreting the narratives of the interviewees and thereby facilitates psychological discovery. The details of methodology are described in the Materials and Methods section [36].

Although every analytical process has its advantages and drawbacks, we chose the Listening Guide, since its intent is to capture the layers of perception and experiences of trauma and stress [37,38] that might otherwise remain unnoticed, thereby broadening the understanding of traumatic situations. By applying this methodology, we aimed to expand knowledge—and to generate new knowledge—regarding the mechanisms and the strategies used by hospital workers to cope with the COVID-19 pandemic. Specifically, we sought to reveal the overt and covert voices [39] emerging from the experiences of front-line workers and to examine how these workers describe and experience their traumas in their struggle to treat patients with COVID-19.

## 2. Materials and Methods

As mentioned above, being aware of the necessity to address the needs of the hospital's HCWs, we conducted a study aimed to expand the knowledge of these needs during the first phase of the covid-19 outbreak. Specifically, the goal of the study was to enable the contrapuntal voices emerging from the hospital staff's experiences to be "heard" and thereby to provide recommendations as to how to meet their changing needs as the crisis unfolded. By paying close attention the narratives of the hospital staff, we were able to address an additional aim, namely, to initiate the establishment of a data-based foundation for both immediate and future interventions, thereby expanding knowledge regarding the psychological mechanisms and strategies that front-line personnel use to cope with exposure to traumatic situations.

The first step of the research comprised the establishment of a focus group of HCWs, all trauma experts, in the hospital. On 18 March 2020, the focus group met to exchange thoughts about the evolving crisis conditions in the hospital and to find the means to evaluate—and indeed alleviate—the situation. The focus group comprised an Emergency Department physician and five medical-social workers with specializations in mental health and trauma. The decision to constitute a focus group of experts as the opening stage for this research project derived from the perspective that the novel coronavirus poses an unmet challenge to medical treatment and hence challenges HCWs in many unknown ways. The focus group was charged with delineating the new situation in the hospital and with deciding on the exact research methodology and research tool(s) that could be applied in a research project and possibly later as adjunct practical tools for dealing with the evolving crisis. The final question put to the focus group was: What is the next step in the research project?" The participants pointed out the need to conduct bottom-up interviews at all levels and sectors in the hospital as a means of understanding the emerging needs of the hospital personnel [40]. The focus group suggested conducting a survey based on semi-structured in-depth interview based on Maslow's Pyramid of Needs' model. The survey and results are described below.

The interview protocol involved questions designed to capture the HCWs' ways of describing their unique experiences. To this end, the interview protocol comprised a standard set of questions, beginning with a request to share with the interviewer thoughts on what it meant to be an HCW in a hospital at the time of the corona crisis. This opening question was followed by encouragement to share thoughts on personal needs. More specific questions were used to clarify the stories as the interviews proceeded, such as: What do you need—physically or anything else? What distresses you at work? What would make you feel safer? What helps you to feel better? What helps you to know that you are appreciated? What motivates you to get up every morning for work? In a year from now, what do you think will have changed in your family, at the hospital, within yourself? Do you have something that you think important to add to the body of knowledge specifically about coronavirus or about trauma situations in general?

Four hundred and fifty semi-structured interviews were conducted in three waves: the first in the middle of March 2020, before recognition of COVID-19 as a global pandemic (163 staff members; 36.2% of the interviewees); the second, two weeks later (157; 34.9%); and the third in the middle of May 2020, at the end of Israel's national lockdown (130; 28.9%). Interviews were conducted with personnel [87 (20.2%) men and 344 (79.8%) women] serving a variety of functions in the hospital: physicians, nurses, pharmacists, respiratory therapists, department supervisors, laboratory technicians, social workers, and administrative workers from various sectors and with different levels of seniority (average 16.6 years; median 15 years, range:0-50) The breakdown of sectors and departments is given in Table 1. The interviews each lasted approximately 20 min to 1 h and were conducted under conditions of assured confidentiality.

As we indicated above, the interview narratives (in audio form and as transcripts) were analyzed by applying Gilligan's Listening Guide methodology [41], which is comprised of four stages, as follows [37,42].

1.  In the first stage, "Listening to the Plot," attention is paid to the whole story of the interviewee. The researchers' goal in this stage is to analyze the story in its context, similar to the analysis of an unfolding plot of a novel. The researchers identify recurring images and words, key metaphors, and dominant themes. The Guide also requires that the researchers document their own reflexive emotional and intellectual responses, thoughts, and feelings, as a means to better recognizing how their responses to the interviewee might affect their understanding of the narrative and the subsequent analysis. This stage is similar to the analysis modes in several of the qualitative thematic methods described in the literature [43].

2.  The second stage, "I Poems," is unique to the Listening Guide method. The second-stage listening and transcript analysis follows the use of the first-person pronoun "I." Within a passage in the transcript, scholars underline every use of "I" together with the attendant verb and any seemingly important accompanying words and then paste these "I Voice" phrases together to compose an "I Poem." This composite traces how the interviewee views herself/himself and the most prominent themes that preoccupy her/him.

3.  The third step, "Listening for Contrapuntal Voices," concentrates on how the interviewee talks about her/his or relationships with others. In this phase, scholars identify the multiple aspects of the story being told, often in multiple voices, with each voice (e.g., "You Voice," "The Voice of Trauma and Stress"; see below) being underlined in a different color. The transcript thus provides a visual way of examining how the different voices change in relation to one another.

4.  In the fourth and final step, "Composing an Analysis," an interpretation of the interviews is developed that synthesizes what has been learned during the entire process by assembling the evidence drawn from the different instances of listening as the basis for composing the analysis. A summary analysis is then constructed [42,43].

The reported study conformed to internationally accepted ethical guidelines and relevant professional ethical guidelines and was approved by the institutional review board (IRB) of Kaplan Medical Center, Rehovot, Israel. To assure confidentiality, each participant was identified by a pseudonym.

**Table 1.** Distribution of the interviewees according to sector and Department.

| Sector | Medicine | Nursing | Admin | Paramed * | Other * | |
|---|---|---|---|---|---|---|
| N = 433 | 72 (16.6%) | 169 (39%) | 41 (9.5%) | 26 (6%) | 125 (28.9%) | |
| Dept. | Corona | Internal med | Surgery | OBGYN ** | Pediatrics | Other ** |
| N = 381 | 50 (13.1%) | 105 (27.6%) | 13 (3.4%) | 54 (14.2%) | 34 (8.9%) | 125 (32.8%) |

Note: Values in the table are number (%) of interviewees that answered the specific question. The numbers do not add up to 450, because, in some cases, interviewees refrained from answering. * Paramed includes social workers, diotitians, physiotherapists, etc. Other includes kitchen workers, pharmacists, security personnel, housekeeping workers, etc. ** Department—OBGYN—Obstetrics and Gynecology; other includes hospital kitchen, pharmacy, security, housekeeping, administration, data and computing, etc.

## 3. Results

### 3.1. The First Step: "Listening to the Plot"

The narratives of the healthcare professionals covered a description of their experiences, explaining in detail their engagement with corona patients and their families and their relationships with members of the hospital staff. Two main themes emerged from the analysis of listening to the plot, the first of which was *preparing for war* in that the participants compared the situation in the hospital to the experience of preparing for war. Sara, for example, said: *"In our country, we know what a war is, and in the healthcare system we know how to function in the hospital during times of war, but still, this is a new war, a war that we have never handled, an invisible enemy, and it is frightening all of us."* In similar vein, Doron said: *"to be significant, to be at the front is important. Before it was the army that was at the front, now it is the turn of the healthcare system to be at the front."*

The second theme to emerge was that of *security and insecurity*, which related to two main aspects of the situation—fear of contamination and uncertainties derived from the assignment of HCWs to new teams and departments whose purpose and essence were unfamiliar. The idea of working in new and unfamiliar teams distressed the personnel, and their narratives reflected their feelings of insecurity. The changing reality was reflected by Dikla a nurse: *"In the Internal Medicine Department, I have been working for the past 18 years with my team, physicians, nurses, secretary—we have a common language. I felt especially secure in those days. How I will be able to use, in an efficient way, a new situation and new staff? This is ridiculous."* Similarly, Tania, a social worker, said: *"This will increase the feeling of insecurity . . . think that the entire situation is new and scary; so, what will I do without my friends who I have been working with for years?"* In particular, the need to be protected during shifts was pronounced. As Sara told us: *"In order to continue to come here, I need to feel that someone is taking care of me. I do not care who in charge of that in the hospital, but I need to feel safe; it is essential for me."*

According to the Listening Guide protocol, at this stage of the analysis, the interviewer should document his/her own reflexive emotional responses. We found that listening to the interviewees and reading the transcripts induced emotional concern for and feelings of empathy with our colleagues who were struggling with unbearable situations and ethical dilemmas. We were full of tears when we heard about the burdens of caring for patients and the cost of that caring. The interviews and the transcript readings thus gave rise to a range of emotions, including sadness, anger, compassion, and frustration, but also to pride in the devoted teams and to an appreciation of their devotion.

*3.2. The Second Step: The "I Poem"*

The second phase of the Listening Guide involves composing the "I Poem," which is a core feature of the approach that serves to identify the active self. The process of composing the "I Poem" [44] can best be understood by examining some examples. Let us start by examining the transcript of the interview with Julie, a nurse in the ICU: *"I cannot believe it . . . because of the workload . . . it is only because of the workload . . . I have to tell you that I haven't eaten for whole days . . . I grab something. It is not that there isn't any food, but we don't have the time and the needs of the staff draw you and you can't ignore them; you need to respond to each one. At other times its different, of course. Here you can't say anything to them. It's the mask; it creates wounds on their noses, so I brought them cream. This kind of mask or any other; so, I saw masks in the grocery store and I bought them pink surgical masks so they would feel joy. Every day I am bringing something to make them happy. All the time. Yes, the protective equipment is a problematic issue by itself . . . I understand since I am involved in that; it depends on the equipment that comes to Israel, but it is not always suitable . . . this equipment is insane."* In her "I Voice," Julie demonstrated her personal difficulties in the Corona ICU and her difficulties in taking care of herself, even with regard to basic needs, such as food. However, as a manager in the ICU, her "I Voice," expressed her competency in taking care of her ICU staff, as reflected in her "I Poem":

Julie's "I Poem"
I cannot believe
I have to tell
I haven't eaten
I grab something [to eat]
I brought them cream
I saw face masks
I bought them masks
I am taking out [something to make them happy]
I understand
I am involved

Michal, a nurse, said: *"I love my job, and I love the feeling of contributing. People around me, outside the hospital, talk about us [the HCWs]. I am in the frontline. It is pleasant and heartwarming."*

Michal's "I Poem" demonstrates her need to feel meaningful. In her "I Voice," she expressed her caring and empathy for patients and fellow team members. She also expressed strength and resilience.

Michal's "I Poem"
I love [my job]
I love [the feeling of contributing]
I am [in the frontline]

Dina said: "I believe this will continue ... . I discovered the richness of family and personal life, which reinforced things that I knew about myself and my [hospital] family—we are sturdy and dedicated and we cope well. I am filled with appreciation for the Infection Management Department that created a safe environment." Her "I Voice" emphasizes her hope and capacity to keep the hard work as a result of teamwork.

Dina's "I Poem"
I believe
I discovered
I knew [about myself]
I am filled [with appreciation]

One can see that, in general, the "I Poems" reflect a layer of self-confidence and feeling of empowerment. This voice stands in contrast to other voices in the narratives that describe difficulties, stressors, and chaos, as described below.

*3.3. The Contrapuntal Voices of Healthcare Workers*

Irrespective of the contrapuntal voices embodied in the narratives, all the HCWs who worked in the hospital's Corona Department stated that *the daily routine in the Corona Department was more complicated than that in the rest of the hospital*. The corona team concurred that the reasons for the differences were both physical and emotional and that acknowledgment should be given to the unique characteristics of the department. Sharon, a nurse, summed up this opinion very succinctly as: *"Corona—it is not extra work, it is completely different work."* Against the background of this commonly held perspective, the third stage of the Listening Guide analytic technique nonetheless enabled us, the interviewers, to identify multiple voices that revealed different aspects of HCWs' experiences and needs, including their attitudes towards the coronavirus pandemic, the staff and the hospital, and their own needs. The Listening Guide analysis of the focus group and the interviews identified five different contrapuntal voices—Trauma and stress, Security, Knowledge, Attachment, and Meaningfulness—intertwined not only with one another but also with the "I Voice" and the "You Voice" (see below). The contrapuntal voices and examples of quotes from the transcripts that represent these voices are described below.

3.3.1. The Voice of Trauma and Stress

The hospital workers' narratives reflected their direct exposure to traumatic events and the pervading presence of death in the hospital, as particularly manifested in the agony of seeing people dying without their families beside them and in the procedures for preparing the deceased for burial by special, double wrapping of the dead body as a precaution against contagion. For example, in response to the interviewer's question "but you are accustomed to the death of patients, what is the difference?" Golda, a nurse, shared her feelings about traumatic moments after a patient's death:

> A deceased is a deceased but the separation from the family is extremely difficult, the
> wrapping process is a different from what you normally do in the internal ward. In addition
> to the regular wrap we put them in a nylon wrap and that is horrifying. A really unpleasant
> sight. It is like you put your patients in a plastic bag and you close it with a zipper. And then

you cover with another bag but from the opposite side. An unpleasant wrapping of a patient since it is supposed to be isolated.

Similarly, Marina, a senior physician in the ICU, shared her difficulties in dealing with professional uncertainty and the absence of definitive information in the medical literature:

Look, the coronavirus is something completely new. A whole new disease that we do not have a clue how to treat, how to behave with it … and the craziest thing [is] that no-one in the world has the knowledge how to treat this disease, no knowledge-based expertise, no medical literature. So, you are constantly calling your colleagues in the country and around the world. Then, you are planning how you will cope with your first coronavirus patient. And then you are planning your second patient and the third. The decisions [as the head of the ICU] are just on your shoulders. They said to me: you are crazy … you are crazy; what are you doing? But I had to listen to myself, my instincts, and I said I have to go with my feelings and intuition. The decision is all yours. And what is most crazy is that you do not know what will happen next. Now it [the patient's condition] is fine and five minutes later the patient can die and there is no-one to consult with because no-one knows [anything] about COVID-19.

In discussing the traumatic nature of their work in the ICU and the Coronavirus Department, Marina and Golda used the personal pronoun "you" in the masculine form when they spoke about taking decisions about life-and-death issues. We note here that in Hebrew, this usage of "you" in the masculine form is a generic usage that does not refer to the gender of the user. According to the Listening Guide methodology, the use of the masculine "you" hints at Marina's and Golda's difficulties in connecting emotionally to their traumatic experience of treating corona patients in the ICU [12,45]. Harel-Shalev and Daphna-Tekoah [45] have defined this Voice as the "You Voice," a voice that enabled the HCW's to distance themselves from recurring exposure to traumatic and painful experiences. It might represent a symptom of dissociation from traumatic events, not as a dissociative disorder, but rather as a coping mechanism allowing them to keep functioning as professionals.

Experiences such as these during routine work in unfamiliar situations were balanced by feelings of competency and an ambition to fight and succeed in the mission to conquer the novel coronavirus.

### 3.3.2. The Voices of Security and Knowledge

These two voices—Security and Knowledge—are presented together since they are intimately intertwined. At the beginning of the crisis, the HCWs expressed their need for security and safety, primarily physical safety, and their need for crucial information and knowledge as a means to help them to feel more secure. With the progression of the pandemic, the HCWs became less anxious about physical safety and medical protection, as the hospital management met these basic needs and as the HCWs acquired the knowledge about how to protect themselves against contracting the disease. However, they still expressed the fear that, in the future, there could be a lack of equipment. According to Maslow, the most fundamental human needs are physiological, namely, air, water, food, shelter, and sleep. For medical personnel, Hale and his colleges [28] extended this level to include the basic determinants of good physical and mental health and safety.

Orr, a nurse in the Corona Department, shared the following thoughts with us:

At the beginning of the corona outbreak, there was a lack of food, protective gear, and clothes and shielding eyeglasses to protect ourselves. We had to shower between the shifts, and there was a shortage of showers in the hospital, and we had to fight for the basic needs to be protected, especially during the weekends. It was horrible. Everyone was terrified. There was a lack of food in the Corona Department. At the beginning, I did not have what to eat during the day. I felt broken and choked … . There were shifts that I did not eat for almost 12 h.

Similarly, Sara, a single mother who moved to the Corona Department and worked 12 h shifts, said:

I did not have a life except the work at the hospital these past few weeks. I did not have a private life at all. I did not meet my family. I am tired all the time, I just want to sleep like a human being, to eat, to be away from the hospital and from the Corona that is all over; these 12-h shifts killed me. I am a single mother and I have a daughter. My daughter was all by herself at our house. It is unbearable; she was all by herself for all those days of the corona, and I was here taking care of other people.

The fear of being infected and of infecting others inside and outside the hospital, especially family and friends, was expressed vividly by the HCWs, mainly those working in the Corona Department. Dorit, a nurse, said:

There was constant anxiety and fear that we would infect others; we [at the Corona Department] felt like lepers … and then the isolation from my family since I was so afraid that I would infect them. I was isolated like a leper. My children could not go out to play with other children because I was terrified that I would infect my children and that they would infect their friends with coronavirus. At the beginning of the coronavirus, my daughter was so stressed out from this crazy situation.

The interviewees tied knowledge to the feeling of security and protection. Dan shared his feelings with us: *"The Head of the Department is constantly updating us … I do not feel detached … I feel secure, knowing where I stand."* In contrast, Avi, an administrator, shared with us that: *" … a lack of communication and information about what is happening at the hospital at the general level and not at the sector level bothers me. I am worried."* In answer to the question, what helps you feel better? Yoav responded: *"Uncertainty concerns me—assessments of the situation and updates by my immediate supervisor would help me."* And Ruth stated: *"I feel like I'm in the dark and don't know what's going on."*

### 3.3.3. The Voice of Attachment

As could be predicted from Maslow's hierarchy of needs, the dominant needs for security and knowledge were replaced by the needs for belonging, love, recognition, and respect from supervisors and the hospital management as the pandemic progressed. The HCWs stated that a sense of family in the departments and the departments and friendship within their teams provided the much-needed sense of support during the pandemic. Moreover, the HCWs stated that without friendship, comradery, and a sense of belonging to a larger family, they would not have been able to work under such difficult conditions. For example, Sisi, a secretary at the hospital, said:

We were all a big family helping each other. I felt so close to all my peers; working together in such a tough time was different from what I had known in the last 26 years that I have been working in the hospital. As a team, we have become closer to each other, and I have discovered additional angels in my team … . In our department there is a sense of "togetherness" and comradery. Professionally, there will be changes; there are thoughts about modifying procedures in light of the current pandemic … . Relating to each other, currently feeling that we are a united and cohesive group.

Tova, a nurse in the ICU, said:

This period is a mixture of emotions. The reality is that everything is so new and unfamiliar. Nevertheless, the staff are so devoted to each other and struggling to do their best to help each other and changing shifts due to the lack of nurses. Sometimes they asked about treatment and I did not have an adequate answer. How I will say it? This is the period that we are re-inventing the protocols and rules of treatment. I am telling them that I am so sorry but there are no guidelines yet.

In addition, the interviewees expressed a need for recognition—appreciation and reinforcement— from their direct supervisors. For example, Tomer said: *"A good word, a compliment, and a positive attitude made me feel valued and . . . reassured."* Orna said: *"A kind word makes my day . . . . It is essential for me to get feedback on my work and to know that I am doing my job well."*

Another narrative relating to attachment was the need for a managerial presence, manifested as "managing by walking around." This practice is considered one of the most important ways to build good manners and performance in the workplace and emphasizes the importance of interpersonal contact, open appreciation, and recognition [46]. The HCWs did indeed voice their desire for appreciation in the form of the need to meet management representatives in the various departments. Alma, for example, said: *"The presence of management in all departments and during all shifts made the staff aware that there was someone with them."* She added: *"Personal appreciation by the management increases motivation and reduces concern . . . .I would like to see more direct communication with management . . . in my team, I feel appreciated. I don't feel I'm getting feedback from management."* Lili said that when management came into the Corona Department to visit the staff and the patients, they asked her personally how she felt, and this was what helped her to feel valued. The HCWs were in agreement in their approval of the actions taken by the hospital management, judging the management's conduct during the crisis to be appropriate and effective. Opinions of the following type were expressed: *Management worked well during the crisis; I want to thank the management for the adaptions that were made by mobilization of staff and change of policies and for taking the time to listen;* and *in my opinion, the hospital and management are doing well.*

### 3.3.4. The Voice of Meaningfulness

The importance of feeling meaningful was verbalized by Hanna: *"Patients with coronavirus helped me to feel valued and meaningful, [especially] the conversations with the patients and the phone conversations with their families out there in their homes, so worried about their loved ones. I was there for the patients and their families, and it allowed me to feel meaningful and to want to continue treating patients."*

## 4. Discussion

The global COVID-19 pandemic has challenged scholars and practitioners to find the means to alleviate stress and to treat the trauma experienced by members of the healthcare professions. Our study was designed to examine the experiences of HCWs during the first weeks of the COVID-19 pandemic, which may be considered as a massive traumatic event. The HCWs in this study, like other medical professionals caring for COVID-19 patients [40], have found themselves in a battle on two fronts: as hospital-based professionals fighting for their patients' lives, giving rise to their perspective of themselves as combatants fighting on the frontline of a war, and as family members fighting to protect their families from exposure to the virus and paying the price for fighting the "new war." As mentioned in the Introduction and Results sections, the image of war has become a common motif in discussions about the COVID-19 pandemic. Medical experts have even suggested that military strategies to be applied to outbreak management and have highlighted the importance of prioritizing healthcare staff capacity, as is done in military scenarios [18]. By documenting knowledge about HCWs, we thus contribute to a scholarly assessment and understanding of various elements of the new war—that against COVID-19. By seeking a dialogue with HCWs and, particularly, by engaging hospital staff in a genuine dialogue that deepens our understanding of the new battlefield and the "new health combatants," we are now in a position to raise questions about "conventional wisdom" in the health system and to expand the knowledge about understudied topics in the new war [12,46].

We believe that to produce a deeper understanding of the experiences of frontline workers, whoever they may be, we should listen to them attentively [47]. Thus, the Listening Guide methodology provides a tool that can capture subconscious expressions through investigation of voices that are not usually otherwise revealed. By implementing the Listening Guide method in this study, we were able to explore more deeply the ways in which the HCW's represent themselves and others—the ways in which they tell their story of the situation. In addition, we suggest that this methodology be integrated into the methods utilized in the healthcare arena and should be further explored in additional healthcare contexts.

During the interviews, the HCWs emphasized the high level of emotional intensity associated with long hospital shifts, the constant fear of death and of exposure to new and unfamiliar traumatic events, and the constant feeling of insecurity. The HCWs indicated that to cope with these emotional facets of their working environment, they needed a secure base. Issues of security and insecurity were revealed in different ways at different stages of the evolving crisis: At the beginning of the interviews, a clear majority of the staff emphasized the need for physical protection and the need to fulfill basic requirements, such as adequate food, a place to rest between long shifts, protective equipment, and showers in the Corona Department for use after their shifts. Another level of insecurity was emphasized by the need for personal recognition from direct supervisors and from hospital management. We found, however, that these needs changed as the pandemic progressed. At the beginning of the pandemic, when work environments were subject to change and scheduling remained uncertain, basic needs and physical safety were emphasized. In particular, respondents noted a shortage of protective equipment. The rapidly changing situation and the lack of supplies at the beginning of the pandemic crisis increased feelings of insecurity and intensified the importance of basic needs. However, as the crisis evolved, the need for security at the physical level was supplanted by a basic craving for security at the psychological and spiritual levels: The respondents focused on interpersonal relationships with their peers and their supervisors and their need for appreciation from their colleagues within their Departments and beyond and from management. The focus thus transitioned from personal health and well-being to a sense of social belonging, a need for respect and appreciation, and even a sense of personal and professional self-fulfillment as predicted by Maslow's theory of needs [27].

Horesh and Brown [48] have encouraged trauma researchers "in the age of COVID-19" to employ all methods of scientific practice, including unique study designs and creative collaborations between disciplines with the aim to deepen the understanding of the health implications of the global coronavirus crisis. In particular, they indicated the need to develop novel methods for empowering and supporting medical personnel, as was done in the current study. Our status as researchers in the field of trauma and health and our particular, and perhaps unique, insider/outsider status as hospital personnel may raise questions about our specific situation and positionalities with regard to this study [46]. In response to such questions, we note that fieldwork, by its very nature, situates researchers among the community that they are researching, either as active participants or as observers or as a combination of the two [49]. As "researchers from within" [48], who are also HCWs, we felt obliged to study the experiences of HCWs in this unpredictable crisis. We were surprised by the high volume and the intensity of the traumatic experiences reported by the hospital personnel. We did not anticipate that HCWs who are accustomed to treating patients in the healthcare system would experience such insecurity and vulnerability. Importantly in this regard, we found that the anonymous qualitative interview—being conducted by skilled social workers  also served as a therapeutic tool and as a proactive means of communication with staff about their needs. The interviews enabled the HCWs to express their vulnerability and to acquire a sense of visibility and value. Brown [49] claims that vulnerability is the source of resilience and that vulnerability allows us to feel the emotions that we really crave—the need for human connection and the ability to belong and to "be seen" is something that every human being wants and needs. Brown holds that for us to be seen, we need to let others see us in a vulnerable state. Thus, the study framework—by enabling HCWs to express their concerns about "not been seen by friends or management" during the battle against the coronavirus—instilled a sense of confidence in the personnel with regard to their ability to communicate their vulnerability, needs, and concerns, particularly the need for personal and psychological security [49]. Bowlby [50], in his influential book "A Secure Base," expands on the need for psychological security when he states that a basic component of human nature is the need for intimate emotional bonds and attachment. By enabling the hospital personnel to give voice to their needs and their insecurities, the interview itself became an intimate and emotional tool and a route of communication that ultimately allowed management to tailor various interventions to the needs of the employees and to increase the feeling of security in an extremely nonsecure situation.

Following the analysis of the data, the following practical interventions were developed by the hospital's Social Work Department:

1.  Ad-hoc meetings aimed at strengthening and supporting staff in transition (in that their departments had changed location and/or function to corona-related locations/functions) were arranged. COVID-19-dedicated teams were approached immediately before or after transition, and a focused, short intervention was conducted with all available staff members.
2.  Telephone support for teams put in isolation after exposure to the coronavirus was established. 140 calls were made to support employees who were in isolation following exposure to patients infected with coronavirus.
3.  Targeted short interventions were initiated for HCWs experiencing anxiety symptoms, and various relaxation techniques, such as eye movement desensitization and reprocessing (EMDR) treatment, were offered for trauma treatment.
4.  Basic information was made available to employees exposed to patients hospitalized for COVID-19. Using the current research results, we created a brochure, in question and answer format, designed to provide information on employee health and rights, workplace guidelines, and the procedure that should be followed after an unwitting exposure to a patient with COVID-19.
5.  A 24/7 hotline was opened for consultations and questions concerning mental or emotional distress.

The rationale for these interventions is embodied in the ideas of Santarone, McKenney and Elkbuli [51] that "maintaining the mental resilience of frontline workers involves offering solutions that allow them to perform their duties." The study showed that the interventions reinforced the concept of the hospital as a protective organization, learning from the knowledge and experience of its staff rather than making assumptions to define these needs. We note that the interventions were not the main intention of the research but evolved from the needs of the HCWs, as expressed in the interviews. The need to generate immediate solutions to an acute crisis informed our decision to conduct a qualitative narrative analysis study. Nonetheless, as a "side-benefit" we have accumulated rich data, which we are now analyzing in greater depth in a mixed-methods study.

## 5. Conclusions

In light of the findings of this study, we recommend that leaders in the health care system should identify hospital HCWs as first responders (similar to combatants in the traditional military environment). Listening to hospital staff and exposing the price of the battle against the novel coronavirus may provide knowledge about new and understudied topics. This knowledge generated by narrative research, such as the current study, may, in turn, assist in providing suitable support programs for professionals fighting the new war in the health battlefield. Moreover, this new knowledge may be invaluable information and contribute to organizational resilience and coping strategies. To conclude, in this era, health care management should exhibit leadership by listening to HCW's needs and adapting suitable interventions aimed to meet those needs.

**Author Contributions:** S.D.-T. participated in designing the work, data collection, took part in data analysis and interpretation, drafted the article and gave her final approval of the version to be published. T.M.B. helped with conceptualizing the work and with data collection and gave her final approval of the version to be published. E.S. contributed to drafting the article, gave critical revisions of the work and and gave his final approval of the version to be published. U.B. participated in conceptualizing the work, took part in data analysis and interpretation, assisted in drafting the article and gave his final approval of the version to be published. All authors have read and agreed to the published version of the manuscript.

**Funding:** This research received no external funding.

**Acknowledgments:** The authors would like to thank Sarit Avishai-Eliner, Eitan Lavon and Lion Poles for the productive discussions and their immense efforts in supporting the hospital's staff in the battle during the coronavirus. A special appreciation to Inez Mureinik for her inspiring remarks and careful editing.

**Conflicts of Interest:** The authors declare no conflict of interest.

## References

1. Ho, C.S.; Chee, C.; Ho, R. Mental health strategies to combat the psychological impact of coronavirus disease 2019 (COVID-19) beyond paranoia and panic. *Ann. Acad. Med. Singap.* **2020**, *49*, 1–6.
2. Tam, C.W.; Pang, E.P.; Lam, L.C.; Chiu, H.F. Severe acute respiratory syndrome (SARS) in Hong Kong in 2003: Stress and psychological impact among frontline healthcare workers. *Psychol. Med.* **2004**, *34*, 1197–1204. [CrossRef]
3. Mohammed, A.; Sheikh, T.L.; Poggensee, G.; Nguku, P.; Olayinka, A.; Ohuabunwo, C.; Eaton, J. Mental health in emergency response: Lessons from Ebola. *Lancet Psychiatry* **2015**, *2*, 955–957. [CrossRef]
4. Krystal, J.H.; McNeil, R.L. Responding to the hidden pandemic for healthcare workers: Stress. *Nat. Med.* **2020**, *26*, 639. [CrossRef]
5. Maunder, R.G.; Leszcz, M.; Savage, D.; Adam, M.A.; Peladeau, N.; Romano, N.; Rose, M.; Schulman, R.B. Applying the Lessons of SARS to Pandemic Influenza. *Can. J. Public Health* **2008**, *99*, 486–488. [CrossRef] [PubMed]
6. Maunder, R.G.; Lancee, W.J.; Balderson, K.E.; Bennett, J.P.; Borgundvaag, B.; Evans, S.; Fernandes, C.M.; Goldbloom, D.S.; Gupta, M.; Hunter, J.J.; et al. Long-term Psychological and Occupational Effects of Providing Hospital Healthcare during SARS Outbreak. *Emerg. Infect. Dis.* **2006**, *12*, 1924–1932. [CrossRef] [PubMed]
7. Mak, I.W.C.; Chu, C.M.; Pan, P.C.; Yiu, M.G.C.; Ho, S.C.; Chan, V.L. Risk factors for chronic post-traumatic stress disorder (PTSD) in SARS survivors. *Gen. Hosp. Psychiatry* **2010**, *32*, 590–598. [CrossRef] [PubMed]
8. Ho, S.M.; Kwong-Lo, R.S.; Mak, C.W.; Wong, J.S. Fear of Severe Acute Respiratory Syndrome (SARS) Among Health Care Workers. *J. Consult. Clin. Psychol.* **2005**, *73*, 344–349. [CrossRef] [PubMed]
9. Da Silva, J.V.; Carvalho, I. Physicians Experiencing Intense Emotions While Seeing Their Patients: What Happens? *Perm. J.* **2016**, *20*, 15–229. [CrossRef]
10. Du Plooy, L.; Harms, L.; Muir, K.; Martin, B.; Ingliss, S. "Black Saturday" and its Aftermath: Reflecting on Postdisaster Social Work Interventions in an Australian Trauma Hospital. *Aust. Soc. Work.* **2014**, *67*, 274–284. [CrossRef]
11. Joubert, L.; Hocking, A.; Hampson, R. Social Work in Oncology—Managing Vicarious Trauma—The Positive Impact of Professional Supervision. *Soc. Work. Health Care* **2013**, *52*, 296–310. [CrossRef] [PubMed]
12. Daphna-Tekoah, S. On the front lines: Narratives of social workers in hospitals. *Qual. Psychol.* **2020**, in preparation.
13. Maslach, C. *Burnout: The Cost of Caring*; Malor Books: Cambridge, MA, USA, 2003.
14. Daphna-Tekoah, S.; Halevi-Sheriki, E. *Symptoms of Distress and Growth among Social Workers in Hospitals in Israel*; Social Work in Health Care In Israel: Tel Aviv-Yafo, Israel, 2019.
15. Liu, N.; Zhang, F.; Wei, C.; Jia, Y.; Shang, Z.; Sun, L.; Wu, L.; Sun, Z.; Zhou, Y.; Wang, Y.; et al. Prevalence and predictors of PTSS during COVID-19 outbreak in China hardest-hit areas: Gender differences matter. *Psychiatry Res. Neuroimaging* **2020**, *287*, 112921. [CrossRef] [PubMed]
16. Kang, L.; Li, Y.; Hu, S.; Chen, M.; Yang, C.; Yang, B.X.; Wang, Y.; Hu, J.; Lai, J.; Ma, X.; et al. The mental health of medical workers in Wuhan, China dealing with the 2019 novel coronavirus. *Lancet Psychiatry* **2020**, *7*, e14. [CrossRef]
17. Greenberg, N.; Docherty, M.; Gnanapragasam, S.; Wessely, S. Managing mental health challenges faced by healthcare workers during covid-19 pandemic. *BMJ* **2020**, *368*, m1211. [CrossRef]
18. Maxwell, D.N.; Perl, T.M.; Cutrell, J.B. "The Art of War" in the Era of Coronavirus Disease 2019 (COVID 19). *Clin. Infect. Dis.* **2020**. [CrossRef]
19. Naqvi, S.H.R.; Fatima, M.; Tun, H.N. Short Message to All Healthcare Providers about Coronavirus Infectious Disease-2019 (COVID 19). *Acta Sci. Microbiol.* **2020**, *3*, 119–122. [CrossRef]
20. Chen, Q.; Liang, M.; Li, Y.; Guo, J.; Fei, D.; Wang, L.; He, L.; Sheng, C.; Cai, Y.; Li, X.; et al. Mental health care for medical staff in China during the COVID-19 outbreak. *Lancet Psychiatry* **2020**, *7*, e15–e16. [CrossRef]

21. Adams, J.G.; Walls, R.M. Supporting the Health Care Workforce During the COVID-19 Global Epidemic. *JAMA* **2020**, *323*, 1439–1440. [CrossRef]

22. Lai, J.; Ma, S.; Wang, Y.; Cai, Z.; Hu, J.; Wei, N.; Wu, J.; Du, H.; Chen, T.; Li, R.; et al. Factors Associated With Mental Health Outcomes Among Health Care Workers Exposed to Coronavirus Disease. *JAMA Netw. Open* **2020**, *3*, e203976. [CrossRef]

23. Marchand-Senécal, X.; Kozak, R.; Mubareka, S.; Salt, N.; Gubbay, J.B.; Eshaghi, A.; Allen, V.; Li, Y.; Bastien, N.; Gilmour, M.; et al. Diagnosis and Management of First Case of COVID-19 in Canada: Lessons applied from SARS. *Clin. Infect. Dis.* **2020**. [CrossRef] [PubMed]

24. McAlonan, G.M.; Lee, A.M.; Cheung, V.; Cheung, C.; Tsang, K.W.T.; Sham, P.C.; Chua, S.E.; Wong, J.G.W.S. Immediate and sustained psychological impact of an emerging infectious disease outbreak on health care workers. *Can. J. Psychiatry* **2007**, *52*, 241–247. [CrossRef] [PubMed]

25. Naushad, V.A.; Bierens, J.J.; Nishan, K.P.; Firjeeth, C.P.; Mohammad, O.H.; Maliyakkal, A.M.; Chalihadan, S.; Schreiber, M.D. A Systematic Review of the Impact of Disaster on the Mental Health of Medical Responders. *Prehosp. Disaster Med.* **2019**, *34*, 632–643. [CrossRef]

26. Hodges, C. Basing Action and Structures on Values in a Post-Corona World. *SSRN Electron. J.* **2020**, 3589690. [CrossRef]

27. Maslow, A.H. A Theory of Human Motivation. *Psychol. Rev.* **1943**, *20*, 20–35. [CrossRef]

28. Hale, A.J.; Ricotta, D.N.; Freed, J.; Smith, C.C.; Huang, G.C. Adapting Maslow's Hierarchy of Needs as a Framework for Resident Wellness. *Teach. Learn. Med.* **2019**, *31*, 109–118. [CrossRef]

29. Thielke, S.; Harniss, M.; Thompson, H.; Patel, S.; Demiris, G.; Johnson, K. Maslow's Hierarchy of Human Needs and the Adoption of Health-Related Technologies for Older Adults. *Ageing Int.* **2012**, *37*, 470–488. [CrossRef]

30. Maslow, A.H. *Toward a Psychology of Being*, 2nd ed.; Van Nostr and Company: New York, NY, USA, 1968.

31. Mahalakshmy, T.; Kalaiselvy, A.; Saya, G.K. Strategies to promote psycho-social wellbeing among health care workers during COVID-19 pandemic. *Int. J. Health Syst. Implement. Res.* **2020**, *4*, 11–16.

32. Woodcock, C. The Listening Guide: A how-to approach on ways to promote educational democracy. *Int. J. Qual. Methods* **2016**, *15*, 1609406916677594. [CrossRef]

33. Gilligan, C. The Listening Guide method of psychological inquiry. *Qual. Psychol.* **2015**, *2*, 69–77. [CrossRef]

34. Woodcock, C. The silenced voice in literacy: Listening beyond words to a "struggling" adolescent girl. *J. Authentic Learn.* **2005**, *2*, 1.

35. Hutton, M.; Lystor, C. The listening guide: Voice-centred-relational analysis of private subjectivities. *Qual. Mark. Res. Int. J.* **2020**. [CrossRef]

36. Petrovic, S.; Lordly, D.; Brigham, S.; Delaney, M. Learning to listen: An analysis of applying the listening guide to reflection papers. *Int. J. Qual. Methods* **2015**, *14*, 1609406915621402. [CrossRef]

37. Mauthner, N.S.; Doucet, A. Reflexive Accounts and Accounts of Reflexivity in Qualitative Data Analysis. *Sociology* **2003**, *37*, 413–431. [CrossRef]

38. Brown, L.M.; Gilligan, C. Listening for voice in narratives of relationship. *New Dir. Child Adolesc. Dev.* **1991**, *1991*, 43–62. [CrossRef]

39. Tekoah, S.D.; Harel-Shalev, A. "Living in a movie"—Israeli women combatants in conflict zones. *Women's Stud. Int. Forum* **2014**, *44*, 26–34. [CrossRef]

40. Shanafelt, T.; Ripp, J.; Trockel, M. Understanding and Addressing Sources of Anxiety Among Health Care Professionals During the COVID-19 Pandemic. *JAMA* **2020**, *323*, 2133. [CrossRef]

41. Senot, C.; Chandrasekaran, A.; Ward, P.T. Role of Bottom-Up Decision Processes in Improving the Quality of Health Care Delivery: A Contingency Perspective. *Prod. Oper. Manag.* **2015**, *25*, 458–476. [CrossRef]

42. Gilligan, C.; Eddy, J. Listening as a path to psychological discovery: An introduction to the Listening Guide. *Perspect. Med. Educ.* **2017**, *6*, 76–81. [CrossRef]

43. Gilligan, C.; Spencer, R.; Weinberg, M.K.; Bertsch, T. On the Listening Guide: A voice-centered relational method. In *Qualitative Research in Psychology: Expanding Perspectives in Methodology and Design*; Camic, P.M., Rhodes, J.E., Yardley, L., Eds.; American Psychological Association (APA): Washington, DC, USA, 2003; pp. 157–172.

44. Serrat, O. Managing by Walking Around. In *Knowledge Solutions*; Springer: Singapore, 2017; pp. 321–324.

45. Harel-Shalev, A.; Daphna-Tekoah, S. Bringing Women's Voices Back In: Conducting Narrative Analysis in IR. *Int. Stud. Rev.* **2016**, *18*, 171–194. [CrossRef]

46. Daphna-Tekoah, S.; Harel-Shalev, A. Beyond binaries: Analysing violent state actors in Critical Studies. *Crit. Stud. Terror.* **2017**, *10*, 253–273. [CrossRef]

47. Kook, R.; Harel-Shalev, A.; Yuval, F. Focus groups and the collective construction of meaning: Listening to minority women. *Women's Stud. Int. Forum* **2019**, *72*, 87–94. [CrossRef]

48. Horesh, D.; Brown, A.D. Traumatic stress in the age of COVID-19: A call to close critical gaps and adapt to new realities. *Psychol. Trauma Theory Res. Pr. Policy* **2020**, *12*, 331–335. [CrossRef]

49. Harel-Shalev, A.; Daphna-Tekoah, S. *Breaking the Binaries in Security Studies*; Oxford University Press (OUP): Oxford, UK, 2019.

50. Bowlby, J.A. *A Secure Base. Parent-Child Attachment and Healthy Human Development*; Basic Books: New York, NY, USA, 1988.

51. Santarone, K.; McKenney, M.; Elkbuli, A. Preserving mental health and resilience in frontline healthcare workers during COVID-19. *Am. J. Emerg. Med.* **2020**, *38*, 1530–1531. [CrossRef]

International Journal of
*Environmental Research*
*and Public Health*

*Article*

# Sense of Coherence, Burnout, and Work Engagement: The Moderating Effect of Coping in the Democratic Republic of Congo

Jeremy Mitonga-Monga and Claude-Hélène Mayer *

Department of Industrial Psychology & People Management, University of Johannesburg,
Johannesburg 2006, South Africa; jeremym@uj.ac.za
* Correspondence: claudemayer@gmx.net; Tel.: +277-6372-1263

Received: 24 February 2020; Accepted: 14 May 2020; Published: 10 June 2020

**Abstract:** Research on coping, sense of coherence, burnout, and work engagement is well documented in western countries. However, a void of studies exists on how coping mechanisms can moderate the relationship among sense of coherence, burnout, and work engagement in a manufacturing company in the Democratic Republic of Congo (DRC). The objective of this research was to examine the moderating effect of coping (COP) in the relationship between sense of coherence (SOC), burnout (BO), and work engagement (WE). The study employed a quantitative research approach, while participants were recruited through convenience sampling. A total of 197 employees ($n = 197$; females 40%) who are permanently employed in a manufacturing organisation in the DRC participated in the study voluntarily. The results indicate that coping related positively to a sense of coherence. Moreover, the results indicate that sense of coherence and work engagement related negatively to burnout. Furthermore, the results show that coping acted as a moderator in the relationships between variables. The study adds value to the WE theory by suggesting that an employee who has a high level of COP, high SOC, low level of BO, will positively engage, perform, and be productive.

**Keywords:** sense of coherence; employee participation; burnout; work engagement; Democratic Republic of Congo (DRC)

---

## 1. Introduction

Work engagement (WE) is a topic of popular interest in the field of management and industrial and organisational psychology, internationally, as well as in Sub-Saharan African contexts [1–3]. WE is described as a positive, fulfilling work-related state of mind that is characterised by rigor, dedication, and absorption [4]. Previous research on WE reveals that WE decreases levels of occupational stress [5,6] and brings about organisational and financial success [7]. Work engagement refers to an energetic state in which the employees are devoted to excellent performance at work, whilst being confident about their own effectiveness [8]. It relates to work-related outcomes, promoting employee health and well-being [9], productivity and flexibility [10], individual morale, and extra-role and organisational performance [11]. Previous research established that a high-level sense of coherence (SOC) influences employees' perceptions of their leader's behaviour [12], work engagement [13], coping strategies [14], and relate negatively to burnout [15].

This research study probes the moderating effect of coping in the relationship between sense of coherence (SOC), burnout (BO), and work engagement (WE). This article investigates the above-mentioned relationships of COP, SOC, BO, and WE in a developing country setting, particularly in the Democratic Republic of Congo (DRC), where organisations remain comparably ineffective, and need to learn how to cope with the demands of the changing work environment in the context of

WE. This study, therefore, investigates, how employees of a manufacturing organisation in the DRC perceive the COP, SOC, and BO to affect their level of WE.

## 2. The Democratic Republic of Congo's Work Context

The DRC is a developing country, with considerable economic potential because of its vast mineral and natural wealth [16]. However, political and economic instability have resulted in high levels of inflation, unemployment, and liquidation of companies, retrenchments, corruption, and the under-development of infrastructure [17]. Security and human rights within workplaces are limited [18], which has resulted in the country ranked below 7% on all six indicators [12], with the lowest scores on government effectiveness, rule of law, political stability, and control of corruption [16]. The country's manufacturing sector lacks basic infrastructure, while employees perceive poor leadership, which impacts production negatively. A previous study by Mitonga-Monga, Coetzee and Cilliers [19] found that EP is predicted by leadership style, SOC, WE and BO. In their study on BO in the DRC, Wolf, Torrente, McCoy, Rasheed and Aber [20] reported that years of experience would influence the association between BO and cumulative risk. Literature and previous research on how COP influences the association between SOC, BO, and WE in the DRC is limited; therefore, hardly any previous research that was done in this context can be presented here. This study aims to fill the void of organisational research in the DRC.

## 3. Theoretical Background

### 3.1. Salutogenesis and Sense of Coherence (SOC)

Salutogenesis is the science of the development of health [21]. It is based on the SOC, which refers to a global life orientation that expresses the extent to which one has a pervasive, enduring, dynamic feeling of confidence, that one's internal and external environments are structured, both predictably and explicably, and that there is a high probability that tasks can be managed, and are worth managing [21–23].

The three SOC components are defined as follows [21,24,25]: (1) comprehensibility—this refers to the extent to which individuals find or structure their world in order to be understandable, meaningful, orderly and consistent instead of chaotic, random and unpredictable; (2) manageability—this refers to the extent to which individuals experience events in life as situations that are endurable or manageable and, which can even be seen as new challenges; and (3) meaningfulness—this refers to the extent to which one feels that life makes sense on an emotional level and not merely on a cognitive level, and that life's demands are worthy of commitment.

Research on SOC [14,21,22,26–28] has shown this construct to be an important component of individuals' health and well-being. It acts as an effective coping strategy [4,29]. It manifests as a readiness and willingness to utilise resources at their disposal [30] to appraise, understand, and make sense of their complex reality and environment, and to choose appropriate strategies to deal with stressors and anxiety in spite of the adversity [31]. Previous research established that a higher level of COP, a strong SOC, and a low level of BO predict WE and performance [1,4].

For the DRC context, research shows that a high SOC relates positively to high levels of education, high income, and positive social relationships, and inversely correlates with cumulative exposure to violence, depression and PTSD symptoms [32]. Mitonga-Monga and Hlongwane [13] found in subsequent research on SOC in a manufacturing company that high levels of SOC perceptions influenced the relationship between leadership style and WE.

### 3.2. Burnout (BO)

BO refers to a persistent, negative work-related state of mind (or syndrome), which is characterised by an array of physical, psychological, and attitudinal symptoms [8,33]. It is a chronic, negative, affective response, with fatigue and emotional exhaustion as major symptoms [34].

The three dimensions of BO are defined as follows [35]: (1) exhaustion refers to the depletion or draining of emotional resources and feelings of being overextended, whilst experiencing distress, a sense of reduced effectiveness, decreased motivation and the development of dysfunctional attitudes and behaviours at work; (2) cynicism refers to interpersonal behaviour that manifests as a negative, callous or excessively detached response to various aspects of the job; and (3) professional efficacy refers to self-evaluation behaviour, which manifests as a feeling of competence, productivity, and achievement at work.

BO develops gradually among individuals who experience crises in their relationships with work, but not necessarily in their relationships at work [33]. It manifests as a persistent, negative, work-related state of mind, which is mediated by self-efficacy beliefs and emotional stability [36]. Researchers have indicated that the development of BO is also characterised by a lack of proper promotion possibilities, policy, and the inability of employees to achieve career goals [19]. Research in the DRC shows that the experience of higher job risks relates to lower motivation and higher BO levels and, therefore, decreased mental well-being [37]. Research in the DRC shows that the experience of higher job risks relates to lower motivation and higher BO levels and, therefore, decreased mental well-being [20]. However, research on BO in industrial work-related contexts are hardly to be found in the DRC. Research in the DRC shows that the experience of higher job risks relates to lower motivation and higher BO levels and, therefore, decreased mental well-being [20]. However, research on BO in industrial work-related contexts are hardly to be found in the DRC. Previous research by Mitonga-Monga, Coetzee, and Cilliers [19] reported that BO related negatively to SOC, EP, and WE.

## 3.3. Work Engagement (WE)

WE refers to a positive, fulfilling work-related state of mind [9,38]. Rothman et al. [27] indicate that it is not momentary or specific, but rather a more persistent and pervasive affective-cognitive state that is not focused on a particular object, event, individual, or behaviour. Employees who are strongly engaged in their daily work display intrinsic motivation through dedication to their jobs and are described as persistent and involved in their work [39]. WE has been frequently linked to work-related outcomes, including health and well-being, productivity and reduced turnover, and stress [40]. The three dimensions of WE include [10,41,42]: (1) vigour–refers to high levels of energy and mental resilience while working, as well as a willingness to exert effort and perseverance even during difficult times; (2) dedication–refers to a sense of significance in terms of one's work, feeling enthusiastic, inspired and proud, and viewing one's job as a challenge; and (3) absorption–refers to a satisfactory state of complete emersion in one's work, as well as focused attention, time distortion, loss of self-consciousness, effortless concentration, absolute control and intrinsic enjoyment. WE relates reciprocally to self-efficacy, positive affect and enthusiasm at work [15]. Work engaged employees with a strong SOC are further likely to be involved in decision-making processes and exhibit low levels of BO [40]. Previous study by Mitonga-Monga and Hlongwane [13] indicates that work engaged employees display intrinsic motivation through dedication to their jobs, work persistence, while they focus on their task performance.

## 3.4. Coping Strategies

Coping strategies have been increasingly researched by scientific scholar during the past decade as important resources in challenging life situations [13]. To comprehend how people positively face adversity is crucial and important to know the factors that may contribute to or promote resilience [44]. Coping has been described as a person's efforts to alleviate, reduce, or manage menacing events that are appraised as challenging or stressful [45]. Prior research endeavours have clustered coping mechanism into two factors, namely problem-focused and emotion-focused [45]. The latter, is aimed at regulating distress and negative emotion rather than trying to change the events themselves, using strategies such as escape support seeking or avoidance. Problem-focused comprises addressing the problem causing the distress the effective problem-focused contribute to positive psychological state by permitting individuals to experience some personal control and a sense of achievement [46].

The literature on coping has further distinguished between active and avoidant coping strategies [47]. Active coping strategies are perceived as behavioural or psychological reactions intended to change the nature of the stressor themselves or how one thinks about them. Whereas, avoidant coping strategies lead individuals into activities, such as alcohol use or destructive mental states, such as withdrawal that prevent them from addressing the stressor directly [47].

Previous research has found that individual who cope with stress by seeking social support or voicing their feeling (emotion-focused) are likely to experience negative outcomes than individual who address the experienced stressor immediately by working on find solutions to their problems [47]. A study by Pisula and Kossakowska [48] on SOC and COP in a sample of mothers and fathers of children with autism, found that a high level of SOC related positively to seeking social support and self-controlling and negatively with accepting responsibility. A previous study by Rothmann and Jorgensen [49] found that WE related positively to problem-focused coping, positive reinterpretation, and growth. High level of COP was found to play an important role in dealing with occupational stress [50]. The following section discusses inter-linkages of the four constructs described above.

### 3.5. Sense of Coherence, Burnout and Work Engagement Relationships

Researchers are investigating the link between COP, SOC, BO, and WE [3,26,49]. Several studies demonstrate that SOC relates positively to employees' level of WE, work involvement and stress [20], and influences the ability to mobilise and generate social resources in the workplace [51]. SOC predicts WE in different cultural groups in South Africa [3]. A strong SOC positively relates to WE, and negatively to the exhaustion and cynicism dimensions of BO [27]. Employees with a weak SOC and a low level of WE tend to develop BO and be less involved and engaged in their work. May, Gilson, and Harter [50] point out that WE relates particularly positively to meaningfulness, which is the third component of SOC. Employees who have developed a strong SOC show WE, are more involved, and actively participate in their work [1,4,20]. WE and BO are conceptualised to be opposing constructs on the ease-disease continuum, while vigour and dedication are the direct opposites of exhaustion and mental distance (cynicism) [3]. Previous research has found that WE and BO relate closely to work-associated well-being [8,52], while managing job burnout prevents ill-health outcomes [1,41,53].

Research has established that BO relates negatively to WE, as it does to SOC [8]. In order to facilitate WE and prevent BO, organisational contexts should foster environments where employees feel enthusiastic, energised, and motivated [54]. However, WE does not always lead to high performance, nor does high performance always indicate WE [55]. This means that employees may show initiative and take responsibility; not because they feel engaged, but rather because they fear redundancy and want to prove their capability. Conversely, employees might fail to show initiative; not because they are unengaged, but rather because constraints in the environment inhibit them from displaying their initiative [55].

### 3.6. COP as Moderator

Several empirical studies have examined the influence of COP in the association between occupational stress, SOC, BO, and WE [10,48], stress mindset, and psychological stress response [56]. However, the findings of these studies were divergent. For example, Rothmann, Jorgensen and Marais [49] reported that high problem-focused, seeking social support, turning to religion and low ventilation of emotions predicted work engagement. Thus, Rodrigues et al. [57] found that coping and stress appraisals do not seem to predict work engagement. Although some authors argue that COP and SOC constructs are likely to reduce individual level of stress [44]. Furthermore, Basson and Rothmann [58] found that both SOC and COP predicted emotional exhaustion, depersonalisation, and personal accomplishment (BO). A study by Van der Colff and Rothmann [4] reported that SOC and ventilation of emotion, low seeking emotional/social support coping predicted emotional exhaustion. Individuals with a strong SOC, high level of confronting COP low burnout are likely to demonstrate a higher level of work engagement [58]. Pisula and Kossakowska [48] studied the relationship between SOC and coping with stress and they found that SOC related negatively to COP. Although a great deal has been learned about the association between SOC, BO, and WE in Western countries [59], little has been learned in African contexts on how COP may moderate the relationship among these variables [3,26,35]. This void is addressed here for the DRC context.

## 4. Purpose and Aim of the Study

The purpose of this study is to examine how COP moderate the relationships between SOC, BO, and WE in a manufacturing organisation in the DRC. This purpose is informed by the void in research, exploring inter-linkages of the four constructs in African contexts, particularly in the DRC. The results of this study contribute to the body of knowledge of COP, SOC, BO, and WE, and can be used to increase organisational health and well-being, WE, whilst decreasing BO. The study addresses the void of research in Industrial and Organisational Psychology within developing countries in Central Africa, and particularly in the DRC.

The following research question guides the investigation and the presentation of the results:

- How do employees' levels of COP influence their level of SOC and BO?
- How do employees' levels of COP influence their level of SOC and WE?

## 5. Research Methodology

### 5.1. Research Paradigm and Design

This study is framed in the positivist paradigm [60] to achieve objective truths, facts, and laws by using a quantitative methodology. It makes use of a non-experimental, quantitative approach, comprising a range of different methods, which aim to describe the relationship between constructs by testing any causal relationships between them [60].

### 5.2. Sample and Setting

A convenience sample of employees ($n$ = 197; females = 40%) in a manufacturing company in the DRC was used (see Table 1 for the demographics). Table 1 indicates that participants were predominantly males (60%) who have a university degree education (61%) and are in their establishment career age (40–55 years). The minority of the participants were proportionally working in human resources, sales, technical, project management, and exploitation management (17%).

### 5.3. Measures

Four standardised questionnaires were used, namely one each for the a.m. constructs, as well as a biographical survey.

Coping Strategies Scale (CSC) [61] consists of 30 items self-reported instrument, measuring problem-focused and seeking support. It is scored on a four-point Likert-type-scale (1 = never, 4 = always). Examples of the items includes the following: problem-focused (tried to solve the problem); and seeking support (went to a friend for advice on how to change the situation), avoidance (Avoided being with people in general). This study obtained a Cronbach alpha coefficient of 0.65 for problem-focused COP, and 0.66 for seeking support COP (See Table 2).

**Table 1.** Sample profiles.

| Characteristic | Category | Frequency | Percentage (%) |
|---|---|---|---|
| Gender | Male | 118 | 60% |
| | Female | 79 | 40% |
| Age | Less than 25 years | 12 | 6% |
| | 25–40 years | 82 | 42% |
| | 40–50 years | 93 | 47% |
| | 55 and older | 10 | 5% |
| Education | Primary school | 7 | 4% |
| | Secondary school | 30 | 15% |
| | University bachelor & honours | 118 | 60% |
| | Masters and doctorate | 42 | 21% |
| Functional department | Human Resources | 33 | 16% |
| | Financial management | 33 | 17% |
| | Distribution & sales | 33 | 17% |
| | Technical management | 33 | 17% |
| | Project management | 33 | 17% |
| | Exploitation management | 32 | 16% |

Source: Own data.

**Table 2.** Descriptive Statistics (Mean and Standard Deviation).

| Variables | Mean | SD |
|---|---|---|
| Coping (COP) | 2.90 | 0.31 |
| Problem-Focused COP | 2.69 | 0.55 |
| Seeking support COP | 3.13 | 0.49 |
| Sense of coherence (SOC) | 4.64 | 0.72 |
| Comprehensibility (SOC) | 5.22 | 1.25 |
| Burnout (BO) | 3.71 | 0.72 |
| Cynicism (BO) | 2.20 | 1.66 |
| Work engagement (WE) | 4.54 | 0.81 |
| Vigour (WE) | 4.86 | 1.12 |
| Dedication (WE) | 4.21 | 1.20 |

The Sense of Coherence (SOC) [30] was used to measure the sense of coherence. The SOC consists of 29 items, using a seven-point Likert-scale ranging from 1 (very often) to 7 (very seldom or never). Example of the items included the following: comprehensibility (do you have the feeling that you are in an unfamiliar situation and don't know what to do?); manageability (has it happened that people whom you counted on disappointed you); and meaningfulness (until now your life has had: no clear goals or purpose at all- very clear goals and purpose). This study obtained a Cronbach alpha coefficient of 0.61 for comprehensibility, and 0.78 for SOC (see Table 2).

The Maslash Burnout Inventory General Survey (MBI-GS) [62] consists of 16 items in the self-report instrument, which measured cynicism, exhaustion, and professional efficacy. It is scored on a seven-point Likert-type-scale (0 = never, 6 = every day). Examples of the items included the following: cynicism (I have become less enthusiastic about my work); exhaustion (I feel used up at the end of a working day); and professional efficacy (In my opinion, I am good at my job. This study obtained a Cronbach alpha coefficient of 0.61 for exhaustion, and 0.64 for cynicism, and 0.71 for total burnout BO (See Table 2).

The Utrecht Work Engagement Scale (UWES) [10] consists of 17 items in the self-report instrument, measuring vigour, dedication, and absorption. It is scored on a seven-point Likert- type-scale (0 = never, 6 = every day). Examples of the items included the following: vigour (I am bursting with energy in my work); dedication (I find my work full of meaning and purpose); and absorption (When I am working, I forget everything else around me. This study obtained a Cronbach alpha coefficient 0.64 for dedication, 0.77 for vigour, and 0.76 for total work engagement (WE) (See Table 2).

The researcher decided to exclude avoidance COP, manageability SOC and meaningfulness SOC, Exhaustion BO and professional efficacy BO, and absorption WE from the interpretation because of the low reliability.

### 5.4. Procedure

Permission to conduct the research was obtained from both the management of the manufacturing company involved in the study, as well as the Ethics Research Review Committee of the overseeing academic institution (No.11/40- AO22/SD-Form/2013). The research assistant distributed research packages amongst the participants, and these comprised of the following: the participant consent form; an invitation letter indicating the aim of the study; both the university and management's approval letter; confirmation of the safekeeping and confidentiality of the responses; instructions on how to complete the instruments; and the actual three instruments, all in hard copy. On completion, each individual was requested to sign the consent form and include this with the completed instruments in an appropriate envelope. The envelope then had to be returned to the research assistant who, in turn, mailed it to the researcher.

### 5.5. Data Analysis

The researchers conducted the statistical analysis with the aid of SPSS program (SPSS Inc., Chicago, IL, USA) [52,63,64]. They investigated the multivariate outliers with Mahalonibis distance using the distribution function for Chi-square. After investigation, three cases did not satisfy the conditions of ($p \leq 0.01$) (62). The three cases were considered to have presence of outliers; therefore, the researchers decided the exclude them from the analysis.

The researcher used the descriptive statistics to explore the data. They calculate the internal consistency of the measuring instruments using item analysis if Cronbach alpha deleted [65]. Because of the low reliability on avoidance COP, Manageability SOC and meaningfulness SOC, Exhaustion BO and professional efficacy BO and absorption WE sub-scales the researchers decided to exclude them from the interpretation [66]. The researchers used Pearson correlation coefficients to determine the relationships between the variables (Problem-focused, seeking support COP, comprehensibility SOC, vigour and dedication WE, and cynicism BO). The researchers used effect size [67] to determine the practical significance of the findings. They set a cut-off alpha value of 95% confidence interval level ($p \leq 0.05$) and a practical effect size of r $\geq$ 0.11 (small effect size) to r $\geq$ 0.31 (medium effect size) were implemented.

The researchers conducted hierarchical multiple regression analyses to determine whether (1) COP moderate the relationship between sense of coherence and Burnout; (2) COP moderated the relationship between SOC and work engagement. The interactions were explored using a simple slope test and the value of the moderator at the −1SD mean +1SD, as well as standard deviations above and below the mean [66]. In order to counter the probability of type I errors, the significant value was set at the 95% confidence interval level ($p \leq 0.05$). For the purpose of this study, the practical significance of $R^2$ values was determined by calculating effects sizes ($f^2$) [68].

## 6. Results

### 6.1. Descriptive Statistics: Means and Standard Deviations

Table 2 presents descriptive statistics for the variables. As shown in Table 2, the participants obtained relatively high scores for the seeking COP (M = 3.13; SD = 0.49) and low scores on problem-focused COP (M = 2.69; SD = 0.55). In terms of the sense of coherence, participants obtained relatively high scores for comprehensibility SOC (M = 5.22; SD = 1.25), sense of coherence SOC (M = 4.64; SD = 0.72.)

As shown in Table 2 above, the participants obtained high scores burnout BO (M = 3.71; SD = 0.72), and cynicism BO (M = 2.20; SD = 1.66). In terms of work engagement, the participants obtained relatively high scores on vigour WE (M = 4.86; SD = 1.12), work engagement WE (M = 4.54; SD = 0.81) and dedication WE (M = 4.12; SD = 1.20).

*6.2. Correlational Analysis*

Table 3 also presents the significant correlation coefficients that were identified between the COP, SOC, BO, and WE variables. The inter-correlations ranged from r ≤ −0.14 (small practical effect size) to r ≥ 0.82 (large practical effect size). These results indicate that the zero-order correlations were below the threshold level of concern (r ≥ 0.90) of multi-collinearity. Problem-focused and seeking support COP positively related SOC and vigour WE and negatively related to and cynicism BO. SOC negatively and significantly related to BO variable. SOC positively and significantly related to vigour and absorption WE variables. BO related negatively and significantly to vigour and dedication WE (the *p* values ranged between *p* ≤ 0.001 and *p* ≤ 0.005).

Table 3. Correlational analysis.

| Variables | Copping COP | Problem-Focused COP | Seeking Support COP | Sense of Coherence SOC | Comprehensibility SOC | Burnout BO | Cynicism BO | Work Engagement WE | Vigour WE | Dedication WE |
|---|---|---|---|---|---|---|---|---|---|---|
| Coping (COP) | 1 | 0.45 ** | 0.53 *** | 0.70 *** | 0.13 | −0.15 * | −0.11 | 0.13 | 0.13 | 0.14 |
| Problem-Focused (COP) | | 1 | 0.04 | 0.24 * | 0.15 | −0.15 * | −0.15 * | 0.06 | 0.16 * | −0.2 |
| Seeking Support (COP) | | | 1 | 0.11 * | 0.03 | 0.03 | 0.13 | 0.03 | 0.02 | 0.05 |
| Sense of Coherence (SOC) | | | | 1 | 0.73 *** | −0.47 ** | −0.40 | 0.09 | 0.22 * | 0.22 * |
| Comprehensibility (SOC) | | | | | 1 | −0.30 * | −0.28 ** | −0.10 | 0.01 | 0.01 |
| Burnout (BO) | | | | | | 1 | 0.82 *** | −0.25 * | −0.32 ** | −0.36 |
| Cynicism (BO) | | | | | | | 1 | −0.04 | −0.14 * | −0.17 * |
| Work Engagement (WE) | | | | | | | | 1 | 0.82 *** | 0.82 *** |
| Vigour (WE) | | | | | | | | | 1 | 0.54 ** |
| Dedication (WE) | | | | | | | | | | 1 |

$n = 197$; *** $p \leq 0.01$, ** $p \leq 0.02$, * $p \leq 0.05$.

### 6.3. Hierarchical Regression Analysis

Table 4 indicates the moderating effect results.

**Table 4.** Hayes' Process Regression Matrix for Moderating effect of the coping (COP) on the relationship between sense of coherence (SOC) and (burnout (BO) ($n = 197$).

| Variables | $B$ ($SE_s$) | $t$ | $p$ | 95% Confidence Interval | | $R$ | $R^2$ |
| --- | --- | --- | --- | --- | --- | --- | --- |
| | | | | LLCI | ULCI | | |
| Constant | 7.33 (3.80) | 1.92 | 0.06 | −0.18 | 14.84 | 0.47 | 0.22 |
| SOC | −1.01 (0.84) | −1.20 | 0.23 | −2.66 | 0.64 | | |
| COP | −0.64 (1.32) | −0.50 | 0.62 | −3.25 | 1.94 | | |
| Interaction_1 | 0.14 (0.29) | 0.47 | 0.64 | −0.43 | 0.70 | | |

Note: $B$ = Unstandardized coefficients; SEs = standard errors; *LLCI = lower level of confidence interval; ULCI = upper level of confidence interval.*

As indicated in Table 4 below, in terms of the main effects, total SOC did not act as a significant predictor of the BO. (F (3; 193) = 18.66; $p \leq 0.05$), (B = −1.01; $SE_B$ = 0.84; 95%CI = (−2.66; 0.64); $p = 0.23$), denoting that SOC was not associated with a decrease in the percentage of the BO. The interactions were explored using a simple slope test and by graphing the interactions using the value of the moderator at the mean, as well as standard deviations above and below the mean [66]. As shows in Table 4, COP did not act as a moderator in the relationship between SOC and BO. (F (3; 193) = 18.66; $p \leq 0.05$), (B = 0.14; $SE_B$ = 0.29; 95%CI = (−0.43; 0.70); $p = 0.64$).

As indicated in Table 5 below, in terms of the main effects, SOC acted as a significant predictor of the WE. (F (3; 193) = 4.39; $p \leq 0.05$), (B = 2.64; $SE_B$ = 0.85; 95%CI = (0.96; 4.31); $p < 0.05$, denoting that SOC was associated with an increase in the percentage of the WE. Furthermore, COP acted as a significant predictor of the WE. (F (3; 193) = 4.39; $p \leq 0.05$), (B = 4.30; $SE_B$ = 1.34; 95%CI = (1.66; 9.94); $p < 0.05$), denoting that COP was associated with an increase in the percentage of the WE. The interactions were explored using a simple slope test and by graphing the interactions using the value of the moderator at the mean, as well as standard deviations above and below the mean [66] As illustrated in Figure 1, the relationship between SOC and WE was stronger for individuals with high level of COP than individual with low level of COP. The participants who scored high on COP also achieved significantly higher scores than their counterpart participants on the WE.

**Table 5.** Hayes' Process Regression Matrix for Moderating effect of the COP on the relationship between SOC and WE ($n = 200$).

| Variables | $B$ ($SE_s$) | $t$ | $P$ | 95% Confidence Interval | | $R$ | $R^2$ |
| --- | --- | --- | --- | --- | --- | --- | --- |
| | | | | LLCI | ULCI | | |
| Constant | −8.19 (3.87) | −2.11 | 0.04 | −15.82 | −0.55 | 0.25 | 0.06 |
| SOC | 2.64 (0.85) | 3.10 | 0.02 | 0.96 | 4.31 | . | |
| COP | 4.30 (−0.89) | 3.21 | 0.00 | 1.66 | 6.94 | | |
| Interaction_1 | −0.89 (0.29) | −3.03 | 0.02 | −1.46 | −0.31 | | |

Note: $B$ = Unstandardized coefficients; SEs = standard errors *LLCI = lower level of confidence interval; ULCI = upper level of confidence interval.*

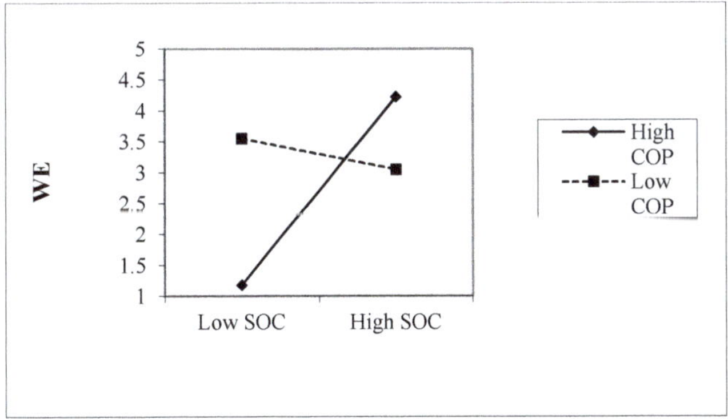


**Figure 1.** Interaction effect between COP, SOC, and work engagement (WE).

## 7. Discussion

Overall, the results suggest that participants' perceptions of problem-focused COP relate significantly and positively to their perception of SOC and vigour WE. Moreover, their perceptions of problem-focused COP related negatively and significantly to their perceptions of cynicism BO. In addition, participants' perceptions of seeking support COP related positively and significantly to their perception of SOC. Participants' perceptions of SOC related significantly and negatively to their perceptions of cynicism BO. Furthermore, participant's perceptions of SOC relate significantly and positively to their levels of vigour and dedication WE. These findings are consistent with those of prior research [23], which reported that a strong SOC and high problem-focused and high seeking support COP are important to foster the abilities and competences of employees to cope with diverse work-related challenges, whilst positively impacting work-related health and well-being. The results are likely to be explained by the fact that participants with a high SOC will likely reciprocate with a higher level of vigour WE [20].

The results suggest that problem-focused and seeking support COP related positively to SOC. This implies that individuals with a strong SOC and proper coping strategies are likely to overcome challenging events or stressful situations posed by their working environment. The results are likely to be explained by the fact that, participants with a strong SOC, who apply positive coping strategies, such as seeking support and problem-focus, are more likely to COP with stressful work environment. In contrast, participants with low level of SOC, are likely to experience threating situations, as they usually perceived stressors as a threat [43,44].

In addition, a low level of comprehensibility SOC relate to a higher level of BO. This implies that participants with a low comprehensibility SOC are likely to experience depletion of emotional resources, demonstrate cynical attitudes. In contrast, participants with a high comprehensibility SOC are less likely to have feelings of depletion of resources at work. These findings mirror those by Van der Colff and Rothmann [4] who found SOC to be negatively related with BO.

Further, the results suggest that high levels of BO relate to low levels of WE. This could possibly be explained by the fact that participants who have feelings of depletion, and who distance themselves emotionally and cognitively from their work are less likely to be energetic, enthusiastic, proud and engrossed in their work tasks. This study's results support previous results for example, Van der Colff and Rothmann [4], which showed low level of WE related to higher levels of BO. These findings are consistent with those by Rožman et al. [8] who found WE to be negatively associate with BO. These results are particularly important in the context of the DRC, which aims to increase health and well-being amongst employees, whilst generally increasing performance in the manufacturing industry through promoting SOC, COP, WE and prevent BO. The present study revealed the important role of SOC in buffering BO, consistent with previous studies [4]; Participants with a strong SOC are likely to view a greater number of events as having coherence. This perceptual seem to be restrained: it influences individual's perceptions of a stressful event, but it does so without their conscious awareness [4].

The results on the effects of COP on SOC and WE revealed that, participants with a strong SOC and high levels of WE perceived a high level of COP, then their counterparts with a low COP. This might be explained by the fact that when participants have higher level of COP, they might respond with high level of SOC and WE [4]. In other words, Participant with high-level of COP strategies and strong SOC are likely to cope with challenging situations and demonstrate high-level of energy, be enthusiastic, proud, and perform in their daily work. These findings mirror the ones of previous studies by Van Colff and Rothman [4], who found that participants with high COP and strong SOC are likely to seek emotional or social support when dealing with occupational stressors in a positive problem-focused manner. Participants with high level of COP and strong SOC are likely to demonstrate higher levels of WE, which in turn, will influence their well-being and ultimately, enhance its performance [10].

*7.1. Limitations of the Study*

The study comes with limitations. Conceptualisation of the study is limited by the fact that hardly any data is available concerning manifestation of positive psychology functioning in any middle, north, or West African country such as the DRC. This means that no comparisons with previous context-specific results could be conducted. In terms of psychometric procedures, translation of the instruments was potentially problematic in terms of experienced confusion about ideas and constructs from Western cultures implemented in culture-specific contexts such as the DRC. The sampling method (not being random) and low reliability prevented generalisation of the results from being applied beyond this organisation's population.

*7.2. Conclusions and Recommendations for Theory and Practice*

The results suggest that employees become work-engaged and dedicated (high level of WE) when they perceive their world of work as being organised and structured. They demonstrate high levels of participation in their work, and are able to cope with their work, and see the meaning in their work (overall high SOC). These results are extremely important within the DRC-context since research regarding this context often focuses on negative aspects and pathogenetic approaches rather than on positive aspects and coping. This study therefore contributes to the positive psychology and positive organisational psychology literature with regard to the Central African context. If employees have a strong SOC, using positive coping strategies, and experiencing low levels of burnout, they would likely be work engaged. This study adds to the growing body of knowledge on COP, SOC, BO in the context of WE in Central African contexts and organisations and supports international studies on SOC and coping in generally challenging work situations. Results of this study support mainly Western-based literature and results from previous studies, however it might be assumed that the culture-specific motivations and the contextual influences and effects differ from Western research settings.

Research-related recommendations, therefore, include that Industrial and Organisational Psychologists should focus their research on constructs such as COP, SOC, BO, and WE in culture-specific contexts, compare them across countries (Pan-African research), and study them, particularly within African organisations in different sectors with mix-method approaches. Further, researchers should develop culture-specific quantitative research instruments to explore culture and language adequate concepts, and not only lean on Western research instruments. It is recommended that positive psychological functioning and its effect on work behaviour of employees and, particularly leadership, should be researched to further predict the way forward for employees and organisations in Central Africa into the fourth industrial revolution.

In terms of practical recommendations, Industrial and Organisational Psychologists and Human Resources Practitioners should become aware of the inter-linkages among COP, SOC, BO, and WE within this specific cultural and organisational context, and focus on positive psychological constructs and employee and organisational functioning, since in the past industrial research in African contexts focused mainly on problems and challenges. Programs that focus on the increase of SOC in terms of mental health and well-being within organisations should be developed with culture-specific backgrounds, particularly in a challenging context like the manufacturing sector in the DRC. The COP, SOC, and WE within organisations should be fostered to counteract BO and should contribute to an overall healthier and empowering work environment.

**Author Contributions:** J.M.-M. was the project leader and was responsible for the conceptualisation and research methodology. C.-H.M. conducted the literature review and contributed to the writing-up of the manuscript. All authors have read and agreed to the published version of the manuscript.

**Funding:** This research received no external funding.

**Conflicts of Interest:** The authors declare no conflict of interest.

## References


1. Coetzee, C.F.; Rothmann, S. Job demands, job resources and work engagement of employees in a manufacturing organization. *S. Afr. Bus. Rev.* **2007**, *11*, 17–32.
2. Kotze, M. How job resources and personal resources influence work engagement and burnout. *Afr. J. Econ. Manag. Stud.* **2018**, *9*, 148–164. [CrossRef]
3. Ramasodi, S.E. Work-Related Sense of Coherence: Demographical Differences and Its Relationship with Work Engagement in a Motor Retail Organization in Gauteng. Master's Thesis, UNISA, Pretoria, South Africa, 2016. Available online: http://uir.unisa.ac.za/bitstream/handle/10500/20983/dissertation_ramasodi_se.pdf?sequence=1&isAllowed=y (accessed on 15 February 2020).
4. Van der Westhuizen, S.C. Incremental validity of work-related sense of coherence in predicting work wellness. *S. Afr. J. Ind. Psychol.* **2018**, *44*. [CrossRef]
5. Beukes, I.; Botha, E. Organisational commitment, work engagement and meaning of work of nursing staff in hospitals. *SA J. Ind. Psychol.* **2013**, *39*, 1–10. [CrossRef]
6. Schaufeli, W.; Salanova, M. Work engagement: On how to better catch a slippery concept. *Eur. J. Work Organ. Psychol.* **2011**, *20*, 39–46. [CrossRef]
7. Bhuvanaiah, T.; Raya, R.P. Employee engagement: Key to organizational success. *SCMS J. Indian Manag.* **2014**, *11*, 61–71.
8. Rožman, M.; Treven, S.; Cingula, M. The impact of behavioural symptoms of burnout on work engagement of older employees: The case of Slovenian Companies. *Našgaspodarstvo/Our Econ.* **2018**, *64*, 3–11. [CrossRef]
9. Schaufeli, W.B.; Bakker, A.B. Job demands, job resources and their relationship with burnout and engagement: A multi-sample study. *J. Organ. Behav.* **2004**, *25*, 293–315. [CrossRef]
10. Scott, D.; Bishop, J.W.; Chen, X. An examination of the relationship of employee involvement with job satisfaction, employee cooperation, and intention to quit in U.S. invested enterprise in China. *Int. J. Organ. Anal.* **2003**, *11*, 3–19. [CrossRef]
11. Bailey, C.; Madden, A.; Alfes, K.; Fletcher, L. The meaning, antecedents and outcomes of employee engagement. A narrative synthesis. *Int. J. Manag. Rev.* **2015**, *19*, 31–53. [CrossRef]
12. Van der Colff, J.J.; Rothmann, S. Occupational stress, sense of coherence, coping, burnout and work engagement of registered nurses in South Africa. *S. Afr. J. Ind. Psychol./Suid-Afrik. Tydskr. Vir Bedryfsielkunde* **2009**, *35*, 1–10. [CrossRef]
13. Mitonga-Monga, J.; Hlongwane, V. Effects of employee's sense of coherence on leadership style and work engagement. *J. Psychol. Afr.* **2017**, *27*, 351–355. [CrossRef]
14. Mittelmark, M.B.; Sagy, S.; Eriksson, M.; Bauer, G.; Pelikan, J.M.; Lindström, B.; Espnes, G.A. *The Handbook of Salutogenesis*; Springer International Publishing: Cham, Switzerland, 2017.
15. Laguna, M.; Razmus, W.; Zalinski, A. Dynamic relationships between personal resources and work engagement in entrepreneurs. *J. Occup. Organ. Psychol.* **2017**, *90*, 248–269. [CrossRef]
16. Mitonga-Monga, J.; Flotman, A.P.; Cilliers, F. Workplace ethics culture and work engagement: The mediating effect of ethical leadership in a developing world context. *J. Psychol. Afr.* **2016**, *26*, 326–333. [CrossRef]
17. Mitonga-Monga, J.; Cilliers, F. Ethics culture and ethics climate in relation to employee engagement in a developing country setting. *J. Psychol. Afr.* **2015**, *25*, 242–249. [CrossRef]

18. Börzel, T.A.; Hönke, J. Mining Companies and the Voluntary Principles on Security and Human Rights in the Democratic Republic of Congo. SFB 700. Governance in Areas of Limited Statehood. Berlin. Deutsche Forschungsgemeinschaft. 2011. Available online: http://edoc.vifapol.de/opus/volltexte/2012/3600/pdf/wp25 (accessed on 22 March 2020).
19. Mitonga-Monga, J.; Coetzee, M.; Cilliers, F. Perceived leadership style and employee participation in a manufacturing company in the Democratic Republic of Congo. *Afr. J. Bus. Manag.* **2012**, *6*, 5389–5398. [CrossRef]
20. Vogt, K.; Hakanen, J.J.; Jenny, G.J.; Bauer, G.F. Sense of Coherence and the Motivational Process of the Job-Demands–Resources Model. *J. Occup. Health Psychol.* **2015**. [CrossRef]
21. Antonovsky, A. *Health, Stress, and Coping*; Jossey-Bass: San Francisco, CA, USA, 1979.
22. Erikson, M.; Mittelmark, M.B. The sense of coherence and its measurement. In *The Handbook of Salutogenesis*; Springer: Berlin/Heidelberg, Germany, 2017; pp. 97–106.
23. Mayer, C.H. *The Meaning of Sense of Coherence in Transcultural Management*; Internationale Hochschulschriften Series; Waxmann: Münster, Germany, 2011.
24. Antonovsky, A. Studying Health vs. Studying Disease. Presented at the Congress for Clinical Psychology and Psychotherapy, Berlin, Germany, 19 February 1990.
25. Antonovsky, A. The structure and properties of the sense of coherence scale. *Soc. Sci. Med.* **1993**, *36*, 725–733. [CrossRef]
26. Maturure, T. Burnout, Work Engagement, and Sense of Coherence in Nurses Working at a Central Hospital in Kwa-Zulu-Natal. Master's Thesis, UNISA, Pretoria, South Africa, 2016. Available online: http://uir.unisa. ac.za/bitstream/handle/10500/20159/dissertation_maturure_t.pdf?sequence=1&isAllowed=y (accessed on 20 April 2020).
27. Rothmann, S.; Steyn, L.J.; Mostert, K. Job Stress, Sense of Coherence and Work Wellness in an Electricity Supply Organization. *S. Afr. J. Bus. Manag.* **2005**, *36*, 55–63.
28. Viljoen, P.J.; Rothmann, S. Occupational stress, ill health and organizational commitment of employees at a university of technology. *S. Afr. J. Ind. Psychol./Suid-Afrik. Tydskr. Vir Bedryfsielkunde* **2009**, *35*, 67–77. [CrossRef]
29. Redelinghuys, F.J.; Rothmann, S. Koherensiesin, coping, uitbranding en begeestering in die bediening (Sense of coherence, coping, burnout and engagement in the ministry). Presented at the 2nd South African Work Wellness Conference, Potchefstroom, South Africa, 25–26 March 2004.
30. Antonovsky, A. *Unraveling the Mystery of Health: How People Manage Stress and Stay Well*; Jossey-Bass: San Francisco, CA, USA, 1987.
31. Saks, A.M. Antecedents and consequences of employee engagement. *J. Manag. Psychol.* **2006**, *21*, 600–619. [CrossRef]
32. Pham, P.N.; Vinck, P.M.; Kinkodi, D.K.; Weinstein, H.M. Sense of coherence and its association with exposure to traumatic events, post-traumatic stress disorder, and depression in eastern Democratic Republic of Congo. *J. Trauma. Stress* **2010**, *23*, 313–321. [CrossRef]
33. Maricutoiu, L.P.; Sulea, C.; Lancu, A. Work engagement or burnout: Which comes first? A meta-analysis of longitudinal evidence. *Burn. Res.* **2017**, *5*, 35–43. [CrossRef]
34. Maslach, C.; Leiter, M.P. *Stress: Concepts, Cognition, Emotion, and Behavior*; Handbook of Stress Series; Fink, G., Ed.; Academic Press: Amsterdam, The Netherlands, 2017; pp. 351–357.
35. Bezuidenhout, A.; Cilliers, F.V.N. Burnout, work engagement and sense of coherence in female academics in higher-education institutions in South Africa. *SA J. Ind. Psychol.* **2010**, *36*, 1–10. [CrossRef]
36. Alessandi, G.; Perinelli, E.; de Longis, E.; Schaufl, W.B.; Theodorou, A.; Borgogni, L.; Caprara, G.V.; Cinque, L. burnout: The contribution of emotional stability and emotional self-efficacy beliefs. *J. Occup. Organ. Psychol.* **2018**, *91*, 823–851. [CrossRef]
37. Wolf, S.; Torrente, C.; McCoy, M.; Rasheed, D.; Aber, J.L. Cumulative Risk and Teacher Well-Being in the Democratic Republic of the Congo. *Comp. Educ. Rev.* **2015**, *59*, 717–742. [CrossRef]
38. Park, J.H.; Ono, M. Effects of workplace bullying on work engagement and health: The mediating role of job insecurity. *Int. J. Hum. Resour. Manag.* **2017**, *28*, 3202–3225. [CrossRef]
39. Andrews, M.C.; Kacmar, K.M.; Valle, M. Surface acting as a mediator between personality and attitudes. *J. Manag. Psychol.* **2016**, *31*, 1265–1279. [CrossRef]

40. Pocnet, C.; Antonietti, J.P.; Massoudi, K.; Györkös, C.; Becker, J.; de Bruin, G.P.; Rossier, J. Influence of individual characteristics on work engagement and job stress in a sample of national and foreign workers in Switzerland. *Swiss J. Psychol.* **2015**, *74*, 17–27. [CrossRef]

41. Csikszentmihalyi, M. *Flow: The psychology of Optimal Experience*; Harper & Row: New York, NY, USA, 1990.

42. Rothmann, S.; Joubert, J.H.M. Job demand, job resources and burnout and work engagement of managers at a platinum mine in the North West Province. *S. Afr. J. Bus. Manag.* **2007**, *38*, 49–61. [CrossRef]

43. Braun-Lewensohn, O.; Abu-Kaf, S.; Al-Said, K. Women in Refugee Camps: Which Coping Resources Help Them to Adapt? *Int. J. Environ. Res. Public Health* **2019**, *16*, 3990. [CrossRef]

44. Meneghel, I.; Martinez, I.M.; Salanova, M.; de Witte, M. Promoting academic satisfaction and performance: Building academic resilience through coping strategies. *Psychol. School* **2019**, *56*, 875–890. [CrossRef]

45. Lazarus, R.S.; Folkman, S. *Stress, Appraisal, and Coping*; Springer: New York, NY, USA, 1984.

46. Duraku, Z.H.; Hoxha, L. Self-esteem, study skills, self-concept, social support, psychological distress, and coping mechanism effects on test anxiety and academic performance. *Health Psychol. Open* **2018**, *5*, 1–9. [CrossRef]

47. Holahan, C.J.; Moos, R.H. Personality, coping and family resources in stress resistance: A longitudinal analyses. *J. Personal. Soc. Psychol.* **1986**, *51*, 389–395. [CrossRef]

48. Pisula, E.; Kossakowa, Z. Sense of coherence and coping with stress among mothers and fathers of children with autism. *J. Autism Dev. Psychol.* **2010**, *40*, 1485–1494. [CrossRef]

49. Rothmann, S.; Jorgensen, L.; Marais, C. Coping strategies and work engagement. *S. Afr. J. Bus. Manag.* **2011**, *37*, 1–11. [CrossRef]

50. May, D.R.; Gilson, R.L.; Harter, L.M. The psychological conditions of meaningfulness, safety and availability and the engagement of the human spirit at work. *J. Occup. Organ. Psychol.* **2004**, *77*, 11–37. [CrossRef]

51. Feldt, T.; Kivimäki, M.; Rrantala, A.; Tolvanen, A. Sense of coherence and work characteristics: A cross-lagged structural equation model among managers. *J. Occup. Organ. Psychol.* **2010**, *77*, 323–342. [CrossRef]

52. Sonnentag, S. Research on work engagement is well and alive. *Eur. J. Work Organ. Psychol.* **2011**, *20*, 29–30. [CrossRef]

53. De Beer, L.T.; Pienaar, J.; Rothmann, S. Job burnout, work engagement and self-reported treatment for health conditions in South Africa. *Stress Health* **2014**, *32*, 36–46. [CrossRef]

54. Bakker, A.; Albrecht, S.; Leiter, M. Key questions regarding work engagement. *Eur. J. Work Organ. Psychol.* **2011**, *20*, 4–28. [CrossRef]

55. Parker, S.K.; Griffin, M.A. Understanding active psychological states: Embedding engagement in a wider nomological net and closer attention to performance. *Eur. J. Work Organ. Psychol.* **2011**, *20*, 60–67. [CrossRef]

56. Horiuchi, S.; Tsuda, A.; Aoki, S.; Yoneda, K.; Sawaguchi, Y. Coping as a mediator of the relationship between stress mindset and psychological stress response: A pilote study. *Psychol. Res. Behav. Manag.* **2018**, *1*, 47–54. [CrossRef]

57. Rodrigues, S.; Sinval, J.; Queirós, C.; Marôco, J.P.; Kaiseler, M. Transitioning from recruit to officer: An investigation of how stress appraisal and coping influence work engagement. *Int. J. Sel. Manag.* **2019**, *27*, 152–168. [CrossRef]

58. Basson, M.J.; Rothmann, S. Sense of coherence, coping and burnout of pharmacists. *S. Afr. J. Econ. Manag. Sci.* **2002**, *5*, 35–62. [CrossRef]

59. Evans, W.; Davis, B. Exploring the relationship between sense of coherence and historical trauma among American Indian Youth, American Indian and Alaska. *Nativ. Ment. Health Res.* **2018**, *25*, 1–26.

60. Transparency International. Corruption Index for the Democratic Republic of Congo. 2018. Available online: https://www.transparency.org/country/COD (accessed on 21 February 2020).

61. Amirkhan, J.H. A factor analytically derived measure of coping: The coping strategy indicator. *J. Personal. Soc. Psychol.* **1990**, *59*, 1066–1074. [CrossRef]

62. Schaufeli, W.; Leiter, M.; Maslach, C.; Jackson, S. Maslach Burnout Inventory-General Survey. In *The Maslach Burnout Inventory: Test Manual*; Maslach, C., Jackson, S.E., Leiter, M.P., Eds.; Consulting Psychologists Press: Palo Alto, CA, USA, 1996.

63. Pallant, J. *SPSS Survival Manual*, 6th ed.; A Step by Step Guide to Data Analysis Using IBM SPSS; Open University Press: New York, NY, USA, 2016.

64. SPSS Inc. *SPSS 25.0 for Windows*; SPSS Inc.: Chicago, IL, USA, 2020.

65. Tabachnick, B.C.; Fidell, L.S. *Using Multivariate Statistics*, 6th ed.; Pearson: Boston, MA, USA, 2014.

66. Cohen, L.; Manion, L.; Morrison, K. *Research Methods in Education*, 6th ed.; Routledge: Abingdon, UK, 2011.
67. Cohen, J. *Statistical Power Analysis for the Behavioral Sciences*, 2nd ed.; Lawrence Erlbaum Associates: Hillsdale, NJ, USA, 1988.
68. Cohen, J. Quantitative methods in psychology: A power primer. *Psychol. Bull.* **1992**, *112*, 155–159. [CrossRef]

International Journal of
*Environmental Research
and Public Health*

*Article*

# Sense of Coherence, Compassionate Love and Coping in International Leaders during the Transition into the Fourth Industrial Revolution

**Claude-Hélène Mayer [1,2] and Rudolf M. Oosthuizen [3,*]**

[1]  Department of Industrial and Organisational Psychology, University of Johannesburg, Johannesburg 2006, South Africa; claudemayer@gmx.net
[2]  Institut für therapeutische Kommunikation und Sprachgebrauch, European University Viadrina, 15230 Frankfurt (Oder), Germany
[3]  Department of Industrial and Organisational Psychology, University of South Africa, Pretoria 0002, South Africa
*  Correspondence: oosthrm@unisa.ac.za

Received: 13 February 2020; Accepted: 17 April 2020; Published: 20 April 2020

**Abstract:** Contemporary workplaces are influenced by rapid changes, high levels of competition, increasing complexities and internationalisation processes. At the edge of the Fourth Industrial Revolution (4IR), insecurities and anxieties are high, and leaders are encouraged more than ever to lead employees with meaningful vision and prudence in order to make use of employees' strengths, and ensure mental health and well-being. The aim of this article is to present new insights into salutogenesis, particularly sense of coherence (SOC), compassionate love (CL), and coping (C) in leaders with different cultural backgrounds. This study strengthens the idea that CL is a coping mechanism. This coping mechanism can be used by leaders to establish a resilient and salutogenic organisations. This article explores the subjective perspectives of 22 international leaders from five different countries and their views regarding SOC, CL and C through a qualitative research approach, using a qualitative online questionnaire for data collection and content analysis for data analysis. The findings on the perspectives of leaders provide new and original insights into how SOC, CL and C are connected, and how these concepts contribute to healthy organisations which are on their way to the transition into the 4IR. Conclusions are drawn. Recommendations for future research and practice are given.

**Keywords:** salutogenesis; sense of coherence (SOC); compassionate love (CL); leadership; coping (C); 4IR workplaces; societal and cultural challenges

## 1. Introduction

The Fourth Industrial Revolution (4IR) is characterised by rapid changes on socio-economic, political and cultural levels. It brings increasing technology-based human–machine interaction, growing digitalisation, and increasing use of smart technologies. These changes also bring about a change in workplaces cultures [1]. Individuals are challenged with understanding complexities, managing them constructively and redefining the meaning of work [2]. Rapid changes, however, cause stress and negative emotions, such as anxiety, frustration and sadness, as well as disorientation, insecurities and ambiguities [3]. Leaders are encouraged to provide guidance and leadership to address new challenges and support employees to cope positively and constructively [4]. Positive emotions in leadership can support salutogenesis and coping [5].

Salutogenesis is concerned with developing, maintaining and increasing the health of individuals [6–8]. The main construct of salutogenesis, sense of coherence (SOC), is a resource and life orientation which supports individuals as a coping mechanism and fosters health, even in stressful situations [9]. SOC consists of three components [6] that will be explored in this study: comprehensibility, manageability and meaningfulness.

Coping (C) can be portrayed as the capacity with which leaders handle an unpleasant occasion [10]. Leaders perceive SOC as a useful and functional guide to behaviour, which gives SOC a prominent role in C in the 4IR context. SOC directs leaders to focus on a specific set of stimuli out of all the possible stimuli with which they could possibly cope. By emphasising important elements of the environment, emotions motivate cognitive and behavioural reactions. Leaders are only likely to perceive a 4IR challenge when emotions are evoked at the same time. Additionally, SOC and compassionate love (CL) prepare leaders for action in response to stimuli [11,12].

The concept of compassionate love (CL) has gained interest in leadership studies [13], highlighting that the concept impacts positively on empowerment, authenticity, and stewardship, providing direction to employees [14]. Compassionate love also increases empathy for others in addition to authentic listening, nurturance and caring skills [15–17]. CL embodies and enacts the qualities of love, altruism, integrity, humility and wisdom combined with an appreciation and empowerment of others [18]. CL is core to the development of a compassionate and person-centred organisation and requires senior leaders to clearly articulate the core values and vision of the organisation and to ensure that they resonate in all the self-organising groups within the system [18]. CL supports leaders by helping them cope with job demands, manage stress and conflicts, set pro-social goals and connect with their employees and stakeholders [17].

The findings of Lloyd [19] indicate that higher levels of CL and SOC were both associated with lower stress. The findings also indicate that CL is positively associated with adaptive C strategies, and both CL and SOC are negatively associated with avoidance-oriented strategies. SOC and maladaptive C emerged as significant predictors of perceived stress in subsequent regression analyses. Interventions or support mechanisms that enhance SOC and reduce reliance on maladaptive C may decrease vulnerability to stress in leaders.

### 1.1. Aim of This Article

The aim of this article is to present the perspectives of leaders on SOC, CL and C and interrelationships that exist therein with regard to their professional work and leadership during transition into the 4IR in different cultural contexts.

Since the literature on the 4IR often focuses on the negative side—technological challenges, fears and rapid, incomprehensible complexities—this study aims to focus on positive aspects in terms of how leaders across the globe cope with these challenges. Leaders were asked about SOC, C and CL in terms of the transition into the 4IR. The leading research question for the findings presented here is: how do SOC and CL support C in leaders during transition into the 4IR? The contribution of this study is to investigate SOC and CL in relation to C, and thereby close a void in the literature.

The relationship between SOC and C is well established in the literature, as well as between self-compassion, SOC, and C strategies, but SOC and CL have not been investigated in relation to C in leaders. Thus, the authors will present the core concepts of this study in more depth.

### 1.2. The Transition into the Fourth Industrial Revolution (4IR)

Leaders face workplace changes and rapid development, increased technologisation, digitalisation, smart-system use, artificial intelligence, questions about the meaning of work, innovation and creativity in 4IR contexts [20,21]. These changes pose challenges to organisations and societies [22]. Magubane [23] suggests that love needs to be part of the transition into the 4IR. Illouz [24,25] sees changes regarding love within the 4IR taking on modern shapes, such as alternate conceptual and relationship changes that are mediated through technological interactions or robotics.

Chandsoda and Salsing [26] emphasise that the 4IR requires people to incorporate sympathy and participation into human relations, work environments and leadership. Van der Hoven [27] emphasises that individuals ought to re-learn how to detach from the advanced world and how to reconnect compassionately with self, others and nature. CL means reconnecting with one's "true self" and nature, collaborating with others, disengaging from the computerised world, tuning into one's inward voice and achieving a state of flow and imagination [27]. Tu and Thien [28] recommend that the fundamental standards of Buddhism offer assistance to leaders through the 4IR, by promoting mindfulness, contemplation, love and trust as basic values.

Sun [29] suggests that organisations ought to utilise "love competition". "Love competition" is a competitive attitude based on CL, shrewdness and market-orientated social skills. Fabritius [30] contends that stress levels, potential conflict, and work insecurity, alluding to "fear of unemployment and trouble of reemployment," are rising [31,32]. Different trends in the 4IR, such as instable working conditions, automation, and fear of loss of work lead to poor health and negative feelings, such as anxiety [33–35]. How do leaders in different cultural contexts cope with the transition into the 4IR?

*1.3. Salutogenesis, Sense of Coherence and Leadership*

Leaders increasingly focus on promoting health and well-being in employees in order to improve the performance of organisations [36]. Salutogenesis has become a favourable approach to increasing health and well-being (source), exploring factors affecting and implementing programmes supporting employee health [37–42]. According to Antonovsky [6], health and well-being are developed through SOC. SOC is a global life orientation that supports individuals in comprehensibility, manageability and meaningfulness—leading to consistent congruence and harmony within an individual. The more pronounced an individual's SOC, the healthier an individual is [43].

Comprehensibility refers to one's understanding of the world based on the ability to process familiar and unfamiliar stimuli as ordered, structured and consistent; manageability, relates to how one copes with challenges and whether one believes that challenges can be solved through the use of resources; meaningfulness relates to how an individual is motivated through the construction of meaning in life and the extent to which life makes sense [7].

SOC supports individuals in focusing on their C strategies and helps leaders to stay healthy in stressful situations [42,44]. SOC prevents mental illness [45] and helps individuals to cope within complex, international and intercultural settings [46]. A high SOC impacts positively on performance, achievement, success and the ability to manage conflicts and intercultural communication [43].

Individuals with a high SOC are more likely to perceive the leader as a good listener, and as an engaging and encouraging intuitive thinker. They also include employees in decision-making and problem-solving and foster participation. This, in turn, leads to employees who are energetic, enthusiastic, proud, inspired and happily engaged in their work [46,47]. Compassionate leadership impacts positively on employees, encouraging creativity in problem solving and a high level of engagement

Organisations that advance and improve working conditions in terms of physical, mental and social well-being, are more likely to have employees perceiving their leaders' management style as positive (i.e., manages who are supportive, consider their concerns, empower them, and listen, including co-operating with them and providing counselling) [46–48]. These employees demonstrate high levels of psychological connection, engagement and performance [47].

SOC can be applied at different system levels, including at the individual, group, organisational and societal levels [49], and is a source of compassionate leadership that supports the development of a strong SOC [50]. Recent leadership research shows that love—defined as a complex concept of compassion, humanity, care and unconditional, compassionate love—impacts positively on leaders in terms of coping with work challenges, such as changes faced during the transition into the 4IR. Harry [51] provides insights, showing the relationship between individuals' wellness attributes (sense of coherence, emotional intelligence and burnout) and their resiliency capacities (career adaptability and hardiness). Organisations need to understand the complex processes involved in coping during transition into the 4IR [52] and the importance of fostering SOC.

### 1.4. Coping and Compassionate Love in Leaders

Leaders could potentially perceive a 4IR challenge, appraise it as such and cope with it without necessarily experiencing intense feelings or emotions. C with emotions includes the idea of C with the leaders' own emotions beyond emotion regulation. Emotion regulation includes conscious efforts to modulate the intensity of leaders' emotions (for example by avoiding emotion-eliciting 4IR situations or by reappraising the meaning of a 4IR situation) [53]. In contrast, C with emotions can include behaviours elicited by an emotion, regardless of whether the behaviours are intended to regulate the emotional experience [54].

Leaders respond differently to emotions [54,55]. The differences in how leaders cope with emotional states can have implications for their long-term psychological and physical health [54]. Machin, Adkins, Crosby, Farrell, and Mirabito [56] postulate that C efforts are actions taken to protect, maintain, or restore well-being. Future-focused and action-oriented challenges promote anticipatory and preventative C strategies. Rather than reacting to immediate 4IR stressors, future-oriented C strategies instead seek to protect the leader before harm occurs. Understanding the goal behind a C tactic also helps in judging coping effectiveness, a troubling issue in contemporary research [57].

Love has been ignored in Western writing on leadership for a long time [58]. However, Barsade and Gibson [59] mention the exceptional role of CL in working environments: CL affects the demeanour of employees, their accomplishments, the organisational culture and the employee relationships. Positive feelings have an immense and positive impact on employees and their performance [60–62]. Several researchers [63,64] have indicated that compassionate love is associated with care, concern und thriving in connection with others. CL is connected to kindness, affection, open-mindedness, caring and kind-heartedness [63]. CL further energises an attitude of humbleness, gratitude, forgiveness and selflessness in leaders within the work environment [14]. CL needs new research and detailed descriptions in organisational contexts [65].

Van Dierendonck and Patterson [14] describe CL as inspiring, meaningful and optimal social functioning in organizations. CL fosters encouragement, genuineness, stewardship and leadership. Barsade and O'Neill [66] state a universal conviction that employment relationships do not contain adequate deepness to be named "love" relationships. Other researchers counter-argue that social relationships at work transfer in-depth emotive encounters and are full of meaning [67]. Savickas [68] suggests that CL at work can significantly affect careers and happiness at work, while O'Neill [69] suggests that CL is a critical factor in employment relationships, which include caring, compassion and gentleness. Emotions perform an exceptional role at work [66] and CL shields negative emotions, such as fear [70]. CL is often referred to in work contexts with regard to compassionate leaders [71].

Rapport [72] states that love within the work environment can be depicted as a frame of moral engagement which goes past culture and signifies a collective culture of morals. CL is often based on values which cultivate natural inspiration—such as spiritual well-being or a calling—compassion and joy, meaning making and sense making, hope, faith and altruistic love (including care, concern, and appreciation for both organisational and employee needs) [73]. CL leads to the removal of fears related to anger, failure, selfishness, guilt or worry [74] and increases vitality and the feeling of being lively within the working environment [75,76].

CL encourages meaningful and optimal human functioning, impacting on the leader's propensity towards virtues, such as humility, gratitude, forgiveness and altruism. The salutogenic approach includes the effectiveness of C processes. It concentrates on what differentiates "successful copers," even in the most stressful of situations, from others and, thus, seeks out personality traits and protective factors that are related to successful C in stressful situations. Work within the framework of this approach mainly involves SOC [7]. Leaders comprehensibility could to a large extent be constructed by their own thoughts and theories. Manageability can be achieved by active information-seeking strategies, social support and C, including positive reinterpretation of the 4IR context. Meaningfulness may be central for quality of life and is achieved through CL and close relations, as well as by work.

SOC integrates essential parts of the stress/C model (comprehensibility, manageability) and of CL (meaning).

## 2. Research Methodology

### 2.1. Research Paradigm

This study is anchored in the qualitative research paradigm and uses a hermeneutical and interpretative framework for analysis and interpretation [77,78]. Hermeneutics are defined as the philosophy and interpretation of meaning [79], focusing on the meaning of a text or a text analogue [80].

### 2.2. Sampling

During the sampling process, purposeful sampling and snowball sampling were used [81]. The researchers used purposeful sampling to ensure that the sample can relate to the research topics. Participants were asked to respond to the questionnaire and refer to their own experiences with regard to the selected concepts. Information about the participants is given in the Findings section. The sample consisted of 22 individuals—9 females and 13 males. The age ranged between 33 and 80 years. Interviewees referred to their religion/belief as follows: seven were Christians; four Roman Catholics; two Protestants; one each atheist, agnostic, atheist/agnostic, Buddhist, Muslim, Jew, and Jesuit; one indicated no religious affiliation.

The sample was classed as "international" due to the fact that leaders came from various countries, and national and cultural backgrounds and all acted in international projects, cooperations and networks. Ten of the participants spoke German as their first language, five English, two Afrikaans and one participant each spoke Japanese, Tswana, Hebrew and Romanian. In terms of nationality, the sample consisted of eight Germans, five US citizens, two Japanese, two South Africans, and one participant each with White, German-Iranian, Israeli, Romanian and Bavarian origin (Bundesland (provincial state) in Germany). In terms of educational background, the participants held the foillowing degrees: nine doctoral degrees, six master's degrees, three "Diploma" degrees (Diplom degrees are equivalent to master's degrees in Germany), two post-doctoral degrees, one national diploma, and one high school certificate. The sample further included seven professors, three academics, three directors, two consultants, one professor/entrepreneur, one entrepreneur, one executive manager, one project manager, one teacher, one psychologist, and one educational officer. All of the leaders held leadership positions in educational fields and had held them for at least two years. All leaders find themselves at "the edge of the 4IR" as organisations in which the leaders' work aim towards a transition into the 4IR using digitalisation and smart systems. Some parts of the work are automated and several organisations are starting to collect and analyse big data—all signs of a transition into the 4IR.

### 2.3. Data Collection and Analysis

Data were collected through structured questionnaires [82] that were sent out to the participants via email. Participants filled in the online questionnaires and sent the filled-in questionnaires back to the researcher. The questionnaires included questions referring to sense of coherence, comprehensibility, manageability and meaningfulness, coping, leadership and compassionate love (see Appendix A). The interviewees indicated the lengths of time they took to fill in the questionnaires. The time ranged from 90 to 240 min.

Data were analysed through the five-step process of content analysis by Terre Blanche, Durrheim, and Painter [83]: (1) familiarisation and immersion, (2) inducing themes, (3) coding, (4) elaboration, and (5) interpretation and checking to ensure data quality. Interviews were coded and categorised by using a deductive research interpretation process, focusing on the constructs of SOC, CL and C. Researchers used an intersubjective validation approach, comparing and discussing the process and codes [84].

The authors considered the usual quality criteria for qualitative research, such as conformability, credibility, transferability and dependability, as well as rigor [85]. These quality criteria were addressed through all the phases of research. In terms of ethical considerations, all participants participated voluntarily, were informed about this study, its aim and purpose and consented to this study.

*2.4. Ethical Considerations*

Ethical considerations included: open discussion of the rights of the participant, respectful communication, respect, the creation of informed consent, confidentiality, anonymity and transparency [86]. Ethical approval for this research study was given by the Institut for Therapeutic Communication and Language Use, Europa Universität Viadrina, Frankfurt (Oder), Germany.

## 3. Results

In the following section, the findings on sense of coherence, coping and compassionate love in leaders are presented. Direct quotations are only provided regarding the three most frequently mentioned categories, and reference to other categories is only provided by description.

*3.1. Salutogenesis and Compassionate Love in Leadership*

When leaders were asked how compassionate love and salutogenesis are connected, 19 out of 23 leaders responded that compassionate love is the basis for salutogenesis. P10, a Japanese female leader, 59 years, emphasises that compassionate love and salutogenesis are strongly interwoven concepts:

"Yes, they are connected tightly. Mental health and well-being provide us with energy to embrace and love others, and vice versa."

P15, a female Israeli leader, also recognises the interconnectedness of CL and health:

"People who have true love and are in good relationship, are physically and mentally healthier. They live longer, less depressed, report fewer pains and more happiness."

Eight participants said that CL is their resource to cope at work. Altogether, seven participants mentioned that compassionate self-love is their core C resource that fosters mental health. P1, a German female leader, emphasises a combined perspective:

"Love is in my view a really crucial factor for happiness, well-being. Although I always used to think that love must be a partner, today I continue to think about it and I enjoy the many loves in my life. . . . Love is a very important factor for my personal well-being and my health. It is not just about being loved, but above all about loving oneself. In me is an ocean of love, I have to share, otherwise I may drown in myself;-) At the same time, the world needs more love, it is a resource . . . "

Further the findings (Table 1) show that seven leaders highlight that CL is a universal emotion and part of "being human". "Being human" is very important in leadership (P1). For six other participants, CL is a general source of happiness and. as such, it is connected to salutogenesis. For five other individuals, CL is the force that creates their inner being, but also the outer balance and harmony they need in their demanding position. CL determines how they feel about people and situations, and how they act and respond. Four individuals highlighted that, if they have an intra-personal feeling of CL towards the world, they are content and happy with their leadership. Happiness further induces salutogenic feelings being at ease with the world. Finally, two individuals view CL as a base for their inner safety and salutogenesis.

**Table 1.** Connection between love and mental health.

| Frequency | Category | Participants |
|---|---|---|
| 19 | Base for well-being | P1, P3, P4, P5, P7, P8, P10, P11, P12, P13, P14, P15, P16, P17, P18, P19, P20, P21, and P22 |
| 8 | Love is a coping resource | P4, P5, P7, P9, P11, P14, P15, and P22 |
| 7 | Self-love improves mental health | P1, P7, P10, P12, P14, P21, and P22 |
| 7 | Love makes us human | P2, P5, P12, P14, P15, P19, and P21 |
| 6 | Base for happiness | P1, P3, P7, P8, P15, and P20 |
| 5 | Love creates balance and harmony | P1, P16, P18, P19, and P20 |
| 4 | Love for the world improves well-being | P1, P5, P7, and P21 |
| 2 | Base for safety | P3 and P5 |

In the following section, the findings are presented with regard to the three components of SOC as aspects which promote and foster mental health and well being when integrated.

### 3.2. How do You Understand the World of Work? (Comprehensibility)

Altogether, 13 leaders say that they can cope best when they understand the world in its complexity and by applying an attitude of CL (Table 2). P19, a Romanian female leader, points out:

"I believe love transcends culture. It is above it and above everything else. Love is a bridge. It is universal, even if expressed differently across cultures. We understand love even if we belong to different cultural backgrounds. This understanding is important to be together and work together. Love is love. We all know when we are being loved. And it helps us to understand each other."

Further, eight leaders highlight that attitude is not only relevant in terms of comprehensibility but also brings appreciation. P8, a male German-Iranian leader, emphasises:

"Appreciation is a main key to open yourself for other people. If you show appreciation, people will open up and understand that a transcultural faux pas was not an intended insult."

P9, a German male leader, mentions:

"Love brings about an appreciation for self and others through its positive energy and strengths. When we love, we feel energized and healthy. And when we appreciate the other we might understand him better."

According to seven leaders, comprehensibility is fostered through a willingness to learn about others. P14, a male South African leader, emphasises:

"In leadership and co-operation, we need to show interest, objectivity, concern, justice, empathy, listening skills, support, willingness to go the extra mile, encouragement, openness, warmth, honesty, a smile, little positive deeds, positive action, word of honour. We need to stay willing to learn, be open. Then cooperation across cultures will work out. But probably it is not valued enough."

Finally, listening with patience and kindness—aspects which are described as part of CL—brings a deeper comprehensibility of complex, transcultural situations which are experienced as demanding or challenging.

**Table 2.** Comprehensibility and love.

| Frequency | Overall Category | Category | Participants |
|---|---|---|---|
| Comprehensibility through ... (31) | | | |
| 13 | Attitude | Understand transcultural experiences with a loving attitude | P1, P2, P5, P7, P9, P10, P13, P14, P16, P18, P19, P20, and P21 |
| 8 | Appreciation | Appreciate others across cultures through an open mindset | P1, P6, P12, P18, P12, P8, P16, and P20 |
| 7 | Learning | Learn about others to increase understanding | P4, P5, P8, P11, P14, P18, and P19 |
| 3 | Listen | Listen with patience and kindness | P2, P17, and P22 |

*3.3. How do You Manage Your Work? (Manageability)*

Participants refer to three major categories which are most helpful to manage and lead across cultures: positive behaviour, positive attitude and positive emotions (Table 3).

**Table 3.** Manageability and love.

| Overall Category | Frequency | Category | Participants |
|---|---|---|---|
| Positive behaviour (33) | 9 | Build an interpersonal connection | P7, P16, P8, P16, P20 P17, P22, P12, and P20 |
| | 8 | Be respectful and trusting | P2, P3, P6, P9, P10, P14, P18, and P22 |
| | 7 | Open communication | P1, P2, P9, P7, P9, P14, and P18 |
| | 6 | Show compassion through behaviour | P2, P9, P15, P16, P21, and P22 |
| | 3 | Challenge prejudices and racism (verbally and through actions) | P5, P11, and P20 |
| Positive attitude (26) | 13 | Actively understand, accept, respect and value the perspective of others | P2, P3, P4, P5, P6, P8, P10, P11, P13, P14, P16, P18, and P22 |
| | 8 | Use knowledge from different transcultural solutions | P1, P7, P8, P9, P14, P18, P19, and P20 |
| | 5 | Show love for humanity | P5, P7, P16, P19, and P21 |
| Positive emotions (13) | 13 | Focus on positive emotions in transcultural cooperation | P1, P7, P9, P10, P13, P18, P19, P20, P21 P2, P5, P14, and P16 |

Building an interpersonal connection is part of managing relationships across cultures. This connection should be built on CL and can support coping with challenging transcultural work situations. P1, a German, female leader, says:

> "Love gives me the strengths to build bridges and connect to other people, no matter where they come from. I believe in the good of the other and love helps me to do that. Love, for me, is a resource to be together, cooperate and build bridges."

Generally, respectful behaviour and trust are foundations to manage transcultural and global work relationships (8 participants). P11, a German male leader, highlights:

"When we behave in a loving way, respectful and kind, it is easy to cooperate in a diverse workforce. Then it is really enjoyable."

Further, eight individuals use knowledge from previous transcultural experiences in which solutions were found. Another 13 participants emphasise that to manage transcultural and global cooperation, they have to understand, respect and value others from different backgrounds. P12, a male Black African leader, highlights both aspects in his response to managing global work relationships:

"We must always see to learn as much as the own and other culture. When we use all this knowledge of humankind in a loving way, we will collaborate well and in a good spirit."

Finally, five leaders mention that a general love for humanity, a kind of world love, which is often called "agape" is important to manage transcultural situations and individuals.

With regard to positive behaviour, open communication (7), showing compassion (3) through behaviour and verbally challenging prejudices and racism (3) are mentioned.

Leaders expressed different views on how to manage global work relationships. In total, 13 participants feel that they need to focus on positive emotions to cope with transcultural cooperation. P4, a male German leader, says:

"When you feel positive feeling towards the other it is so much easier to respect, to appreciate and to learn from the other. When I focus on the positive feelings, I can cope much easier with any transcultural situation."

The majority of leaders found coping easier when positive thoughts, feelings and behaviours were executed or experienced.

*3.4. What Makes Your Work Meaningful? (Meaningfulness)*

Altogether, 51 statements were made with regard to meaningfulness. Many statements refer to meaningfulness and its creation through loving (work) relationships (Table 4). Meaningfulness is strongly constituted through a loving human connection (13 participants). CL represents an overall meaning for 11 participants. This experienced meaningfulness is connected to fulfilment, vocation and life purpose. The following statements refer to CL and meaning in this regard. However, four statements also refer to the negative aspects in the case of an absence of CL, which is associated with negative health and depression. P6, a US-American male leader, points out:

"I can devote myself to meaningful work because I have the security that comes with being embedded in relationships of mutual love. And even though I have collaborated with other scholars who I do not particularly know, or sometimes do not know at all, most of my collaborations have been in what I would call loving relationships. And the work we have done or are doing is meaningful both because we are jointly putting our knowledge and abilities to the project and because we like working with one another."

P18, a female US-American leader, states:

"I have worked for institutions that see their employees and students as fungible cogs in a wheel. These employers use their employees only for the monetary value that can be gained, as if the only value in an institution of higher learning is the bottom line. I am now in a place that recognizes the importance of vocation—both with the faculty and staff and with the students."

P7, a German female leader, emphasises:

"Love is an energy that transforms people for the better, for peace, sustainability, and humanity, and positivity in its original form."

CL is viewed as an overall positive concept that is only negative when it is intentionally or unintentionally misused or absent. It is seen as a universal positive source which can connect individuals of different cultures.

**Table 4.** Meaningfulness and love.

| Overall Category | Frequency | Category | Participants |
|---|---|---|---|
| Love creates meaningfulness (51) | 13 | Meaningfulness creation through loving work relationships and social connection at work, transformation of relationships | P4, P5, P6, P7, P8, P10, P11, P13, P15, P18, P19, P20, P21, and P22 |
| | 11 | Love has an overall meaning | P1, P3, P4, P5, P6, P7, P8, P18, P19, P20, P21, and P22 |
| | 10 | Love constitutes fulfilment, vocation and purpose in life | P3, P5, P6, P7, P15, P18, P19, P20, P21, and P22 |
| | 6 | Love creates meaning to know what is good for me | P7, P9, P10, P15, P18, and P19 |
| | 5 | Love creates health and happiness | P4, P10, P13, P15, and P19 |
| | 4 | No love brings depression, health risks | P3, P4, P15, and P16 |
| | 2 | The love to God gives meaning | P19 and P20 |

### 3.5. Coping Mechanisms

In terms of C mechanisms at work, the participants point out that CL builds a base for salutogenesis and is a major general C mechanism (Table 5). Altogether, 10 statements refer to the idea that loving one's work and working with people creates happiness and well-being, and thereby acts as a strong social C mechanism. P1, a female German leader, says:

**Table 5.** Love as a coping mechanism.

| Overall Category | Frequency | Category | Participants |
|---|---|---|---|
| Love as a general coping mechanism (40) | 18 | My love builds a base for well-being and coping with challenges | P1, P3, P4, P5, P7, P8, P10, P11, P12, P13, P14, P15, P16, P17, P18, P19, P20, and P21 |
| | 8 | Love is a resource | P4, P5, P7, P9, P11, P14, P15, and P22 |
| | 7 | Love is a human approach | P2, P5, P12, P14 P15, P19, and P21 |
| | 5 | Love creates balance | P16, P19, P20, P1, and P18 |
| | 2 | Safety | P3 and P5 |
| Love as a social coping mechanism (15) | 10 | Love my work and working with people is a base for happiness and well-being | P1, P3, P7, P8, P15, P20, P1, P5, P7, and P21 |
| | 5 | Loving social support improves mental health | P8, P15, P19, P20, and P22 |
| Self-love as a coping mechanism | 7 | Self-love improves mental health | P1, P7, P10, P12, P14, P21, and P22 |

"Working within loving relationships makes me happy."

An absence of love results in poor health. P3, a male, South African leader, states the following:

"Love relationships impact strongly on my mental health and well-being. I am loved by my family; thus I feel loved and are very happy. However, my organisation is not always in love with me, which impact strongly on my well-being at work. I am not loved by my organisation due to my race, and political agendas. New policies are now recommended that white people not be considered for promotion, to advance black people. This impacts very negatively on my sentiments for my organisation."

For this leader, the organisation shows love if it acts non-discriminatory. However, this leader does not feel loved because he is structurally discriminated against with regard to his race. This structural discrimination in South Africa is founded in compensation policies and mechanisms stemming from socio-historical discrimination. Therefore, the attitude of this leader could be judged as "non-loving" in a way, because he does not appreciate that the policies are made to compensate for previous injustice. He only sees the policy as being non-loving, since advantages are given to members of other race groups in this case.

Seven individuals view CL as an intra-personal resource—a self-coping mechanism. P22, a German female leader, emphasises:

"There is a close connection between love and mental health. Without love—not to love and to feel, not to be loved—makes you sick. Persons get isolated. But for mental health, it is important, to love yourself to cope with all the challenges. Love is a big resource for myself. It is part of me."

Further statements regarding CL refer to general CL concepts, seeing CL as a human approach to deal with challenges in life and at work. Again, seven individuals highlight self-love as a key to C, while five people emphasise that loving social support improves mental health. Finally, a few participants view CL as a balancing approach and two people highlight that CL creates a feeling of safety and acts as a coping mechanism.

*3.6. How Is CL Valued in Your Organisational Leadership Culture?*

Leaders provide statements on their views on how CL impacts on their national, socio-cultural and leadership culture. Altogether, 13 individuals say that CL is not a priority in their culture, referring to German, US-American and South African culture explicitly. Four statements mention that love is usually hidden in leadership culture in general. P18, a female, US-American leader, takes a critical view regarding love and her US-American culture:

"In the US love is not valued in leadership culture and barely at all in culture as a whole. Our patriarchal capitalist system is such that it seeks to undermine the individual in order to control her—to summarize Carol Gilligan, patriarchy separates men from women, men from men, and everyone from everyone else. It stunts emotion. There is a false narrative that rationality is the proper focus of attention for a leader, but this rationality rarely is founded in actual logical principles but in a wish to dominate."

A male, US-American leader, P5, contributes:

"Again, I don't think most of our political leaders are acting out of tough love, they appear to be acting out of narcissism, which I suppose you could call self-love, or narcissistic love. In America we generally believe that only narcissists are motivated to pursue leadership. My sense is that given the choice, Americans would rather be their own boss, rather than be someone else's boss and those who aspire to power and rank tend to be those least deserving of it. I suspect this stems from the idealization of the independent family farmer at the time of the US's initial formation, an ideal that has remained powerful into the 20th century. My ancestors immigrated to rural Minnesota from Europe to farm or start small businesses (barber) and I suspect that small family business ethos continues to permeate the American psyche. Although leaders elsewhere, like Korea, in which hierarchy runs deep through all social relations, do not seem any less narcissistic, loving or noble than anywhere else."

Ten statements emphasise that CL is of high value in the leadership culture, referring to aspects of CL in the US, German, Black African, Afrikaans and Japanese leadership culture. P9, a German male leader, says that there is CL in the German leadership culture, but there is also a need to increase it:

"I think there can be love in the leadership culture and also one can love their work in the sense of a general love. However, this is rather seldom. We need more love in the workplaces in future to overcome all the challenges."

Finally, P10, a Japanese female leader, says:

"In our leadership culture, jintokku (benevolence) is valued. Leaders should be competent, but competence only does not necessarily serve as a requirement for a great leader. Of course, leaders should be competent. But at the same time, they have to be kind and considerate towards others."

According to several participants (Table 6), CL is not necessarily reflected in their national culture or their organisational culture but is often part of their individual or socio-cultural group culture. Several leaders highlight that they would wish for a CL leadership culture and some leaders connect CL with other outstanding values, such as benevolence for the Japanese leadership context.

**Table 6.** Love in leadership culture.

| Overall Category | Frequency | Category | Participants |
|---|---|---|---|
| Love is not a priority in leadership culture (13) | 5 | In German culture—the word love is not mentioned in German leadership culture | P4, P6, P7, P8, P13, P16, P19, and P22 |
| | 4 | Generally hidden in leadership culture | P3, P4, P8, and P11 |
| | 3 | In US culture | P5, P16, and P18 |
| | 1 | In South African culture | P14 |
| High value of love . . . (10) | 5 | In US culture, through understanding and freedom and "tough love" | P5, P6, P11, P12, and P16 |
| | 2 | German leadership culture | P9 and P16 |
| | 1 | Black African culture | P12 |
| | 1 | Afrikaans culture | P3 |
| | 1 | Japan: Jintokku (benevolence) | P10 |

## 4. Discussion

This study contributes to previous research on salutogenesis and SOC, CL and C in leadership and transcultural and global work contexts [8]. The findings support studies on SOC as a C mechanism to manage work stressors and challenges [9]. The findings show that leaders focus particularly on meaning at work and their lifestyle choices [2], emphasising that CL, as a positive emotion, contributes to managing challenges in contemporary and future workplaces. Leaders in this study do not focus on negative factors when facing challenges during the transition into the 4IR [3] but stay in a positive mindset. This might be a side effect of their strong focus on salutogenesis, SOC, and a strong focus on meaningfulness. They use CL as their main resource for coping at work. The findings, therefore, support previous studies such as Li, Hou and Wu [5] and contribute from a positive psychology mindset to the stress management literature [14]. As shown above [15], CL is defined in the context of three of the six key features [17], namely: listening skills, self-growth, and building a (healthy) community. Leaders in this study foster CL to connect with others, create positive social bonds and grow individually [17].

Meaningfulness is very important for the leaders and provides them with motivation and strengths in 4IR workspaces [20]. SOC and CL can be promoted as new skills and actions to promote systemic, human-based leadership qualities [87], which are important for managing C when faced with new 4IR stressors. The findings do not display a critical view of CL and leadership in relation to Illouz's [88] critique on capitalist societies. That no critical views on CL are presented might be a shortcoming of the findings. Rather, the participants follow Chandsoda and Salsing's [26] approach, incorporating sympathy and participating in human relations, based on CL, connecting passionately with the self and other [27,29].

Dooris, Doherty and Orme [36] emphasise that organisations and leaders are increasingly focusing on promoting salutogenesis improve the performance of organisations. The findings show that SOC components can aid leaders [37,38,42]. Further, the findings show that meaningfulness is the most important component, followed by manageability and comprehensibility as SOC, fostering happiness and well-being [46,47], positivity and listening skills [48]. The findings support that CL improves health in leaders, supporting Gray [50]. It is also assumed that CL and SOC are interlinked, including the idea suggested by Eriksson [51] that SOC needs to be applied at different system levels. As in this study, Harry [51] also supports the idea of implementing SOC interventions in organisations. Further, this study supports the idea of fostering positive feelings [63], thereby promoting affection and interest in others [63,64], increasing emotive encounters and meaning [68] through SOC interventions.

As suggested by Winston [89], leaders aim at "doing good" and would like to improve the life of others as well as working "for the better". Leaders do not refer to negative emotions, such as anger, failure, selfishness, guilt or worry [73]. This positive mindset might increase relaxation, other positive feelings, general well-being [75,76], and spirituality. This further fosters a general appreciation of humanity, with a focus on personal and organisational growth, vocation and professional calling [73]. This study supports previous studies on the positive effects of salutogenesis, SOC and CL regarding C in leadership positions.

## 5. Limitations

Limitations of this study include the qualitative nature of the research, which only provides a selected, in-depth insight into a small number of participants. These participants are from specific cultures and middle- and higher-economic backgrounds within their societies. The sample is, therefore, biased. The research study provides insights into subjective, ambiguous views and experiences and cannot be generalised [90]. The findings might be helpful to provide a starting point for mixed-method studies on SOC, CL and C.

## 6. Conclusions

The aim of this article was to present the positive coping mechanisms of leaders for coping with challenges faced in the transition into the 4IR with special regard to SOC and CL, thereby contributing to opening a positive and constructive approach on coping during the transition into the 4IR.

Leaders believe in CL as an important mindset to improve transcultural work relationships and to understand complex situations (comprehensibility). CL helps leaders to focus on the positive, look beyond cultural differences and "see the good" in others. CL helps to be appreciative and open-minded and minimises cultural and religious stereotypes. CL relates further to conscious and mindful listening.

Leaders manage transcultural work relationships by primarily focusing on managing behaviour, attitude and emotions (manageability). Healthy workplaces consist of well-managed resources, balanced relationships and positive humane connections. When leaders maintain positive relationships, they cope better with challenges. Love is highly important as a resource to cope with stress.

Meaningfulness is connected with positive and loving relationships, overall meaning creation, purpose and fulfilment, positive feelings, health and well-being. Two individuals view CL and meaningfulness in relation to God.

In conclusion, leaders see a strong connection between CL and C. They feel that CL is strongly associated with being human, happiness, balance and harmony, love for the world and safety. Leaders also highlight that when they have to make decisions, maintaining a CL perspective is often challenging.

Overall, CL is a strong coping mechanism in leaders of this study. CL is connected with SOC and C (40 statements). CL is a social and intra-personal coping mechanism as well as an important part of the leaders' personal leadership culture, though not necessarily part of the national leadership culture. CL is a concept that needs to be actively implemented into leadership.

**Author Contributions:** Both authors (C.-H.M. and R.M.O.) designed and wrote up this study. C.-H.M. collected and analysed the data. Intersubjective validation of this study was performed by both authors. Both authors have read and agreed to the published version of the manuscript. All authors have read and agreed to the published version of the manuscript.

**Funding:** This research received no external funding.

**Acknowledgments:** We thank the participants for their engagement in this study and for sharing their in-depth insights.

**Conflicts of Interest:** All other authors declare no conflict of interest.

## Appendix A

Biographical data
Sex:
Age:
Mother tongue:
Cultural background:
Nationality:
Religious affiliation/belief:
Highest educational degree:
Profession:
Leadership position:
Marital status:
Marital status with someone from another culture? If yes, which cultures:
Have you have lived outside of your birth country (more than one year)?
If yes, where did you live longer than one year?

1. What is love for you and how to you show love in professional contexts?
2. Please give an example of love in the workplace.
3. How are love and leadership connected?
4. How does love impact on your (work) relationships?
5. How can love support the establishment of positive transcultural interaction?
6. How are love and mental health and well-being connected?
7. How do you understand the world?
8. What is important to understand the world?
9. Which resources do you use to cope with challenging work situations?
10. What makes your life and your work meaningful?
11. Which resources to you use to cope with the transformation into the 4IR?

12.  How is love valued in your culture/in your leadership culture. Please give an example.

## References


1.   Philbeck, T.; Davis, N. The fourth industrial revolution: Shaping a new era. *J. Int. Aff.* **2019**, *72*, 17–22.
2.   Fink, A.; Elisabetta, E. Skill flows and the Fourth Industrial Revolution: Future questions and directions for the ASEAN Economic Community. In *Skilled Labour Mobility and Migration*; Edward Elgar Publishing: Cheltenham, UK, 2019.
3.   Schwab, K. *The Fourth Industrial Revolution*; Crown Business, Penguin Random House: New York, NY, USA, 2017.
4.   Kamaruddin, S.F.; Laura, D.I.; Susan, R.J.; Musa, B.M.; Eng, T.H. The impact of listening and speaking anxieties on the fourth Industrial revolution: What can educators do? *J. Lang. Commun.* **2019**, *6*, 115–129.
5.   Li, G.; Hou, Y.; Wu, A. Fourth Industrial Revolution: Technological drivers, impacts and coping methods. *Chin. Geogr. Sci.* **2017**, *27*, 626–637. [CrossRef]
6.   Antonovsky, A. *Health, Stress and Coping*; Jossey-Bass: San Francisco, CA, USA, 1979.
7.   Antonovsky, A. *Unravelling the Mystery of Health: How People Manage Stress and Stay Well*; Jossey-Bass: San Francisco, CA, USA, 1987.
8.   Bauer, G.F.; Roy, M.; Bakibinga, P.; Contu, P.; Downe, S.; Eriksson, M.; Mana, A. Future directions for the concept of salutogenesis: A position article. *Health Promot. Int* **2019**, in press. [CrossRef] [PubMed]
9.   Saboga-Nunes, L.; Bittlingmayer, U.H.; Okan, O. Salutogenesis and health literacy: The health promotion simplex. In *International Handbook of Health Literacy*; The Policy Press: Bristol, UK, 2019; pp. 649–664.
10.  Wallace, D.; Boynton, M.; Lytle, L. Multilevel analysis exploring the links between stress, depression, and sleep problems among two-year college students. *J. Am. Coll. Health* **2017**, *65*, 187–196. [CrossRef]
11.  Maxwell, J.S.; Davidson, R.J. Emotion as motion: Asymmetries in approach and avoidant actions. *Psychol. Sci.* **2007**, *18*, 1113–1119. [CrossRef]
12.  Maykrantz, S.A.; Houghton, J.D. Self-leadership and stress among college students: Examining the moderating role of coping skills. *J. Am. Coll. Health* **2020**, *68*, 89–96. [CrossRef]
13.  Shuck, B.; Alagaraja, M.; Immekus, J.; Cumberland, D.M.; Honeycutt-Elliott, M. Operationalizing Compassionate Leadership Behavior. In *Academy of Management Proceedings*; Academy of Management: Briarcliff Manor, NY, USA, 2016; Volume 2016, p. 14266.
14.  Dierendonck, D.; Patterson, K. Compassionate love as a cornerstone of servant leadership: An integration of previous theorizing and research. *J. Bus. Ethics* **2015**, *128*, 119–131. [CrossRef]
15.  Reynolds, K. Is Servant-Leadership Gender Bound in the Political Arena? In Proceedings of the Annual Meeting of the Midwest Political Science Association 67th Annual National Conference, Chicago, IL, USA, 2–5 April 2009; Available online: http://www.allacademic.com/meta/p364186_index.html (accessed on 19 April 2020).
16.  Reynolds, K. Servant-Leadership—A Feminist Perspective. *Int. J. Servant Leadersh.* **2015**, *10*, 35–63.
17.  Villiers, C. Boardroom Culture: An Argument for Compassionate Leadership. *Eur. Bus. Law Rev.* **2019**, *30*, 253–278.
18.  De Zulueta, P.C. Developing compassionate leadership in health care: An integrative review. *J. Healthc. Leadersh.* **2016**, *8*, 1. [CrossRef] [PubMed]
19.  Lloyd, J.L. Relationship between Self-Compassion, Sense of Coherence, Coping Strategies and Perceived Stress in Clinical Psychology Trainees. Ph.D. Thesis, Staffordshire University & Keele University, Stoke, UK, 2017.
20.  Alabi, M.; Telukdarie, A.; Van Rensburg, N. Industry 4.0: Innovative solutions for the water industry. In Proceedings of the International Annual Conference of the American Society for Engineering Management, Huntsville, AL, USA, 18–21 October 2019.
21.  Sisodia, R.; Wolfe, D.; Sheth, J.N. *Firms of Endearment. How World Class Companies Profit from Passion and Purpose*; Pearson Prentice Hall: Upper Saddle River, NJ, USA, 2003.
22.  World Economic Forum. The Future of Jobs. Employment, Skills and Workforce Strategy for the Fourth Industrial Revolution. Global Challenge Insight Report. 2016. Available online: http://www3.weforum.org/docs/WEF_Future_of_Jobs.pdf (accessed on 19 April 2020).
23.  Magubane, K. *This Love Thing: A New Age Love Story*; umSinsi Press: Cape Town, South Africa, 2019.

24. Illouz, E. *Warum Liebe Weh Tut*; Suhrkamp: Berlin, Germany, 2011.

25. Illouz, E. Warum endet die Liebe? Sternstunde Philosophie SRF Kultur. 29 March 2019. Available online: https://www.youtube.com/watch?v=3HT3c0rawI0 (accessed on 19 April 2020).

26. Chandsoda, S.; Salsing, P.S. Compassion and cooperation: The two challenging ethical perspectives in the Fourth Industrial Revolution. *J. Int. Bus. Stud.* **2018**, *9*, 101–115.

27. Van der Hoven, J. Using Flow to Create Meaningful Work in the Fourth Industrial Revolution. 22 October 2017. Available online: https://leaderless.co/blog/2017/10/22/the-flow-of-the-fourth-industrial-revolution/ (accessed on 19 April 2020).

28. Tu, T.N.; Thien, T.D. *Buddhism and the Fourth Industrial Revolution*; Vietnam Buddhist University Press, Hong Duc Publishing House: Hanoi, Vietnam, 2019.

29. Sun, Z. *Innovation and Entrepreneurship in the 4th Industrial Revolution*; Joint Workshop on Entrepreneurship: Lae, Papua New Guinea, 2018.

30. Fabritius, S. Ventures for a Better Society; 4th Entrepreneurial Revolution. Master's Thesis, Aalto University School of Science, Aalto, Finland, 2017.

31. Champeau., M.; Arsac, M.; Pineau, P.; Denyset, F. *Non-Standard Employment around the World*; ILO Cataloguing in Publication Data: Geneva, Switzerland, 2016; pp. 1925–1926.

32. Jae, R.N. Research on the employment instability and its causes. *J. Labour Econ.* **2005**, *28*, 111–139.

33. De Witte, H.; Pienaar, J.; De Cuyper, N. Review of 30 years of longitudinal studies on the association between job insecurity and health and well-being: Is there causal evidence? *Aust. Psychol.* **2016**, *51*, 18–31. [CrossRef]

34. Patel, P.C.; Devaraj, S.S.; Hicks, M.J.; Wornell, E.J. County-level job automation risk and health: Evidence from the United States. *Soc. Sci. Med.* **2018**, *202*, 54–60. [CrossRef]

35. Rojas-Rueda, D. Autonomous vehicles and mental health. *J. Urban Ment. Health* **2017**, *3*, 2.

36. Dooris, M.; Doherty, S.; Orme, J. The application of salutogenesis in universities. In *The Handbook of Salutogenesis*; Mittelmark, M.B., Sagy, S., Eriksson, M., Bauer, G.F., Pelikan, J.M., Lindström, B., Espnes, G.A., Eds.; Springer Nature: Cham, Switzerland, 2017; pp. 237–246.

37. Gimpel, C.; Von Scheidt, C.; Jose, G.; Sonntag, U.; Stefano, G.B.; Michalsen, A.; Esch, T. Changes and interactions of flourishing, mindfulness, sense of coherence, and quality of life in patients of a mind-body medicine outpatient clinic. *Forsch. Komplementärmedizin* **2014**, *21*, 154–162. [CrossRef]

38. Krause, D.C.; Mayer, C.H. *Exploring Mental Health: Theoretical and Empirical Discourses on Salutogenesis*; Pabst Science Publisher: Lengerich, Germany, 2012.

39. Mayer, C.H.; Boness, C.M. *Creating Mental Health across Cultures. Coaching and Training for Managers*; Pabst Science Publisher: Lengerich, Germany, 2013.

40. Mayer, C.H.; Krause, C. Promoting mental health and salutogenesis in transcultural organisational and work contexts. In *International Review of Psychiatry 23, December 2011*; Mayer, C.H., Krause, C., Eds.; Taylor & Francis: Oxford, UK, 2016.

41. Mayer, C.H.; Krause, C. 'Editorial', in C.-H. Mayer & C. Krause (eds.), 'Salutogenese in Beratung und Psychotherapy' [Salutogenesis in counselling and psychotherapy]. *Prax. Klin. Verhalt. Und Rehabil.* **2013**, *26*, 92–196. [CrossRef]

42. Tan, K.K.; Chan, S.W.C.; Wan, W.; Vehviläinen-Jukunen, K. A salutogenic program to enhance sense of coherence and quality of life for older people in the community: A feasibility randomized controlled trial and process evaluation. *Patient Educ. Couns.* **2016**, *99*, 108–116. [CrossRef]

43. Mayer, C.H.; Boness, C.M. Interventions to promoting sense of coherence and transcultural competences in educational contexts. *Int. Rev. Psychiatry* **2011**, *23*, 516–524. [CrossRef]

44. Mayer, C.H.; Louw, L.; Von der Ohe, H. Sense of coherence in Chinese and German students. *Health Sa Gesondheid* **2019**, *24*, a1151. [CrossRef] [PubMed]

45. Pretorius, N.; Arcelus, J.; Beecham, J.; Dawson, H.; Doherty, F.; Eisler, I.; Gallagher, C.; Gowers, S.; Isaacs, G.; Johnson-Sabine, E.; et al. Cognitive-behavioural therapy for adolescents with bulimic symptomatology: The acceptability and effectiveness of internet-based delivery. *Behav. Res. Ther.* **2009**, *47*, 729–736. [CrossRef] [PubMed]

46. Mayer, R.E. *Applying the Science of Learning*; Pearson: Boston, MA, USA, 2011.

47. Mitonga-Monga, J.; Cilliers, F. Perceived ethical leadership: Its moderating influence on employees' organisational commitment and organisational citizenship behaviours. *J. Psychol. Afr.* **2016**, *26*, 35–42. [CrossRef]

48.	Mitonga-Monga, J.; Coetzee, M.; Cilliers, F. Perceived leadership style and employee participation in a manufacturing company in the Democratic Republic of Congo. *Afr. J. Bus. Manag.* **2012**, *6*, 5389–5398. [CrossRef]

49.	Eriksson, M. The Sense of Coherence in the Salutogenic Model of Health. 2017. Available online: https://www.ncbi.nlm.nih.gov/books/NBK435812/ (accessed on 19 April 2020).

50.	Gray, D. Developing leadership resilience through a sense of coherence. In *Contemporary Leadership Challenges*; Alvinius, A., Ed.; IntechOpen: London, UK, 2017; Available online: https://www.intechopen.com/books/contemporary-leadership-challenges/developing-leadership-resilience-through-a-sense-of-coherence (accessed on 19 April 2020).

51.	Harry, N. Employee Well-Being in the African Call Centre Digital Workspace. In *Thriving in Digital Workspaces*; Springer: Berlin/Heidelberg, Germany, 2019; pp. 109–129.

52.	Konstam, V.; Celen-Demirtas, S.; Tomek, S.; Sweeney, K. Career adaptability and subjective wellbeing in unemployed emerging adults: A promising and cautionary tale. *J. Career Dev.* **2015**, *42*, 463–477. [CrossRef]

53.	Gross, J.J. Antecedent- and response-focused emotion regulation: Divergent consequences for experience, expression, and physiology. *J. Personal. Soc. Psychol.* **1998**, *74*, 224–237. [CrossRef]

54.	Segerstrom, S.C.; Smith, G.T. Personality and coping: Individual differences in responses to emotion. *Annu. Rev. Psychol.* **2019**, *70*, 651–671. [CrossRef]

55.	Bossuyt, E.; Moors, A.; DeHouwer, J. On angry approach and fearful avoidance: The goal-dependent nature of emotional approach and avoidance tendencies. *J. Exp. Soc. Psychol.* **2014**, *50*, 118–124. [CrossRef]

56.	Machin, J.E.; Adkins, N.R.; Crosby, E.; Farrell, J.R.; Mirabito, A.M. The marketplace, mental well-being, and me: Exploring self-efficacy, self-esteem, and self-compassion in consumer coping. *J. Bus. Res.* **2019**, *100*, 410–420. [CrossRef]

57.	Folkman, S.; Moskowitz, J.T. Coping: Pitfalls and promise. *Annu. Rev. Psychol.* **2004**, *55*, 745–774. [CrossRef]

58.	Hoyle, J.R. *Leadership and the Force of Love. Six Keys to Motivating with Love*; Corwin Press: Thousand Oaks, CA, USA, 2001.

59.	Barsade, S.G.; Gibson, D.E. Why does affect matter in organizations? *Acad. Manag. Perspect.* **2007**, *21*, 36–59. [CrossRef]

60.	Barsade, S.; Brief, A.P.; Spataro, S.E. The affective revolution in organizational behavior: The emergence of a paradigm. In *Organizational Behavior: The State of the Science*; Greenberg, J., Mahwah, N.J., Eds.; Lawrence Erlbaum: London, UK, 2003; pp. 3–52.

61.	Hareli, S.; Rafaeli, A. Emotion cycles: On the social influence of emotion in organizations. In *Research in Organizational Behaviour*; Brief, A.P., Staw, B.M., Eds.; Elsevier: New York, NY, USA, 2008; pp. 35–59.

62.	Robinson, M.D.; Watkins, E.R.; Harmon-Jones, E. *Handbook of Cognition and Emotion*; Guilford Press: New York, NY, USA, 2013.

63.	Eldor, L. Public Service Sector: The compassionate workplace—The effect of compassion and stress on employee engagement, burnout, and performance. *J. Public Adm. Res. Theory* **2017**, *28*, 86–103. [CrossRef]

64.	Underwood, L.G. The human experience of compassionate love. In *Altruism and Altruistic Love*; Post, S.G., Underwood, L.G., Schloss, J., Hurlbut, W.B., Eds.; Oxford University Press: Oxford, UK, 2009; pp. 72–88.

65.	Patterson, K. Servant Leadership and Love. In *Servant Leadership*; van Dierendonck, D., Patterson, K., Eds.; Palgrave Macmillan: London, UK, 2010; pp. 67–76.

66.	Barsade, S.G.; O'Neill, O.A. What's love got to do with it? A longitudinal study of the culture of companionate love and employee and client outcomes in a long-term care setting. *Admin. Sci. Q.* **2014**, *59*, 551–598. [CrossRef]

67.	Fineman, S. *Emotions in Organizations*; Sage: New York, NY, USA, 2000.

68.	Savickas, M.L. The meaning for work and love: Career issues and interventions. *Career Dev. Q.* **1991**, *39*, 291–379. [CrossRef]

69.	O'Neill, O.M. The FACCTs of (work) life: How relationships (and returns) are linked to the emotional culture of companionate love. *Am. J. Health Promot.* **2018**, *32*, 1312–1315. [CrossRef] [PubMed]

70.	Rynes, S.I.; Bartunek, J.M.; Dutton, J.E.; Margolis, J.D. Care and compassion through an organizational lens: Opening up new possibilities. *Acad. Manag. Rev.* **2012**, *37*, 503–523.

71.	Parry, K.; Kempster, S. Love and leadership: Constructing follower narrative identities of charismatic leadership. *Manag. Learn.* **2013**, *45*, 21–38. [CrossRef]

72. Rapport, N. *Cosmopolitan Love and Individuality: Ethical Engagement beyond Culture*; Lexington Books: Lanham, MD, USA, 2019.
73. Fry, L.W.; Matherly, L.L. Spiritual Leadership and Organizational Performance: An Explorative Study. In Proceedings of the Academy of Management Meeting, Atlanta, GA, USA, 11–16 August 2006; Available online: https://www.iispiritualleadership.com/wpcontent/uploads/docs/SLTOrgPerfAOM2006.pdf (accessed on 19 April 2020).
74. Soh, C.; Connolly, D. New Frontiers of Profit and Risk. The Fourth Industrial Revolution's Impact on Business and Human Rights. *New Political Econ.* **2020**, in press. [CrossRef]
75. Cooper, M. *The Compassionate Mind Approach to Reducing Stress*; Constable & Robinson: London, UK, 2013.
76. Gilbert, P. *The compassionate Mind: A New Approach to Life's Challenges*; New Harbinger Publications: Maryland, CA, USA, 2009.
77. Creswell, W. *Qualitative Inquiry and Research Design: Choosing among Five Approaches*; SAGE Publications: New York, NY, USA, 2013.
78. Hassan, I.; Ghauri, P.N. Research Design. *Int. Bus. Manag.* **2014**, *30*, 75–81.
79. Bleicher, J. *Hermeneutics as Method, Philosophy and Critique*; Routledge & Kegan Paul: London, UK, 1980.
80. Willcocks, L.P.; Mingers, J. *Social Theory and Philosophy for Information Systems*; John Wiley & Sons Ltd.: Hoboken, NJ, USA, 2004.
81. Naderifar, M.; Goli, H.; Ghaljaie, F. Snowball Sampling: A Purposeful Method of Sampling in Qualitative Research. *Strides Dev. Med. Educ.* **2017**, *14*, e67670. [CrossRef]
82. Rahi, S. Research design and methods: A systematic review of research paradigms, sampling issues and instruments development. *Int. J. Econ. Manag. Sci.* **2017**, *6*, 1–5. [CrossRef]
83. Terre Blanche, M.; Durrheim, K.; Painter, D. *Research in Practice: Applied Methods for the Social*; UCT Press: Juta, South Africa, 2006.
84. Yin, R.K. *Case Study Research: Design and Methods*; Sage: Los Angeles, CA, USA, 2014.
85. Colepicolo, E. *Information Reliability for Academic Research: Review and Recommendations*; New Library World: Novato, CA, USA, 2015.
86. Roth, W.M.; Unger, H.V. Current Perspectives on Research Ethics in Qualitative Research. *Forum Qual. Soz. Forum Qual. Soc. Res.* **2018**, *19*, 1–12. [CrossRef]
87. Mayer, C.H. Key factors of Creativity and the Art of Collaboration in Twenty-First-Century workplaces. In *Thriving in Digital Workplaces: Innovations in Theory, Research and Practice*; Coetzee, M., Ed.; Springer Nature: Cham, Switzerland, 2019; pp. 147–166.
88. Illouz, E. *Cold Intimacies: The Making of Emotional Capitalism*; Polity Press: Cambridge, UK, 2007.
89. Winston, B.E. *Be a Leader for God's Sake*; Regent University School of Leadership Studies: Virginia Beach, VA, USA, 2002.
90. Atieno, O.P. An analysis of the strengths and limitation of qualitative and quantitative research paradigms. *Probl. Educ. 21st Century* **2009**, *13*, 13–18.

MDPI
St. Alban-Anlage 66
4052 Basel
Switzerland
Tel. +41 61 683 77 34
Fax +41 61 302 89 18
www.mdpi.com

*International Journal of Environmental Research and Public Health* Editorial Office
E-mail: ijerph@mdpi.com
www.mdpi.com/journal/ijerph